SOYBEANS

Volume 2

Genome, Biotechnology, Stress, and Disease

Soybeans: History, Taxonomy, Genetics, Breeding, Diseases, and Biotechnology (2-volume set)
ISBN: 978-1-998511-74-7 (hbk)

Soybeans, Volume 1: History, Breeding, and Product Profile
ISBN: 978-1-998511-70-9 (hbk)
ISBN: 978-1-998511-71-6 (ebk)

Soybeans, Volume 2: Genome, Biotechnology, Stress, and Disease
ISBN: 978-1-998511-72-3 (hbk)
ISBN: 978-1-998511-73-0 (ebk)

SOYBEANS

Volume 2

Genome, Biotechnology, Stress, and Disease

R. Ambika Rajendran, PhD
Ayyagari Ramlal, MSc, MRSB, MSIVB
Sanjay Kumar Lal, PhD
Godfree Chigeza, PhD
Editors

AAP | APPLE ACADEMIC PRESS

First edition published 2026

Apple Academic Press Inc.
1265 Goldenrod Circle, NE,
Palm Bay, FL 32905 USA

760 Laurentian Drive, Unit 19,
Burlington, ON L7N 0A4, Canada

CRC Press
2385 NW Executive Center Drive,
Suite 320, Boca Raton FL 33431

4 Park Square, Milton Park,
Abingdon, Oxon, OX14 4RN, UK

Apple Academic Press exclusively co-publishes with CRC Press, an imprint of Taylor & Francis Group, LLC

Library and Archives Canada Cataloguing in Publication

Title: Soybeans / R. Ambika Rajendran, PhD, Ayyagari Ramlal, MRSB, Sanjay Kumar Lal, PhD, Godfree Chigeza, PhD, editors.
Names: Rajendran, R. Ambika, editor. | Ramlal, Ayyagari, editor. | Lal, Sanjay Kumar, editor. | Chigeza, Godfree, editor.
Description: First edition. | Includes bibliographical references and indexes. | Contents: Volume 2: Genome, biotechnology, stress, and disease.
Identifiers: Canadiana (print) 2025030208X | Canadiana (ebook) 20250302136 | ISBN 9781998511747 (set) | ISBN 9781998511723 (v. 2 ; hardcover) | ISBN 9781998511730 (v. 2 ; ebook)
Subjects: LCSH: Soybean.
Classification: LCC SB205.S7 S69 2026 | DDC 633.3/4—dc23

Library of Congress Cataloging-in-Publication Data

CIP data on file with US Library of Congress

ISBN: 978-1-998511-72-3 (hbk)
ISBN: 978-1-998511-73-0 (ebk)

DOI: 10.1201/9781998511730

About the Editors

R. Ambika Rajendran, PhD

Senior Scientist of Plant Breeding, Division of Genetics, Indian Council of Agricultural Research (ICAR)–Indian Agricultural Research Institute (IARI), Pusa Campus, New Delhi, India

R. Ambika Rajendran, PhD, is a Senior Scientist of Plant Breeding, Division of Genetics, Indian Council of Agricultural Research–Indian Agricultural Research Institute, Delhi, India. She worked at the Directorate of Maize Research (DMR), New Delhi (currently referred to as the Indian Institute of Maize Research [IIMR], Punjab, India) from 2009 until 2016. Later, she joined the Division of Genetics, IARI, as a scientist in a senior scale in 2016. Dr. Rajendran has over 15 years of research experience. She currently works in abiotic stress tolerance and doubled haploidy in soybeans. She worked in maize breeding for high oil along with baby corn. Her significant research includes heterotic grouping and patterning of quality protein maize using molecular markers, *in vitro* mutagenesis in rice, identification of sources of high oil post-*Fusarium* stalk rot resistance genetic stocks in maize, and identification of tolerant sources for pre-germination anaerobic stress tolerance in soybean. She has edited two books, namely *Genetics and Breeding in Pulses* and *Handbook of Cereals*, and has also served as an editor for a book on soybean. She has many peer-reviewed research articles published in journals of international and national repute and book chapters to her credit. She is also serving as a reviewer for various national and international journals. Dr. Rajendran earned her bachelor degree from Tamil Nadu Agricultural University (TNAU), Tamil Nadu, India, her master's degree from Kerala Agricultural University (KAU), Kerala, India, and her doctorate from Tamil Nadu Agricultural University (TNAU).

Ayyagari Ramlal, MSc, MRSB, MSIVB

Plant Biotechnology Laboratory School of Biological Science, University Sains Malaysia, Malaysia; Division of Genetics, Indian Council of Agricultural Research (ICAR)–Indian Agricultural Research Institute (IARI), Pusa Campus, New Delhi, India

Ayyagari Ramlal, MRSB, is a Senior Doctoral Scholar at the School of Biological Sciences, Universiti Sains Malaysia, Malaysia. He completed his bachelor's and master's degrees in Botany from the University of Delhi, Delhi, India, in 2017 and 2020, respectively. He has worked as a Junior Research Fellow at the Division of Genetics, Indian Agricultural Research Institute (ICAR–IARI), Delhi, India, and Banaras Hindu University (BHU), Uttar Pradesh, India. He also worked on an innovation research project initiated by the University of Delhi. He has received young author and the best author awards from various societies and organizations in India. He is the recipient of the prestigious Malaysia International Scholarship (MIS) 2024, Ministry of Higher Education (MOHE), Malaysia, and 2025 Student Travel Award for the 2025 *In Vitro* Biology Meeting, Society for *In Vitro* Biology (SIVB), VA, USA. He has published more than 30 peer-reviewed research and review articles in international journals of repute, along with several book chapters. He served as an editor for the book *Phytohormones in Abiotic Stress* and is currently serving as a reviewer for many national and international journals of repute. Additionally, he is a review editor for the Biodiversity section of *Frontiers in Young Minds*. He is a member of the Royal Society of Biology (RSB), UK and Society for *In Vitro* Biology (SIVB), MD, USA.

Sanjay Kumar Lal, PhD

Principal Scientist, Division of Genetics, Indian Council of Agricultural Research–Indian Agricultural Research Institute, India

Sanjay Kumar Lal, PhD, is a Principal Scientist at the Division of Genetics, Indian Council of Agricultural Research–Indian Agricultural Research Institute, and Coordinator of the Delhi Centre for All India Coordinated Research Project (AICRP) on Soybean. He is also the group/project leader for genetic enhancement of soybean. He has been working as an active researcher and possesses over 25 years of experience in conducting and doing research. His area of specialization includes soybean breeding, tissue culture, and molecular breeding. He has published over 50 original articles in journals of repute. He has released several varieties of soybean including Pusa9712 and Pusa9814. He has also guided many students including master's and supervised doctoral theses.

Godfree Chigeza, PhD
Soybean Breeder, International Institute of Tropical Agriculture (IITA), Lusaka, Zambia

Godfree Chigeza, PhD, is currently associated with the International Institute of Tropical Agriculture (IITA), Lusaka, Zambia, as a Soybean Breeder. Prior to this appointment, he worked as breeder project lead at MRI Zambia, Syngenta Africa and Middle East (AME). He was the projects team lead for maize breeding at the Agricultural Research Council-Grain Crops Institute, Potchefstroom, South Africa, and worked as a field trial and horticultural manager at the Centre for Novel Agricultural Products (CNAP) Department of Biology, University of York, UK. Dr. Chigeza has a PhD in Plant Breeding from the University of KwaZulu Natal, South Africa. He received his MPhil in Agricultural Sciences and graduate diploma in Agricultural Sciences (Plant Breeding and Biotechnology) at the University of Queensland, Australia. He earned his graduate diploma in Seed Technology and BSc in Agricultural Crop Science at the University of Zimbabwe.

Contents

Contributors

Nisha Agrawal
Indian Institute of Soybean Research (ICAR–IISR), Indore, Madhya Pradesh, India

Pawan K. Amrate
Department of Plant Breeding and Genetics, Jawaharlal Nehru Krishi Vishwa Vidyalaya, Jabalpur, Madhya Pradesh, India

Jhilick Banerjee
Department of Plant Breeding and Genetics, Jawaharlal Nehru Krishi Vishwa Vidyalaya, Jabalpur, Madhya Pradesh, India

Akash Barela
Department of Plant Breeding and Genetics, Jawaharlal Nehru Krishi Vishwa Vidyalaya, Jabalpur, Madhya Pradesh, India

Rayees Ahmad Bhat
Department of Botany, University of Delhi, New Delhi, Delhi, India

Pulak Debbarma
Division of Genetics, Indian Council of Agricultural Research (ICAR)–Indian Agricultural Research Institute (IARI), Pusa Campus, New Delhi, India

Chandrasekaran Deepika
Department of Genetics and Plant Breeding, Tamil Nadu Agricultural University, Coimbatore, Tamil Nadu, India

Sourav Singh Deo
Department of Botany, Deshbandhu College, University of Delhi, New Delhi, India

Murugesan Dhandapani
Tamil Nadu Rice Research Institute, Tamil Nadu Agricultural University, Aduthurai, Tamil Nadu, India

Subramanian Divya
Department of Genetics and Plant Breeding, Tamil Nadu Agricultural University, Coimbatore, Tamil Nadu, India

Pranay Gundala
Department of Genetics and Plant Breeding, Indira Gandhi Krishi Vishwavidyalaya, Raipur, Chhattisgarh, India

Preeti Gupta
Independent Researcher, New Delhi, India

Prem Kumar Gupta
Modern Diagnostics and Research Center, Gurgaon, Haryana, India

Amooru Harika
Division of Plant Physiology, ICAR–Indian Agricultural Research Institute (IARI), Pusa Campus, New Delhi, Delhi, India

Shivani Jawarkar
Department of Plant Breeding and Genetics, Jawaharlal Nehru Krishi Vishwa Vidyalaya, Jabalpur, Madhya Pradesh, India

Sumit Kumar Kashyap
Department of Molecular and Cellular Engineering, Jacob Institute of Biotechnology and Bioengineering, Sam Higginbottom University of Agriculture, Technology, and Sciences, Naini, Prayagraj, Uttar Pradesh, India

Fatima Nazish Khan
Department of Biotechnology, Jamia Millia Islamia, New Delhi, India

Raj Vrajesh Kothari
Department of Plant Biotechnology, MGM College of Agricultural Biotechnology, Gandheli, Aurangabad, Maharashtra, India

Rahul Kumar
Division of Genetics, Indian Council of Agricultural Research (ICAR)–Indian Agricultural Research Institute (IARI), Pusa Campus, New Delhi, India; Indian Council of Agricultural Research (ICAR), Research Complex for North Eastern Hill Region, Tripura Center, Lembucherra, Agartala, Tripura, India

Megha Kumari
Department of Botany, Hansraj College, University of Delhi, Delhi, India

Giriraj Kumawat
Indian Institute of Soybean Research (ICAR–IISR), Indore, Madhya Pradesh, India

Sanjay Kumar Lal
Division of Genetics, Indian Council of Agricultural Research (ICAR)–Indian Agricultural Research Institute (IARI), Pusa Campus, New Delhi, India

Ponnuchamy Magudeeswari
School of Crop Improvement, College of Post Graduate Studies in Agricultural Sciences, Central Agricultural University, Imphal, Manipur, India

Rohit Kumar Mahto
Division of Genetics, Indian Council of Agricultural Research (ICAR)–Indian Agricultural Research Institute (IARI), New Delhi, India; School of Biotechnology, Institute of Science, Banaras Hindu University (BHU), Varanasi, Uttar Pradesh, India

Bingi Pujari Mallikarjuna
ICAR–Indian Agricultural Research Institute, Regional Research Center, Dharwad, Karnataka, India

Kalaiarasan Manivelan
Tamil Nadu Rice Research Institute, Tamil Nadu Agricultural University, Aduthurai, Tamil Nadu, India

Ashlesha Manta
Department of Botany, University of Delhi, New Delhi, Delhi, India

Priya Meghwal
Indian Institute of Soybean Research (ICAR–IISR), Indore, Madhya Pradesh, India

Riya Mishra
Department of Plant Breeding and Genetics, Jawaharlal Nehru Krishi Vishwa Vidyalaya, Jabalpur, Madhya Pradesh, India

Balamurugan Mohanapriya
Department of Genetics and Plant Breeding, Tamil Nadu Agricultural University, Coimbatore, Tamil Nadu, India

Satish Nagar
Indian Institute of Soybean Research (ICAR–IISR), Indore, Madhya Pradesh, India

K. Gopalakrishna Naidu
All India Coordinated Research Project (AICRP) on Soybean, University of Agricultural Sciences, Dharwad, Karnataka, India

Shrishti Naidu
Department of Botany, Hansraj College, University of Delhi, Delhi, India

Vennampally Nataraj
ICAR– IISR Indian Institute of Soybean Research, Indore, Madhya Pradesh, India

Aparna Nautiyal
Department of Botany, Deshbandhu College, University of Delhi, New Delhi, India

Byonkesh Nongthongbam
Department of Botany, University of Delhi, New Delhi, Delhi, India

Raja Sekhar Reddy Noru
Kerala Agricultural University, College of Agriculture Vellayani, Thiruvananthapuram, Kerala, India

M. Vani Praveena
Department of Genetics and Plant Breeding, Acharya N.G. Ranga Agricultural University, Andhra Pradesh, India

Raman Pushpa
Tamil Nadu Rice Research Institute, Tamil Nadu Agricultural University, Aduthurai, Tamil Nadu, India

Shivam Raj
Department of Life Sciences, Hansraj College, University of Delhi, New Delhi, Delhi, India

Ambika Rajendran
Division of Genetics, ICAR–Indian Agricultural Research Institute (IARI), Pusa Campus, New Delhi, Delhi, India

Dhandapani Raju
Division of Plant Physiology, ICAR–Indian Agricultural Research Institute, Pusa Campus, New Delhi, India

Ayyagari Ramlal
Division of Genetics, ICAR–Indian Agricultural Research Institute (IARI), Pusa Campus, New Delhi, India; School of Biological Sciences, Universiti Sains Malaysia (USM), Georgetown, Penang, Malaysia

Milind B. Ratnaparkhe
Indian Institute of Soybean Research (ICAR–IISR), Indore, Madhya Pradesh, India

Santoshi Rawat
National Institute of Food Technology, Entrepreneurship, and Management, Thanjavur, Tamil Nadu, India

Purnima Ray
Navinchandra Mafatlal College of Agriculture, Navsari Agricultural University, Navsari, Gujarat, India

Anjana Rustagi
Department of Botany, Gargi College, Delhi University, Delhi, India

Manisha Saini
Division of Genetics, Indian Agricultural Research Institute (ICAR)—Indian Agricultural Research Institute (IARI), Pusa Campus, New Delhi, India

Viswabharathy Sakthivel
Department of Genetics and Plant Breeding, Tamil Nadu Agricultural University, Coimbatore, Tamil Nadu, India

Palanivelu Sathya
Department of Genetics and Plant Breeding, Tamil Nadu Agricultural University, Coimbatore, Tamil Nadu, India

Aravindan Shanmugam
Department of Genetics and Plant Breeding, Tamil Nadu Agricultural University, Coimbatore, Tamil Nadu, India

Deepshikha Sharma
Division of Genetics, Indian Council of Agricultural Research (ICAR)–Indian Agricultural Research Institute (IARI), New Delhi, India; Department of Botany, University of Delhi, New Delhi, India

Divya Sharma
Department of Botany, University of Delhi, New Delhi, Delhi, India

M. Shivakumar
Indian Institute of Soybean Research (ICAR–IISR), Indore, Madhya Pradesh, India

M. K. Shrivastava
Department of Plant Breeding and Genetics, Jawaharlal Nehru Krishi Vishwa Vidyalaya, Jabalpur, Madhya Pradesh, India

Aakash Singh
Department of Botany, University of Delhi, New Delhi, Delhi, India

Archana Singh
Department of Botany, Hansraj College, University of Delhi, Delhi, India; Delhi School of Climate Change and Sustainability, Institution of Eminence, Maharishi Karnad Bhawan, University of Delhi, Delhi, India

Indrakant K. Singh
Department of Zoology, Molecular Biology Research Lab, Deshbandhu College, University of Delhi, Kalkaji, New Delhi, India

Nisha Sogan
Department of Botany, Deshbandhu College, University of Delhi, New Delhi, India

Sreeramanan Subramaniam
School of Biological Sciences, Universiti Sains Malaysia (USM), Georgetown, Penang, Malaysia; Chemical Center Biology (CCB), Universiti Sains Malaysia (USM), Bayan Lepas, Penang, Malaysia; Department of Biology, Faculty of Science and Technology, Universitas Airlangga, Surabaya, Indonesia

P. G. Suresha
DST–Agharkar Research Institute, Pune, Maharashtra, India

Meniari Taku
Division of Genetics, Indian Council of Agricultural Research (ICAR)–Indian Agricultural Research Institute (IARI), New Delhi, India

Akshay Talukdar
Division of Genetics, Indian Council of Agricultural Research (ICAR)–Indian Agricultural Research Institute (IARI), Pusa Campus, New Delhi, India

Manisha Yadav
Department of Botany, Hansraj College, University of Delhi, Delhi, India

Reviewers

Jessica Jayenthi James Antony
Department of Crop Science, Faculty of Agricultural Science and Forestry, UPM Bintulu Sarawak Campus, Malaysia

Pooja Baweja
Department of Botany, Maitreyi College, University of Delhi, Delhi, India

Pezangulie Chakruno
SASRD Nagaland University, Medziphema Campus, Nagaland, India

Godfree Chigeza
International Institute of Tropical Agriculture (IITA) Zambia, Lusaka, Zambia

Iten M. Fawzy
Department of Pharmaceutical Chemistry, Faculty of Pharmacy, Future University in Egypt, Cairo, Egypt

Praveen Gupta
Department of Botany, University of Lucknow, Lucknow, Uttar Pradesh, India

Nita Khandekar
Indian Institute of Soybean Research (ICAR–IISR), Indore, Madhya Pradesh, India

Kuldeep Kumar
Indian Institute of Pulse Research (ICAR–IIPR), Kanpur, Uttar Pradesh, India

Sanjay Kumar Lal
Division of Genetics, ICAR–Indian Agricultural Research Institute (IARI), Pusa Campus, New Delhi, India

Jayanthi Madhavan
Division of Genetics, ICAR–Indian Agricultural Research Institute (IARI), Pusa Campus, New Delhi, India

Sahil Mehta
Department of Botany, Hansraj College, University of Delhi, Delhi, India

Sweta Mishra
Department of Plant Breeding and Genetics, PG College of Agriculture, Dr. Rajendra Prasad Central Agricultural University, Pusa, Samastipur, Bihar, India

Gaurav Pal
Department of Plant and Microbial Biology, North Carolina State University, Raleigh, NC, United States

Ambika Rajendran
Division of Genetics, Indian Agricultural Research Institute (ICAR–IARI), Pusa Campus, New Delhi, India

Shubhra Rajput
Department of Biological Sciences, State University of New York, Old Westbury, New York, United States; Institute of Chemical Biology and Drug Discovery, Stony Brook University, New York, United States

Dhandapani Raju
Division of Plant Physiology, ICAR–Indian Agricultural Research Institute (IARI), Pusa Campus, New Delhi, India

Ayyagari Ramlal
School of Biological Sciences, Universiti Sains Malaysia (USM), Georgetown, Penang, Malaysia; Division of Genetics, Indian Council of Agricultural Research (ICAR)–Indian Agricultural Research Institute (IARI), Pusa Campus, New Delhi, India

Beena M. Ravindran
Saaten-Union Biotec GmbH, Hovedisser Street, Leopoldshöhe, Germany

Vishnu Vardhan Reddy
International Institute of Tropical Agriculture (IITA), Ibadan, Nigeria

Shukla Saluja
Department of Botany, Sri Venkateswara College, University of Delhi, New Delhi, India

Abdel Nasser B. Singab
Department of Pharmacognosy, Faculty of Pharmacy, Ain Shams University, Cairo, Egypt; Center of Drug Discovery Research and Development, Ain Shams University, Abbasia, Cairo, Egypt

Virendra Pal Singh
Division of Plant Physiology, ICAR–Indian Agricultural Research Institute (IARI), Pusa Campus, New Delhi, India

Philips Varghese
Retired Scientist and Soybean Breeder, Maharashtra, India

Abbreviations

2,4-D	2,4-dichlorophenoxyacetic Acid
AA	amino acid
AA	ascorbic acid
AAD-12	aryloxyalkanoate dioxygenase-12
ABA	abscisic acid
AC	activated charcoal
ACC	1-aminoacyclopropane 1-carboxylate synthase
AM	arbuscular mycorrhizal
AMF	arbuscular mycorrhizal fungi
AMV	alfalfa mosaic virus
B5	gamborg medium
B5BKN	B5-BA-Kin-NAA
B5DBIG	B5-2,4-D-IBA-Myoinositol-Glutamine
B5DK	B5-2,4-D-Kin
BAP	6-benzylaminopurine
BBI	Bowman-Birk Protease Inhibitor
bla	β-Lactamase
BPMV	bean pod mottle virus
CaMV	cauliflower mosaic virus
CAT	catalase
CHR	chalcone reductase
Chr.	chromosome
CIM	composite interval mapping
CNV	copy number variation
COR	cold regulator genes
CRISPR	clustered regularly interspaced short palindromic repeat
CRISPRi	CRISPR interference
CVD	cardiovascular disease
DAS	days after sowing
DCP	2,4-dichlorophenol
DE	differentially expressed
DH	doubled haploid
DMO	dicamba monooxygenase
DOE-JGI	Department of Energy Joint Genome Institute
DOM	days to maturity
DSB	double-strand break

DSCA	3,6-dichloro salicylic acid
DTM	days to maturity
EDS1	enhanced disease susceptibility
EMS	ethyl methane sulfonate
EPSPS	5-dnolpyruvylshikimate-3-phosphate synthase
ER	endoplasmic reticulum
ER	estrogen receptor
ERF	ethylene response factor
Erα	estrogen receptor alpha
ERβ	estrogen receptor beta
EST	expressed sequence tag
ETI	effector triggered immunity
FA	fatty acid
FAS	fatty acid synthesis
FDA	Food and Drug Administration
FN	fast neutron
FT-NIR	Fourier Transform Near-Infrared Spectroscopy
GA3	gibberellic acid (A3)
GABA	gamma-aminobutyric acid
Gb	giga base
GC	gas chromatography
GE	genome editing
GEA	genome-environment association
GGA	glutamate glyoxalase aminotransferase
GHG	greenhouse gas
GI	glycemic index
Gln	glutamine
Glu	glutamate
GM	genetically modified
GO	glycolate oxidase
GS	genomic selection
GT	glyphosate-tolerant
GUS	beta-glucuronidase
GWAS	Genome-Wide Association Studies
H_2O_2	hydrogen peroxide
HPLC	high-performance liquid chromatography
HPPD	4-hydroxyphenylpyruvate dioxygenase
hr	hour
HR	hypersensitivity response
HRNMR	high-resolution nuclear magnetic resonance
HT	herbicide tolerance
HTP	high-throughput phenotyping

IBA	indole-3-butyric acid
IFS	isoflavone synthase
IPM	integrated pest management
Kin	kinetin
KTI	Kunitz Trypsin Inhibitor
LC-MS	liquid chromatography-mass spectrometry
LDL	low-density lipoprotein
LEA	late embryogenesis abundant
LG	linkage group
LIS	legume information service
LJ	long juvenile trait
LL	liberty link®
LPL	lipoprotein lipase
LRR	leucine-rich repeat
LTR	long terminal repeat
MABC	marker-assisted back-crossing
MAF	minor allele frequency
MAPK	mitogen-activated protein kinase
MARS	marker-assisted recurrent selection
MAS	marker assisted selection
MBB	mexican beetle
MCIM	mixed-model-based composite interval mapping
MDA	malondialdehyde
MDA	monodehydroascorbate
mg L^{-1}	milligram per liter
MS/MSO	Murashige and Skoog Medium
MSA	multiple sequence alignment
MSM6	Finer and McMullen Medium
MYMIV	mungbean yellow mosaic india virus
NAA	naphthalene acetic acid
NBS-LRR	nucleotide-binding site-leucine rich repeat
NDR1	non-race specific disease resistance-1
NGS	next-generation sequencing
NHEJ	non-homologous end-joining
NHX	na^+-h^+ exchanger
NIL	near-isogenic line
NN	Nitsch and Nitsch Medium
NPBT	new plant breeding technologies
NSPP	number of seeds per pod
P	phosphorus
PAL	phenylalanine ammonia-lyase
PAMP	pathogens associated molecular pattern

PAT	phosphinothricin-n-acetyltransferase
PAV	presence-absence variation
PC	phosphatidylcholine
PCD	programmed cell death
PDCAAS	protein digestibility-corrected amino acid score
PEP	phosphoenol pyruvate
PG	phosphoglycolate
PGA	3-phosphoglycerate
PGen	pegasus genomic
PGR	plant growth regulator
PH	plant height
PI	protease inhibitor
Pic	picloram
PLH	potato leafhopper
PM	plasma membrane
PMN	plant metabolic network
PP	photoperiod
PRI	plant resistance to insect
PRR	pattern recognition receptor
PSV	peanut stunt virus
PTC	plant tissue culture
PTI	PAMP triggered immunity
PVP	plant variety protection
QDR	quantitative disease resistance
QDRL	QDR loci
QTL	quantitative trait loci
QTL	quantitative trait locus
QTN	quantitative trait nucleotide
RBD	randomized block design trial
RFLP	restriction fragment length polymorphism
RGC	ratio of glycinin to β-conglycinin
RIL	recombinant inbred line
RKN	root-knot nematode
RNAi	RNA interference
ROS	reactive oxygen species
RR	roundup ready
RRX	roundup ready 2 xtend®
RWC	relative water content
S3P	shikimate-3-phosphate
SA	salicylic acid
SAR	systemic acquired resistance
SBA	soybean aphid

SbDV	soybean dwarf virus
SBL	soybean looper
SCN	soybean cyst nematode
SCN	soybean cyst nematode disease
SDS	sudden death syndrome
SE	somatic embryogenesis
SGC	sum of glycinin and β-conglycinin
SGE	soy germ extract
SH	seed height
SL	seed length
SM	soybean meal
SMV	soybean mosaic virus
SNP	single nucleotide polymorphism
SOD	superoxide dismutase
SSD	single-seed descent method
SSR	simple sequence repeat
Suc	sucrose
SV	seed volume
SV	structural variation
SVNV	soybean vein necrosis virus
SW	seed width
SW/plant	seed weight per plant
TAG	triacylglycerol
TALEN	transcription activators like effector nuclease
tAPX	ascorbate peroxidase
TCM	Traditional Chinese Medicine
TDZ	thidiazuron
TF	transcription factor
TM	traditional medicine
TRG	glycine betaine, trigonelline
TRSV	tobacco ringspot virus
USDA	U.S. Department of Agriculture
USDA-ARS	U.S. Department of Agriculture-Agricultural Research Service
VBC	velvet bean caterpillar
VNTR	variable number of tandem repeat
WGD	whole genome duplication
WGRS	whole-genome resequencing
Wm82	Williams 82
YS	Yeung and Sussex
YSaa	Yeung and Sussex Amino Acid
ZFN	zinc finger nuclease

Foreword

Soybean (*Glycine max* (L.) Merr.; Fabaceae) is a functional food crop widely used as both a food and feed crop, with a chromosome number of 2n = 2x = 40. It is a self-pollinating, short-day, annual crop that contains high amounts of protein (40%), fats (20%), and oil. It also has a well-balanced composition of macronutrients, essential amino acids, micronutrients, and minerals. Two-thirds of the world's soybean protein is utilized for feeding livestock, while 25% is used in the production of edible oil globally.

Soybeans are widely used for various culinary purposes, including soy paste, soy milk, miso, tofu, and natto (traditional products), along with soybean meal and oil (Western products). Because of its nutritional and economic significance, soybean is often referred to as "gold from the soil" and "cow of the field."

Soybean produces a plethora of compounds with therapeutic applications, including storage proteins (which help in managing hypocholesterolemia and chronic kidney diseases), trypsin inhibitors and lectins (which have anticancer properties), lipids (such as linoleic acid and alpha-linolenic acid), carbohydrates, and other minor constituents. Given its diverse benefits, soybean has great potential to replace many existing food crops. It is also known for its medicinal properties and can be used in treating various diseases, including cardiovascular diseases, contributing to the development of safer drugs.

In cropping systems, soybean is a valuable source of nitrogen, as it has the ability to fix atmospheric nitrogen through *Rhizobium* species. It also possesses other agronomically beneficial traits. However, the crop is vulnerable to various abiotic and biotic stresses. For instance, soybean is highly prone to water stress during germination, emergence, and grain-filling stages.

Soybean offers a wide range of opportunities for genetic manipulation to enhance yield and other agronomic traits. Genetic improvements for yield include lodging resistance, stability across diverse environments, tolerance to both water stress and excess moisture, adaptation of improved cultivars and production methods, increased atmospheric CO_2

concentration, enhanced nitrogen fixation, efficient allocation of assimilates during the seed-filling period, and tolerance to high plant populations, as well as resistance to major pests and pathogens.

Volume 1 emphasizes various aspects of soybean taxonomy, domestication, and evolution, while Volume 2 provides insights into current trends in genomic information, conventional breeding techniques, associated challenges, and the need for modern breeding methods. This book takes a holistic approach, covering a wide range of research areas and topics of interest, ultimately paving the way for future advancements in crop improvement.

—Dr. Viswanathan Chinnusamy
Joint Director (Research),
Indian Agricultural Research Institute (IARI), New Delhi, India

Preface

Plants are continuously exposed to various environmental stresses. Being sessile organisms, they must develop adaptive strategies to overcome harsh climatic conditions and stressors. These stresses limit the full potential of crops to achieve maximum yield and productivity. Crop failures are a major concern, and the need to produce more food to meet the demands of a growing population has increased exponentially. Crops face diverse abiotic stresses due to fluctuating climate conditions, which are, in turn, influenced by anthropogenic activities. Therefore, understanding a crop's physiology and disease resistance is essential for improving productivity and meeting global food demands.

Soybean is often regarded as a "wonder crop" and is also referred to as "gold from the soil." It is both an oilseed and a legume crop of significant economic and nutritional importance.

With these objectives in mind, this book is designed to provide broad coverage of soybean and its utility, including its history and domestication, development of abiotic stress tolerance, genomics, and biotechnology. Efforts have been made to encompass the vast scope of soybean research, though the rapidly increasing knowledge in this field makes it impossible to include all relevant material in an entry-level text. Consequently, in light of current advancements, this book covers a wide range of topics, including molecular biology, environmental physiology, agronomy, and genetics, making it a valuable resource for agriculturalists, plant physiologists, biotechnologists, and, in particular, soybean breeders.

This edition comprises four parts: *Soybean Genomics, Proteomics, and Tools; Improvements in Soybean*; *Soybean Diseases and Biotechnological Interventions*; and *Soybean in Medicine*. Every effort has been made to ensure readability and maintain a structured approach.

—Editors

PART I

Soybean Genomics, Proteomics, and Tools

CHAPTER 1

Soybean Genetics Emphasizing Chromosome and Genome Organization

MENIARI TAKU,[1,*] PULAK DEBBARMA,[1] RAHUL KUMAR,[1,2] MANISHA SAINI,[1] AYYAGARI RAMLAL,[1,3] and AKSHAY TALUKDAR[1]

[1]*Division of Genetics, Indian Council of Agricultural Research (ICAR) – Indian Agricultural Research Institute (IARI), Pusa Campus, New Delhi, India*

[2]*Indian Council of Agricultural Research (ICAR) Research Complex for North Eastern Hill Region, Tripura Center, Lembucherra, Agartala, Tripura, India*

[3]*School of Biological Sciences, Universiti Sains Malaysia (USM), Georgetown, Penang, Malaysia*

*Corresponding author

ABSTRACT

Soybean (*Glycine max* (L.) Merr.) is a tetraploid with partially diploidized characteristics, possessing 20 pairs of chromosomes (2n = 40) resulting from two rounds of genome-wide duplication. Cytogenetically, soybean exhibits diploid-like behavior and showcases disomic inheritance for most genetic loci. The haploid soybean genome, comprising approximately 1 gigabase (Gb), is organized into 20 chromosome pairs, with an estimated 40–60% of the DNA being repetitive. The soybean genome structure has been intricate due to a polyploidy event followed by subsequent genomic rearrangements, expansion of pericentromeric regions, and instances of gene losses and duplications. Notably, the genome is characterized by large pericentromeric regions that are gene-poor yet repeat-dense, covering

Soybeans, Volume 2: Genome, Biotechnology, Stress, and Disease.
R. Ambika Rajendran, Ayyagari Ramlal, Sanjay Kumar Lal, and Godfree Chigeza (Eds.)

DOI: 10.1201/9781998511730-2

around 57% of the genome sequence. The availability of the soybean genome sequence has facilitated the placement of various genomic elements, both internal to soybean and from closely related species. Despite these evolutionary changes, considerable conservation persists between soybean and its cultivated bean relatives, offering insights into the evolutionary history spanning approximately 13 million years since the occurrence of the polyploidy event.

1.1 INTRODUCTION

Soybean (*Glycine max* (L.) Merr.) is an ancient polyploid (paleopolyploid) and is highly valued for its seed protein and oil content. Over time, soybean has experienced multiple genome duplications and subsequent re-diploidization events, accompanied by chromosome rearrangements, marking it as a paleopolyploid species. These genomic complexities have led to the presence of up to 12 occurrences of specific genome regions within soybean (Cannon et al., 2012; Doyle et al., 2010; Petereit et al., 2022; Pfeil et al., 2005; Schmutz et al., 2010). Notably, soybean holds the distinction of being the first legume species to have its complete genome sequenced, serving as a vital reference point for the broader legume family, which encompasses over 20,000 species. Archaeological findings, including charred soybean specimens, strongly suggest that the Yellow River basin in China played a pivotal role in the domestication of soybean (Lee et al., 2011; Zhao, 2004). It is widely accepted that modern cultivated soybean originated from its wild counterpart, *Glycine soja* Sieb. & Zucc., through a domestication process that took place in East Asia around 6000–9000 years ago (Kim et al., 2012). Despite its significance, the cytogenetic understanding of soybean remains relatively underdeveloped compared to other major crops such as maize, barley, rice, wheat, tomato, brassicas, pea, and faba bean. Soybean, being a prized agricultural commodity, is expected to encounter growing demands for enhancement due to the challenges posed by global climate change, similar to other agricultural products. The reference genome assembly of cultivated soybean originates from a northern U.S. cultivar named Williams 82 (Wm82), developed through collaborative efforts between the soybean research community and the Department of Energy Joint Genome Institute (DOE-JGI) (Schmutz et al., 2010). Traditionally, soybean breeders have primarily focused on enhancing soybeans by utilizing landraces and occasionally incorporating

G. soja. However, there has been limited exploration of wild perennial glycine species to broaden the genetic diversity of soybean. Nevertheless, recent advancements in whole-genome sequencing of both cultivated and wild soybeans have provided insights into the structural, functional, and comparative genomics of this vital crop and its relatives. The availability of complete genome sequences has enabled a deeper comprehension of soybean's genome organization, thereby fueling increased research into its structural and functional genomics. Studies on chromosome organization and genome sequences have facilitated an enhanced understanding of the genetic underpinnings of various soybean traits, expediting the development of improved soybean varieties. This chapter will focus on a brief discussion of soybean chromosomes and genome organization.

1.2 CHROMOSOME ORGANIZATION

The soybean genome is around 1.1 giga base pairs in size, which makes it one of the larger plant genomes. Soybean has 20 pairs of chromosomes (2n = 40) that are derived from at least two rounds of genome-wide duplication, with 75% of the genes occurring as multiple copies. The soybean genome is composed of approximately 46,000 protein-coding genes, which is more than twice the number of genes found in the human genome. The genome was sequenced using a whole-genome shotgun method and combined with physical and high-density genetic maps to provide a draft sequence assembly at the chromosomal level. They predict 46,430 protein-coding genes. About 78% of the predicted genes occur in chromosome ends, which comprise less than one-half of the genome but account for nearly all of the genetic recombination (Schmutz et al., 2010). The *G. max* has 40 chromosomes (2n = 40), which are small in size (1.42–2.84 μm) with similar and distinguishing morphology. The genus *Glycine* consists of subgenera – *Soja* and *Glycine*. The subgenus *Soja* consists of two species: *G. max* (2n = 2x = 40), the cultivated species, and *G. soja* (L.) Sieb or *G. ussuriensis* (2n = 2x = 40), a wild species (Poehlman and Poehlman, 1987). These two species show cross-fertilization or alogamy. There are 15 wild species of soybean for which *G. tabacina* and *G. tomentella* have polyploid forms (including 2n = 4x = 80) (Poehlman and Poehlman, 1987). *G. max* has never been found growing wild, and its most probable progenitor among the known species is *G. soja*. About 15,000 *G. max* accessions and a smaller number of other

Glycine species are kept at Urbana, Illinois, and Stoneville, Mississippi, respectively, by the USDA. Soybean accessions are also maintained by the Applied Genetics Department of the Korea Atomic Energy Research Institute in Seoul, Korea, as well as by the National Institute of Agricultural Sciences in Hiratsuka, Japan, and the Asian Vegetable Research and Development Center in Taiwan.

Cytogenetically, soybean behaves like a diploid and has disomic inheritance for most loci, even though it has gone through a few rounds of genome-wide duplication or polyploidy events. This was made possible with the help of a process called diploidization (Walling et al., 2006). Small or large chromosome-level rearrangements, that is, inversions, translocations, and deletions, resulted in diploidization, which led to disomic pairing in soybean (Levy and Feldman, 2002). Cytogenetic investigations in soybean have lagged behind many major crops such as maize, wheat, and tomato, despite its economic importance for human consumption, livestock feed, and industrial use. The main reason is the large and symmetrical number of chromosomes, which lack morphological distinguishing landmarks. Only a pair of nucleolus organizer chromosomes is occasionally visible, and this has been verified by FISH (Krishnan et al., 2001). A few attempts to use mitotic metaphase chromosomes of soybean to construct chromosome maps were not successful. The mitotic and meiotic chromosomes are unsuitable for modern cytogenetic studies because they are quite small, only 4–6 μm in length. However, pachytene chromosomes are much less compacted (Singh and Hymowitz, 1988) and are very useful for molecular cytogenetics. The length of pachynema chromosomes ranges from 39.79 μm (chromosome 1) to 10.63 μm (chromosome 20). Each pair is unique with characteristic landmark features: (i) Seven short arms (chromosomes 5, 7, 10, 13, 18, 19, 20) are heterochromatic; (ii) chromosome 2 contains the least (14.22%) heterochromatin, chromosome 5 contains the shortest heterochromatic short arm and is easy to identify from other chromosomes, chromosome 7 consists of 59.44% heterochromatin, and chromosomes 18 and 19 are difficult to distinguish, but these chromosomes are very easy to separate from the clump of pachynema chromosomes. Chromosome 13 has satellite DNA that is completely embedded in the nucleolus, and the short arm is completely heterochromatic. Genetic mapping in soybean has evolved significantly, progressing from techniques such as restriction fragment length polymorphisms (RFLPs) to simple sequence repeats (SSRs) and single nucleotide polymorphisms (SNPs). Studies have revealed evidence of long-range interchromosomal homology, although not entirely collinear, suggesting intrachromosomal rearrangements that align with the concept of chromosome-level diploidization. Soybean chromosomes predominantly feature highly euchromatic arms with minimal medium to high copy dispersed repeats, which are mostly concentrated in the centromeric and pericentromeric regions. This observation implies that euchromatic regions are selective against high-copy repeats, either through exclusion or efficient elimination mechanisms. In euchromatic regions, the average ratio of genetic to physical distance stands at approximately 1 cM per 197 kb, contrasting with 1 cM per 3.5 Mb in heterochromatic regions. Although classical cytogenetics has associated some linkage groups (LGs) with specific chromosomes, the extent of chromosome coverage by molecular LGs remains unclear. Notably, approximately 80% of predicted genes are located in chromosome ends, accounting for nearly all genetic recombination despite constituting less than half of the genome. The distribution of SSR markers and putative genes indicates a higher concentration near telomeres compared to centromeres. Despite the pivotal role of centromeres in chromosomal segregation during cell division, DNA sequences within these regions exhibit limited conservation, even within a single genus.

1.3 GENOME ORGANIZATION

After the sequencing of the reference cultivar Wm82, the first observation about the genome was its size. The soybean genome, with a haploid size of approximately 1100 Mb and organized into 20 pairs of chromosomes, underwent whole-genome duplication (WGD) around 5 to 13 million years ago (Mya), resulting in a partially diploidized tetraploid state. This genome evolution involved a diploid ancestor (n = 11), which experienced aneuploid loss (n = 10), followed by polyploidization (2n = 40) and subsequent diploidization (n = 20). The evidence of WGD is evident in the soybean genome's comparison with itself, revealing regions of internal synteny where genes are found in the same order and orientation across different parts of the genome. However, these synteny blocks are occasionally disrupted by small insertions, deletions, or inversions, reflecting the numerous rearrangements that have occurred since the WGD event. The soybean genome structure is complicated by duplicated regions that have been segmented and reshuffled, leaving a substantially duplicated genome with nearly 75% of the genes present in numerous copies. According to Shoemaker et al. (1996), hybridization of RFLP probes revealed that the soybean genome contains most probes duplicated an average of 2.6 times, leading to its classification as a "paleopolyploid" genome. This phenomenon of polyploidization, common in genome evolution, is often followed by a diploidization process, as observed in grasses (Levy and Feldman, 2002), which restores diploid-like pairing and disomic genetics. It is speculated that the relatively large genome size and high gene count in soybean may be attributed to recent polyploidy events in its evolutionary history. Chromosome studies conducted in the genus *Glycine* have indicated that soybean possesses approximately twice the number of chromosomes compared to its closest relatives (Doyle et al., 2004). It has been proposed that the presence of two primary types of centromeric heterochromatin repeat sequences within the soybean genome suggests the amalgamation of two cytologically distinct progenitor genomes to form the duplicated *Glycine* ancestral genome, indicating allopolyploidy in *Glycine*. As a result of soybean's duplication history, most genes are present in duplicate or more, even within small gene families. This duplication can complicate standard laboratory procedures, as the paralogs resulting from the duplication are often quite similar. Consequently, gene discovery via knockout methods may prove more challenging in soybean compared to less duplicated plant genomes. Another notable characteristic of the soybean genome is the substantial and distinct pericentromeric regions present in all chromosomes, constituting approximately 57% of the current assembly. These regions are rich in repeats and low in genes, with significantly suppressed rates of recombination. However, despite their gene-poor nature relative to the euchromatic chromosome arms, the pericentromeres collectively harbor a substantial number of genes, comprising more than 21% of the predicted high-confidence genes. Approximately 93% of recombination events occur within the repeat-poor, gene-rich regions of the soybean genome, which represent only 43% of its total size. Interestingly, around 21.6% of confidently identified genes are located in regions abundant in repeats and transposons, primarily situated in the central portions of chromosomes. The soybean genome exhibits a remarkable feature in the preservation of duplicated gene blocks, a consequence of multiple duplication events and subsequent diploidization, along with chromosomal rearrangements. These duplicated gene blocks often involve more than two chromosomes, with an average of 61.4% of homologous genes residing in blocks spanning just two chromosomes. However, a small percentage (5.63%) extends across three chromosomes, while 21.53% traverse four chromosomes. Notably, chromosome 14 displays extensive fragmentation, with matches to 14 other chromosomes, the highest among all chromosomes. Conversely, chromosome 20 shows a high degree of homology with the long arm of chromosome 10, with minimal matches elsewhere in the genome.

1.4 FUNCTIONAL GENOMICS

Around half of the paralogs resulting from the WGD in soybean have experienced sub-functionalization at the expression level, highlighting the regulatory impact of WGD in soybean. Interestingly, despite the occurrence of WGD, genes within highly connected networks demonstrated greater conservation compared to those with lower connectivity. Moreover, WGD genes displayed a decreased frequency of alternative splicing relative to non-WGD genes, which was associated with reduced intron length, exon number, and transcriptional activity. The influence of WGD extended to the evolution of transposable elements as well. DNA methylation profiling analyses revealed hypomethylation patterns affecting the expression of adjacent genes, with CG body-methylated genes being notably abundant in duplicated genes exhibiting higher expression levels than single-copy genes. Additionally, the dynamics of DNA demethylation/methylation were found to play crucial roles in stress responses, including those triggered by continuous cropping (Liang et al., 2019), salinity stress (Song et al., 2012), and cyst nematode infection (Rambani et al., 2020). A recent comprehensive examination of miRNA distribution in the soybean genome, along with their family organization, mature sequence diversity, and target prediction, indicated an active evolution of the soybean genome with continuous functional diversification.

1.5 COMPARATIVE GENOMICS

A comparative analysis of multiple reference genomes representing diverse genetic backgrounds is essential for comprehending the significance of key alleles in the domestication and genetic improvement of vital crops like soybean. The soybean genome spans approximately 1 Gb and contains approximately 64,000 protein-coding genes, which is considerably larger than the Arabidopsis genome (125 Mbp) and nearly two and a half times the size of the genomes of model legumes such as *Medicago truncatula* and *Lotus japonicus*, as well as the grape genome (each approximately 450–470 Mbp). Moreover, the soybean genome is roughly double the size of the common bean and poplar genomes (625 and 550 Mbp, respectively), but less than half the size of the maize genome (2300 Mbp). In terms of predicted coding genes, soybean also exhibits a relatively high count, with approximately 46,400 coding genes, compared to approximately 26,500 in *Arabidopsis*, approximately 30,400 in grape, and approximately 45,000 in poplar. Approximately 57% of the genomic sequence in soybean is located within repeat-rich, low-recombination heterochromatic regions encircling the centromeres. This proportion is similar to that observed in sorghum, where 62% of the sequence is heterochromatic, but differs from rice, where only 15% resides in heterochromatin. The demarcation of these regions is primarily dictated by suppressed recombination and corresponds to shifts in gene density and transposon density. Among transposable elements, long terminal repeat (LTR) retrotransposons are the most prevalent in soybean, constituting 42% of the genome. While this proportion is lower than in sorghum and maize, it surpasses that of rice. The availability of the soybean genome sequence has established a common reference framework for genomic features, including genes, regulatory elements, transposons, and markers, across soybean and related species. Consequently, various genome browsers have been developed, such as SoyBase, Legume Base, Phytozome, and SoyKB, each offering distinct specializations and functionalities. There is notable genome conservation observed between *Glycine* (soybean) and other species within the Phaseoleae family. For instance, soybean exhibits extensive synteny with cowpea, where the entirety of cowpea chromosome 5 aligns with soybean chromosome 14 (Gm14), as well as with homoeologous segments on Gm02 and Gm01. Additionally, common bean displays considerable synteny with soybean. However, certain barriers hinder direct comparisons between soybean and other species within the Phaseoloid clades. One notable difference is the significant expansion of pericentromeric regions in soybean over the past approximately 10 million years, potentially resulting in transpositions or unexpected rearrangements. Despite the occasional loss of synteny in certain regions between soybean and other Phaseoloid genomes, the overall similarity between soybean and other species in this clade remains high enough to readily identify orthologs, irrespective of genomic context. A comparative genetic analysis conducted on three *G. max* accessions: Wm82, the southern U.S. accession Lee (PI 548656), and *Glycine soja* accession PI 483463 revealed a high level of conservation among these accessions, with variations predominantly observed in genomically clustered gene families. Notably, there are approximately 40–42 inversions per chromosome between either Lee or Wm82 and *G. soja*, and roughly 32 inversions per chromosome between Wm82 and Lee. Over the past approximately 10 million years, there has been a significant expansion of pericentromeric regions in soybeans, as evidenced by genomic comparisons. Within these regions, numerous genes have been lost in one or the other homoeologous regions, and the remaining genes have been separated due to transposon insertions. Chang et al. (2013) developed the initial linkage map for *G. latifolia*, a perennial species. Their mapping efforts with the *G. max* genome sequence revealed a 70% failure rate in aligning the reads. While the soybean genome serves as a valuable reference and enables the mapping of perennial sequences to identify a sufficient number of SNPs for future fine-mapping studies, the creation of a specific perennial genome reference sequence would be advantageous.

1.6 PANGENOME ANALYSIS

Although the soybean reference genome (Wm82) is considered one of the most comprehensive plant reference genomes available, it does not fully encompass the genetic diversity present within the species. This limitation has spurred interest in developing a pan-genome for soybeans, which would provide a more comprehensive representation of both structural and nucleotide variation across different individuals. A pan-genome encompasses the complete set of genes within a species and typically includes a core set of genes shared by most or all individuals, as well as variable genes that are only partially shared among individuals. Notably, over 90% of the observed length variation in assembled genomes is attributed to presence-absence variations (PAVs), underscoring the significant role of PAVs in driving genome size variation. Core genes are defined as the total gene set present in all accessions, while softcore genes are shared by over 90% of the collection. Dispensable genes are those present in some accessions, while private genes are exclusive to a single accession. Following a pangenome analysis involving 29 accessions, it was observed that wild soybeans exhibited a higher proportion of private structural variations (SVs) (averaging 22.2%) compared to cultivated soybeans (averaging 6.7%). Examination of the sequence composition of each PAV revealed that 78.5% of the PAVs originated from repetitive DNA, supporting the hypothesis that variations in repetitive sequences largely contributed to genome divergences (Kumar and Bennetzen, 1999). The combined total of dispensable and private gene families constituted a significant portion of the overall gene sets, averaging 19.1% across individual accessions. WGD regions demonstrated a higher proportion of core and soft-core genes, while non-WGD regions exhibited a higher ratio of dispensable and private genes. Nucleotide diversity within the WGD regions from the 29 genomes was notably lower compared to non-WGD regions. Additionally, WGD regions harbored fewer private SVs than non-WGD regions, suggesting that genome duplication not only constrained gene evolutionary rates but also played a crucial role in shaping SV evolution. Thus, the pan-genome serves as a comprehensive resource for understanding genome composition and offers substantial benefits to the soybean breeding and research community.

1.7 CONCLUSION

The comprehension of the soybean genome, comprising billions of nucleotides and their organization, provides opportunities for enhancing crops and expanding genetic diversity through crossbreeding with related species. By unraveling the genetic basis of individual traits and deciphering the regulatory networks governing these traits, we can engineer crops to meet specific objectives. With the global population on the rise, there is a pressing need to accelerate the rate of soybean yield improvement. Chromosomal and genomic data serve as fundamental frameworks for generating extensive new experimental insights, including tissue-specific gene expression patterns and comprehensive genome-wide association studies (GWAS).

KEYWORDS

- **disomic inheritance**
- **genetic diversity**
- **genome-wide association studies**
- **genome-wide duplication**
- **pericentromeric regions**
- **polyploidy**
- **soybean genome**

REFERENCES

Arumuganathan, K., & Earle, E. D. (1991). Estimation of nuclear DNA content of plants by flow cytometry. *Plant Molecular Biology Reporter, 9*(3), 229–241.

Blanc, G., & Wolfe, K. H. (2004). Widespread paleopolyploidy in model plant species inferred from age distributions of duplicate genes. *Plant Cell, 16*, 1667–1678.

Cannon, S. B., & Shoemaker, R. C. (2012). Evolutionary and comparative analyses of the soybean genome. *Breeding Science, 61*, 437–444.

Doyle, J. J., & Egan, A. N. (2010). Dating the origins of polyploidy events. *New Phytologist, 186*, 73–85.

Goldberg, R. B. (1978). DNA sequence organization in the soybean plant. *Biochemical Genetics, 16*, 45–68.

Gurley, W. B., Hepburn, A., & Key, J. L. (1979). Sequence organization of the soybean genome. *Biochimica et Biophysica Acta, 561*, 167–183.

Hymowitz, T. (1970). On the domestication of the soybean. *Economic Botany, 24*, 408–421.

Lee, J. M., Grant, D., Vallejos, E. C., & Shoemaker, R. C. (2001). Genome organization in dicots. II. *Arabidopsis* as a 'bridging species' to resolve genome evolution events among legumes. *Theoretical and Applied Genetics, 103*, 765–773.

Li, Y. H., Zhou, G., Ma, J., Jiang, W., Jin, L. G., Zhang, Z., et al. (2014). De novo assembly of soybean wild relatives for pan-genome analysis of diversity and agronomic traits. *Nature Biotechnology, 32*, 10.

Liu, Y., Du, H., Li, P., Shen, Y., Peng, H., Liu, S., et al. (2020). Pan-genome of wild and cultivated soybeans. *Cell, 182*(1), 162–176.

Ott, A., Trautschold, B., & Sandhu, D. (2011). Using microsatellites to understand the physical distribution of recombination on soybean chromosomes. *PLoS ONE, 6*(7), e22306.

Pfeil, B. E., Schlueter, J. A., Shoemaker, R. C., & Doyle, J. J. (2005). Placing paleopolyploidy in relation to taxon divergence: A phylogenetic analysis in legumes using 39 gene families. *Systematic Biology, 54*(3), 441–454.

Poehlman, J. M. (1987). Breeding soybeans. In J. M. Poehlman (Ed.), *Breeding Field Crops* (pp. 421–450).

Probst, A. H., Judd, R. W., & Caldwell, B. E. (1973). Soybeans: Improvement, production, and uses. In B. E. Caldwell (Ed.), *Agronomy, 1*, 1–15.

Schlueter, J. A., Dixon, P., Granger, C., Grant, D., Clark, L., Doyle, J. J., & Shoemaker, R. C. (2004). Mining EST databases to resolve evolutionary events in major crop species. *Genome, 47*, 868–876.

Schmutz, J., Cannon, S. B., Schlueter, J., Ma, J., Mitros, T., Nelson, W., et al. (2010). Genome sequence of the palaeopolyploid soybean. *Nature, 463*(7278), 178–183.

Sedivy, E. J., Wu, F., & Hanzawa, Y. (2017). Soybean domestication: The origin, genetic architecture and molecular bases. *New Phytologist, 214*, 539–553.

Sherman-Broyles, S., et al. (2014). The wild side of a major crop: Soybean's perennial cousins from down under. *American Journal of Botany, 101*(10), 1651–1665.

Shoemaker, R. C., Polzin, J., Labate, J., Specht, J., Brummer, E. C., Olson, T., Young, N., Concibido, V., Wilcox, J., Tamulonis, P., Kochert, G., & Boerma, H. R. (1996). Genome duplication in soybean (*Glycine subgenus soja*). *Genetics, 144*, 329–338.

Singh, R. J., & Hymowitz, T. (1988). The genomic relationship between *Glycine max* (L.) Merr. and *G. soja* Sieb. and Zucc. as revealed by pachytene chromosomal analysis. *Theoretical and Applied Genetics, 76*, 705–711.

Tian, A. G., et al. (2004). Characterization of soybean genomic features by analysis of its expressed sequence tags. *Theoretical and Applied Genetics, 108*(5), 903–913.

Torkamaneh, D., Lemay, M., & Belzile, F. (2021). The pan-genome of the cultivated soybean (PanSoy) reveals an extraordinarily conserved gene content. *Plant Biotechnology Journal, 19*, 1852–1862.

Turner, M., Oliver, Y., & Subramanian, S. (2012). Genome organization and characteristics of soybean microRNAs. *Genomics, 13*, 169.

Valliyodan, B., Cannon, S. B., Bayer, P. E., Shu, S., Brown, A. V., Ren, L., et al. (2019). Construction and comparison of three reference-quality genome assemblies for soybean. *The Plant Journal, 100*, 1066–1082.

Walling, J. G., Shoemaker, R., Young, N., Mudge, J., & Jackson, S. (2006). Chromosome level homeology in paleopolyploid soybean (*Glycine max*) revealed through integration of genetic and chromosome maps. *Genetics, 172*, 1893–1900.

Yan, H. H., Mudge, J., Kim, D. J., Larsen, D., Shoemaker, R. C., Cook, D. R., & Young, N. D. (2003). Estimates of conserved microsynteny among the genomes of *Glycine max, Medicago truncatula*, and *Arabidopsis thaliana*. *Theoretical and Applied Genetics, 106*(7), 1256–1265.

Zhang, M., Liu, S., Wang, Z., Yuan, Y., Zhang, Z., Liang, Q., et al. (2022). Progress in soybean functional genomics over the past decade. *Plant Biotechnology Journal, 20*, 256–282.

Zou, J. J., Singh, R. J., Lee, J., Xu, S. J., Cregan, P. B., & Hymowitz, T. (2003). Assignment of molecular linkage groups to soybean chromosomes by primary trisomics. *Theoretical and Applied Genetics, 107*, 745–750.

CHAPTER 2

Grain Structure, Composition, and Its Characteristics

RAHUL KUMAR,[1,2,*] MANISHA SAINI,[1] MENIARI TAKU,[1] PULAK DEBBARMA,[1] ROHIT KUMAR MAHTO,[1,3] DEEPSHIKHA SHARMA,[1,4] SANTOSHI RAWAT,[5] PURNIMA RAY,[6] and AKSHAY TALUKDAR[1]

[1]*Division of Genetics, Indian Council of Agricultural Research (ICAR)–Indian Agricultural Research Institute (IARI), New Delhi, India*

[2]*Indian Council of Agricultural Research (ICAR) Research Complex for North Eastern Hill Region, Tripura Center, Lembucherra, Agartala, Tripura, India*

[3]*School of Biotechnology, Institute of Science, Banaras Hindu University (BHU), Varanasi, Uttar Pradesh, India*

[4]*Department of Botany, University of Delhi, New Delhi, India*

[5]*National Institute of Food Technology, Entrepreneurship, and Management, Thanjavur, Tamil Nadu, India*

[6]*Navinchandra Mafatlal College of Agriculture, Navsari Agricultural University, Navsari, Gujarat, India*

[*]Corresponding author

ABSTRACT

The soybean (*Glycine max* (L.) Merr.), originally cultivated in East Asia, is now grown globally for its high protein and oil content. The seed comprises two parts: the seed coat (hull) and the embryo. The embryo,

Soybeans, Volume 2: Genome, Biotechnology, Stress, and Disease.
R. Ambika Rajendran, Ayyagari Ramlal, Sanjay Kumar Lal, and Godfree Chigeza (Eds.)

DOI: 10.1201/9781998511730-3

consisting of two cotyledons and the embryonic axis, accounts for 90% of the seed weight. The embryonic axis includes the radicle, hypocotyl, and epicotyl. During germination, the cotyledons serve as nutrient reserves, while the radicle becomes the primary root. The hypocotyl lifts the cotyledons and epicotyl above the soil, and the epicotyl develops into the stem. Protein and oil bodies are distinct subcellular organelles that store soy protein and oil. These, often surrounded by spherosomes, are found in the cytoplasm of palisade-like cotyledon cells. On average, soybean seeds contain 40–41% protein by dry weight. The USDA germplasm collection includes accessions with oil content ranging from 8.1 to 24.0%. Carbohydrates make up about 35% of the seed's dry weight. Minor components include phytic acid (1.0 to 2.3%) and isoflavones (0.2–4.2 mg/g seed dry weight).

2.1 INTRODUCTION

The soybean (*Glycine max* (L.) Merr.) is an East Asian legume crop that is now planted globally due to its high protein and oil content. In 2021–22, its global output was 372.56 million metric tons (USDA Global Market Analysis, 2022). The increase in soybean output can be linked to higher crop yield and demand to meet the world's growing food and fuel needs (Rajendran et al., 2022; Ramlal, 2022). Brazil is the world's top soybean producer, accounting for 37.3% of total production, followed by the United States (32.4%), Argentina (12.48%), China (4.4%), India (3.19%), Paraguay (2.28%), Canada (1.68%), and others (6.27%) (USDA Global Market Analysis, 2022). Soymilk, soy sauce, soy paste, tempeh, miso, tofu, and natto are all classic Asian cuisines made using soybeans (Yan et al., 2017). Soybeans are mostly processed into soybean meal (SM) and seed oil in Western cultures (Kumar et al., 2023b). Processors smash soybeans to produce full-fat flakes, which are then extracted using organic solvents to produce oil and defatted flakes. Full-fat flakes can also be crushed into full-fat flour for use as a food ingredient or utilized as animal feed components (Cui et al., 2004). After removing the fat, defatted flakes undergo grinding to produce SM, a versatile ingredient utilized in the production of texturized vegetable protein, soy concentrate, and soy isolates. Additionally, it serves as a valuable source of high-quality protein in animal feed formulations (Yan et al., 2017). Approximately 55% of soybean oil produced is utilized as cooking and salad oil, 24% as baking and frying fats and oils, 4% as an ingredient in margarines, 7% for other food and industrial purposes, and 11% as a substrate for biodiesel synthesis (Iowa State University Extension Service, Agricultural Marketing Resource Centre, 2013). Soybean component structure and chemical composition are critical from a processing and nutritional standpoint (Rajendran et al., 2023; Xu et al., 2011). Constant variations in soybean qualities necessitate adjustments to processing conditions and product reformulation, which might result in altered product quality (Tao et al., 2017).

2.2 MORPHOLOGY AND COMPOSITION OF SOYBEAN SEED

Soybean usually contains one to three seeds per pod. These seeds vary significantly in their shapes, sizes, and colors (Hina et al., 2020; Kumar et al., 2022). The shape might range from almost spherical to flat and elongated. Seeds are 5–11 mm in diameter and weigh 120–180 mg each (Li et al., 2008). The shape of mature soybeans is approximately spherical and varies greatly depending on cultivar and growth conditions. The size and shape of the seed are pivotal factors influencing both the weight of the seed and the overall yield in soybean cultivation (Kumar et al., 2023a; Salas et al., 2006; Yan et al., 2017). The seed size and shape represent complex quantitative characteristics influenced by several genes and are highly responsive to environmental factors (Hina et al., 2020). The USDA soybean genome database (SoyBase; http://www.soybase.org) has cataloged over 400 quantitative trait loci (QTLs) associated with seed size and shape. The seed coat (hull) and embryo are the two sections of the soybean seed (Table 2.1). Soybean hulls come in a variety of colors and patterns, including yellow, green, brown, and black. The inheritance of soybean seed color exhibits a moderately complex pattern, governed by multiple loci. Classical genetics studies have identified at least five genetic loci (I, R, T, W1, and O), with most of them implicated in the flavonoid-based pigmentation pathway (Palmer et al., 2004; Yang et al., 2010). Among these loci, three (I, R, and T) are directly involved in pigment biosynthesis, while O and W1 influence pigmentation only in the presence of recessive alleles of i r or i t, respectively (Palmer et al., 2004). The hilum might be black, brown, buff, or pale yellow, and the cotyledons can be yellow or green (Song et al., 2016). The hilum, serving as the point of attachment to the pod, is located within the seed coat. The embryo, which consists of two cotyledons and an embryonic axis, contributes 90% of the seed weight (Medic et al., 2014; Wolf, 1972). The embryonic axis is made up of the radicle, hypocotyl, and epicotyl. The hull holds the dicotyledon seed (two cotyledons) together (Sharma et al., 2014). When the hull is removed, the cotyledons separate, and the germ is dislodged. The commercial soybean includes approximately 8% hull, 90% cotyledons, and 2% hypocotyl (Medic et al., 2014). The cuticle is found within the seed coat, followed by palisade cells, a layer of hourglass-shaped cells, and numerous layers of parenchyma cells. The presence of the hourglass-shaped cells also aids in the identification of soy products in foods. A layer of parenchyma cells and aleurone cells covers the cotyledon. The core of the cotyledon is packed with elongated palisade-like cells (not to be mistaken with the palisade cells in the hull), which hold the majority of the oil and protein in the form of oil and protein bodies (Sharma et al., 2014; Smith et al., 1972).

The cotyledons act as food reserves for the growing plant during germination; the radicle serves as the main root, the hypocotyl lifts the cotyledons and epicotyl above the soil surface, and the epicotyl serves as the plant's stem (Sharma et al., 2014). The cotyledons are the most essential portions of the seed from a commercial standpoint because they store protein and oil. Soy protein and oil are packaged into distinct subcellular structures known as protein and oil bodies. Protein and oil bodies are found in the cytoplasm of palisade-like cotyledon cells, and spherosomes are usually seen around the protein body. Protein bodies are enormous spherical organelles with diameters ranging from 2 to 10 μm, whereas oil bodies are considerably smaller particles with diameters ranging from 0.2 to 0.5 μm (Wolf, 1972). Solvent extraction is used to remove the oil from modern soybean processing, after which the oil is further processed. The high-quality leftover protein meal is utilized in food and animal feed. The hulls of the soybeans can be removed or left in the meal. The hulls include a high concentration of cellulose and hemicellulose, as well as acid-digestible fiber and zinc and iron (Korngay, 1978). The oil extracted through solvent extraction is known as crude oil, and it contains a variety of materials that must be removed through processing before the oil is fit for human use.

TABLE 2.1 Chemical Composition of Soybeans and Their Components (Dry Weight Basis)

Components	Yield (%)	Protein (%)	Fat (%)	Carbohydrates (%)	Ash (%)
Whole soybeans	100.0	40.3	21.0	33.9	4.9
Cotyledon	90.3	42.8	22.8	29.4	5.0
Hull	7.3	8.8	1.0	85.9	4.3
Hypocotyl	2.4	40.8	11.4	43.4	4.4

Source: Adapted from Liu (1997).

2.2.1 PROTEIN

Based on dry weight, soybean seeds typically contain approximately 40–41% protein. The inheritance of protein content in seeds is quantitatively governed and exhibits a negative correlation with both seed oil content and yield (Lee et al., 2019). Through the utilization of accessions with diverse genetic backgrounds, researchers have identified over 300 QTLs regulating soybean seed oil and protein content (Van et al., 2017). Seed proteins are categorized based on their functions into four groups: metabolic enzymes, structural (including ribosomal and chromosomal), membrane, and storage proteins (Krishnan, 2001; Nielsen et al., 1997). Storage proteins account for 65–80% of total seed protein (Murphy, 2008). Glycinin and b-conglycinin are the two primary seed storage proteins, and they belong to the legumin (11S globulins) and vicilin (7S globulins) families of proteins, respectively. Glycinin and b-conglycinin make up roughly 40% and 25% of total embryo protein, respectively (Zarkadas et al., 2007). Their primary function is to provide amino nitrogen to the developing seed (Murphy, 2008; Ramlal et al., 2022a). Glycinin is a hexamer, and A1aB2, A1bB1b, A2B1a, A3B4, and A5A4B3 are the five primary monomeric subunits of glycinin, where A and B stand for acidic and basic polypeptides, respectively (Maruyama et al., 2005). The A5A4B3 subunit differs from the others in that it contains an extra acidic polypeptide in its structure (Liu, 1997). Among soybean protease inhibitors (PIs), the Bowman-Birk trypsin-chymotrypsin inhibitor and the Kunitz trypsin inhibitor (KTI) have received considerable attention in research. In monogastric animals, significant levels of PIs found in soybean seeds (constituting approximately 6% of soybean protein) hinder pancreatic enzymes, leading to reduced protein digestibility and potentially causing pancreatic hypertrophy in certain animals (Krishnan, 2001). Hemagglutinins (lectins) are proteins characterized by their strong capability to agglutinate both red blood cells and intestinal mucosa cells through interactions with the carbohydrates present on their cell surfaces. Soybean lectins are also antinutritional factors because they hinder animal growth by preventing nutrients from being absorbed (Liener, 1994). Lectin concentrations in soybean seeds range between 1% and 2% of seed dry mass (George et al., 2008). Heat treatment of soybeans reduces the action of lectins and PIs, although 10–20% of the residual trypsin inhibitor activity persists (Friedman, 2001). Lipoxygenase enzymes are not antinutritional factors, although they are linked to disagreeable odors ("greeny" and "beany") in soybean diets. They catalyze the oxidation of fatty acids (FAs) with the cis, cis-1,4-pentadiene structure (usually linoleic and α-linolenic acids) to hydroperoxides, which are further degraded to secondary compounds (e.g., hexanal) that cause off-flavors (Gerde et al., 2008).

2.2.2 LIPIDS

The USDA soybean germplasm collection contains accessions with lipid (oil) content ranging from 8.1 to 24.0% on a dry seed basis (Hill et al., 2001, 2005; Peregrine et al., 2008). Oil content in soybean is governed by several QTLs (Bilyeu et al., 2016). Zhang et al. (2020) identified significant QTLs for soybean protein and oil content on chromosome 15, specifically linked to a sugar transporter gene known as GmSWEET39. They observed a strong association between a specific two-nucleotide CC deletion in GmSWEET39, resulting in a truncated C-terminus, and the manifestation of high seed oil content coupled with low seed protein content. This suggests a pleiotropic effect of GmSWEET39 on both protein and oil content in soybeans. Soybean seed lipids serve as energy storage for the plant, membrane constituents, signaling molecules, pathogen defense, and so on. Triacylglycerols (TAGs) are the most common storage lipids found in oil bodies. A TAG molecule is made up of three FAs that have been esterified to a glycerol backbone. Linoleic acid (18:2) is the most prevalent FA in commodity soybean TAGs, followed by oleic (18:1), palmitic (16:0), linolenic (18:3), and stearic (18:0) acids. Because of the selectivity of plant lipases, the distribution of FAs within a single TAG and among various TAGs is not random (Pokorny et al., 2011). According to Liu (1997), saturated FAs such as palmitic and stearic acid, along with long-chain FAs containing more than 18 carbons, are evenly and randomly distributed across positions 1 and 3 on triglycerides. Oleic and linolenic acids are found in all three positions, while any remaining positions are typically occupied by linoleic acid. Oil quality, nutritional value, flavor, oxidative stability, melting point, crystallization form, and other factors are influenced by FA composition and distribution (Yadav, 1996). Soybean oil has a high nutritional value since it contains a lot of unsaturated FAs like oleic, linoleic, and linolenic acids, which have one, two, and three double bonds, respectively. Unsaturated FAs offer numerous health advantages, such as shielding against atherosclerosis, reducing levels of total cholesterol, low-density lipoprotein (LDL) cholesterol, and TAGs in the plasma, and dampening inflammatory responses (Bahrami, 2009). Linoleic and linolenic acids are also necessary FAs that mammals cannot produce and must thus get from their meals.

The quantity of double bonds, on the other hand, is directly proportional to oil stability. Due to its elevated linolenic acid content, soybean oil is inclined to oxidation and the development of off-flavors, rendering it relatively unstable. As a result, soybean oil is frequently hydrogenated in order to reduce the quantity of unsaturated double bonds. Soybean seed membrane lipids are mostly made up of phospholipids (3% of total soybean lipids or 0.74% of dry-seed weight) (Wang et al., 1997), which are similar to TAGs except that the FA on the sn-3 position of the glycerol moiety is substituted by a phosphate group. During the degumming stage of the oil refining process, phospholipids are removed from crude soybean oil. The phospholipid-rich fraction is known as soybean lecithin and is utilized as an emulsifier in the food industry. Phosphatidylcholine (PC) (55.3%), phosphatidylethanolamine (26.3%), and phosphatidylinositol (18.4%) are the three primary phospholipids in soybean lecithin (Wang et al., 1997). In addition to phospholipids, soybean seeds contain sphingolipids as minor membrane elements, the most prevalent of which is glucosylceramide (Gerde et al., 2008). Phytosterols are another family of bioactive lipid components found in soybean seeds that have been intensively researched due to their ability to decrease serum LDL cholesterol (Kritchevsky et al., 2005). Campesterol, stigmasterol, and β-sitosterol are the most abundant phytosterols in soybeans (Gerde et al., 2008). As a by-product of the soybean oil refining process, phytosterols are collected and used to make commercial extracts for the health and nutrition business. Tocopherols are natural plant defenses that protect biological membranes from the oxidative breakdown of polyunsaturated FAs (Sattler et al., 2003). They prevent lipid peroxidation by scavenging active oxygen species and peroxyl radicals, preventing them from reacting with lipids (Kamal et al., 1996). Tocopherol concentrations in soybean seeds typically range from 198 to 278 ppm on a dry basis. Tocopherols in crude soybean oil range between 1205 and 2195 ppm, which contribute to the oil's oxidative stability; however, a significant quantity is lost during the refining process (Gerde et al., 2008). Because there is a positive relationship between oil unsaturation and total tocopherols in soybean seeds, soybean cultivars with low linolenic acid content have lower tocopherol content (Dolde et al., 1999).

2.2.3 CARBOHYDRATES

Carbohydrates are the third most prevalent component in soybean seeds, accounting for approximately 35% of the dry seed weight. They are abundant in the soybean seed coat (86% of the dry weight of the seed coat) but can also be identified in embryonic parenchyma cells. Although the hulls remove some of the seed carbohydrates, SM can still contain up to 40% total carbohydrates (Karr-Lilienthal et al., 2005; Middelbos et al., 2008). About 60% and 40% of carbohydrates are insoluble and soluble sugars, respectively. Soluble sugar in soybean dry matter is comprised of 5% sucrose, 1.5% raffinose, and 3% stachyose. The concentration of soluble sugars, particularly sucrose, in soybean seeds plays a crucial role in determining the taste and flavor of soy-based products like soymilk and tofu (Wilson, 2004). Although limited in number, some QTLs associated with sucrose content have been identified through linkage analysis, as documented in the SoyBase database (www.soybase.org). Kim et al. (2005) found four sucrose content-related QTLs in chromosomes 2, 11, and 19 using the RIL populations derived from a cross between "Keunolkong" × "Shinpaldalkong." Skoneczka et al. (2009) identified one main effect QTL in the Sat_213-Satt643 interval of chromosome 6 through two F2 populations, which could explain 76% of the observed genetic variations in sucrose content. Soybeans contain roughly half of their total carbohydrates as structural carbohydrates, with the other half as nonstructural carbohydrates (Karr-Lilienthal et al., 2005). Cell-wall polysaccharides (cellulose, hemicellulose, and pectins) are structural carbohydrates, whereas non-structural carbohydrates include starch and other mono-, di-, and oligosaccharides (Obendorf et al., 2009). Cell-wall polysaccharides, along with lignin (non-polysaccharide), enzyme-resistant starch, and oligosaccharides, are components of dietary fiber, a nondigestible element of food/feed. The amount of dietary fiber in soybean seeds varies depending on the cultivar and the technique of measurement. The enzymatic-gravimetric method for assessing dietary fiber content is commonly employed in nutritional analysis to ascertain the nutritional value of various foods. Redondo-Cuenca et al. (2008) discovered 24.4% total dietary fiber in raw seeds of a commercial soybean line, while raw black soybeans contained 19.7% total dietary fiber (Lee et al., 2006) on a dry seed basis using the AOAC 991.42 and AOAC 985.29 methodologies, respectively. For soybean trading purposes, the crude fiber analysis (e.g., AACC 32–30, AOAC 978.10) and range between 4% and 8% seed dry weight) is used to determine the fiber content of soybean seeds (AACC, 2000; Brumm, 2004; Naeve et al., 2006).

However, it has been found that crude fiber analysis recovers (captures) only around 20% of hemicellulose and 50–80% of cellulose on average (Van Soest et al., 1973). As a result, Badley crude fiber measurements underestimate cell-wall polysaccharide levels (Badley et al., 1975). The dehulling procedure removes the majority of cell-wall polysaccharides from soybeans. Thus, hulls are a high source of dietary fiber (83% total dietary fiber) that may be advantageous in some food applications (Dust et al., 2004). Non structural Carbohydrates: Soybeans typically contain 11 to 25% soluble carbohydrates, which include 15–20 distinct sugar species (Maughan et al., 2000; Obendorf, 2008). Sucrose, raffinose, and stachyose stand out as the predominant soluble sugars in abundance. Sucrose gives SMs (such as natto and tofu) a nice sweet taste and is a beneficial component in soybean seeds (Taira, 1990). Galactooligosaccharides (raffinose, stachyose, and verbascose) are considered antinutritional factors in humans and non-ruminant animals because they promote flatulence and digestive discomfort (Krause et al., 1994; Liying et al., 2003). Galactooligosaccharides are assumed to play a key physiological role in mature seeds, vegetative organs, and leaves as storage or transport carbohydrates (Dey, 1985). Starch is a polymer that stores D-glucose monomeric units. It is produced in granular form in soybean cotyledon cells. Developing soybean seeds collect starch until 30–40 days after blooming, following which the starch content drops quickly to 1% on a dry seed basis at maturity. Amylose and amylopectin constitute the primary components of starch, with amylose exhibiting comparatively lower susceptibility to enzyme hydrolysis. Stevenson et al. (2006) discovered that starches from immature soybean seeds collected 20 days before harvest had low absolute amylose contents (11.8–16.2% of the starch db) and a short average branch-chain length of amylopectin (20.4–20.9 glucose units), indicating a favorable starch structure for rapid enzyme hydrolysis. According to the findings of Stevenson et al. (2006), Wilson et al. (1978), and Saldivar et al. (2011), starch functions as a temporary reserve material in soybeans that is rapidly hydrolyzed to glucose at later stages of maturity to provide energy for the developing seed.

2.2.4 MINOR CONSTITUENTS

Phytic acid (myo-inositol-1,2,3,4,5,6-hexaphosphoric acid) is the primary phosphorus storage form in legumes and cereals. It contributes 65–80% of the total phosphorus in soybean seeds (Raboy et al., 1993) and varies from 1.0 to 2.3% on a dry seed basis (Anderson and Wolf, 1995). Oltmans et al. (2004) reported that low phytate was controlled by recessive alleles (pha1 and pha2) at two independent loci exhibiting duplicate dominant epistasis. Phytic acid accumulates during the seed development process, with the majority of it localized in the protein bodies of soybean cotyledons (Anderson and Wolf, 1995). Phytic acid phosphorus, unlike inorganic phosphorus, is not available to humans or non-ruminant animals (e.g., pigs and poultry) because they lack the phytase enzyme in their digestive tract (Oltmans et al., 2005). As a result, it is a significant source of phosphorus in manure and an environmental problem (Israel et al., 2006). Soybean seeds contain phytic acid in the form of mixed phytate and phytin salts. Charged phytate salts bind metal cations such as zinc, calcium, magnesium, and iron, rendering them inaccessible to humans and non-ruminant cattle (Israel et al., 2007; Raboy et al., 1984). As a result, phytic acid is regarded as an antinutritional component in foods/feeds. Phytic acid can also bind to soybean storage proteins and influence their functional properties (e.g., isoelectric point and solubility) (Kumar et al., 2005; Kwanyuen and Burton, 2005). Several efforts have been undertaken to create low phytic acid soybean lines in order to alleviate environmental issues caused by phosphorus excretion in animal manure. Soybean mutants LR33 (Hitz et al., 2002) and M153 (Wilcox et al., 2002) reduced phytate content by 50% and 75%, respectively, with an associated increase in inorganic phosphorus content. However, both mutant lines had low seeding rates and yields, making them of limited commercial utility in breeding. Yuan et al. (2007) documented the discovery of two soybean lines, Gm-lpa-TW-1 and Gm-lpa-ZC2, characterized by reduced phytate content (66.6% and 46.3%, respectively), increased levels of inorganic phosphorus (6 times and 1.4 times, respectively), and consistent total phosphorus content in seeds compared to their parent lines. When compared to a commercial cultivar, these seeds demonstrated good seedling emergence in all settings except subtropical, with no yield loss.

Isoflavones, a class of phytoestrogens, have phenolic compounds that are comparable to the estrogen 17b-estradiol, allowing them to bind to estrogen receptors (ERs) and exhibit mild estrogenic activity (Sakai and Kogiso, 2008). Many studies have shown that soybean isoflavones consumed in combination with soy proteins have a cardioprotective effect, primarily by lowering serum LDL cholesterol levels and enhancing the oxidative stability of LDL cholesterol (Kritchevsky and Chen, 2005). Isoflavones supplied in the absence of soy protein had no cholesterol-lowering effects. Soybean isoflavones have also been linked to a reduction in the risk of many malignancies and osteoporosis (Weaver et al., 2005). Isoflavone isomers include aglycons (daidzein, genistein, and glycitein), conjugated glycosides (daidzin, genistin, and glycitin), malonyl glucosides (600-O-malonyl daidzin, 600-O-malonyl genistin, and 600-O-malonyl glycitin), and acetyl (600-O-acetyl daidzin, 600-O-acetyl genistin, and 600-O-acetyl glycitin). Soybeans have a high isoflavone content (0.2–4.2 mg/g seed dry weight) (Hammond et al., 2005), and the content of specific isoflavones varies depending on genotype, growing location, and crop year. Mature soybean seeds do not have a uniform distribution of isoflavones: the hypocotyl has the highest concentration of isoflavones (4–6 times higher than the cotyledons), followed by the cotyledons and the seed coat, in that order (Kim et al., 2007). Isoflavones are lost in substantial amounts during soybean processing: 44% during tofu manufacturing, 53% during alkaline extraction in soy isolate production, and 12% and 49% during the soaking and heat processing processes of tempeh production, respectively. As a result, the isoflavone level in SMs varies greatly, ranging from 1 μg/g in soy sauce to 540 μg/g in tempeh (Murphy et al., 1999). Because these highly polar molecules are eliminated during the oil refining process, soybean oil contains only tiny amounts of isoflavones (Setchell, 1998).

2.3 FUTURE PROSPECTS

Soybean stands among the top five crops that significantly influence global agriculture, alongside rice, wheat, maize, and cotton. Enhancing soybean seed composition traits through breeding presents a complex challenge; yet, the integration of extensive genomic resources and tools offers promising avenues for soybean breeders and researchers. By leveraging both traditional breeding methods and genomic techniques, the identification of genomic loci and haplotypes can be facilitated, aiding in the advancement of seed composition traits and the creation of cultivars with enhanced food-grade attributes.

2.4 CONCLUSION

Soybean seed shape, size, and color vary significantly depending on cultivar and growth conditions. A variety of nutritionally essential components can be found in soybean seed. Soybean's principal constituents include lipids, protein, and carbohydrates, while minor constituents include phytic acid, isoflavones, ash, and others. The composition and quantities of soybean constituents exhibit considerable diversity influenced by genetic characteristics, environmental variables, the stage of plant maturity, and processing parameters.

KEYWORDS

- **cotyledons**
- **embryonic axis**
- **genetic and environmental variability**
- **isoflavones**
- **phytic acid**
- **protein and oil bodies**
- **soybean seed composition**

REFERENCES

AACC. (2000). *Official Methods of Analysis (Crude Fiber in Flours, Feeds, and Feedstuffs)* (10th ed., Sec. 32–20).

Anderson, R. L., & Wolf, W. J. (1995). Compositional changes in trypsin inhibitors, phytic acid, saponins, and isoflavones related to soybean processing. *Journal of Nutrition, 125*, 581S–588S.

Badley, R., Atkinson, D., Hauser, H., Oldani, D., Green, J., & Stubbs, J. (1975). The structure, physical, and chemical properties of the soybean protein glycinin. *Biochimica et Biophysica Acta (BBA)–Protein Structure, 412*, 214–228.

Bahrami, G. (2009). Trans and other fatty acids: Effects on endothelial functions. In R. R. Watson (Ed.), *Fatty Acids in Health Promotion and Disease Causation* (pp. 3–41). AOCS Press.

Brumm, T. J. (2004). Chapter two: Quality and grading factors of IP soybeans. In *Managing Grain After Harvest*. Department of Agricultural and Biosystems Engineering, Iowa State University, Ames.

Chen, S., Kritchevsky, D., & Baer, D. (2007). Trans fatty acid effect on cardiovascular disease: Animal and human studies (pp. 1–37).

Cui, Z., James, A. T., Miyazaki, S., Wilson, R. F., & Carter, T. (2004). Breeding specialty soybeans for traditional and new soyfoods. In *AOCS Press* (pp. 274–332). Champaign, IL, USA.

Dey, P. M., & Dixon, R. A. (1985). *Biochemistry of Storage Carbohydrates in Green Plants*. Academic Press.

Dolde, D., Vlahakis, C., & Hazebroek, J. (1999). Tocopherols in breeding lines and effects of planting location, fatty acid composition, and temperature during development. *Journal of the American Oil Chemists' Society, 76*, 349–355.

Dust, J. M., Gajda, A. M., Flickinger, E. A., Burkhalter, T. M., Merchen, N. R., & Fahey Jr., G. C. (2004). Extrusion conditions affect chemical composition and *in vitro* digestion of select food ingredients. *Journal of Agricultural and Food Chemistry, 52*, 2989–2996.

Friedman, M., & Brandon, D. L. (2001). Nutritional and health benefits of soy proteins. *Journal of Agricultural and Food Chemistry, 49*, 1069–1086.

George, M., Bhide, S., Thengane, R., Hosseini, G., & Manjaya, J. (2008). Identification of low lectin mutants in soybean. *Plant Breeding, 127*, 150–153.

Gerde, J., & White, P. J. (2008). Soybeans: Chemistry, production, processing, and utilization (pp. 193–227). AOCS Press, Urbana.

Hammond, E., Johnson, L. A., Su, C., Wang, T., & White, P. J. (2005). Soybean oil. In F. Shahidi (Ed.), *Bailey's Industrial Oil and Fat Products* (pp. 577–653). Wiley, New York.

Hill, J. L., Peregrine, E. K., Sprau, G. L., Cremeens, C. R., Nelson, R. L., Kenty, M. M., Kilen, T. C., & Thomas, D. A. (2001). Evaluation of the USDA soybean germplasm collection: Maturity groups VI–VIII (FC 03.659–PI 567.235B). *US Department of Agriculture Technical Bulletin No. 1894* (pp. 1–130).

Hill, J. L., Peregrine, E. K., Sprau, G. L., Cremeens, C. R., Nelson, R. L., Orf, J. H., & Thomas, D. A. (2005). *Evaluation of the USDA Soybean Germplasm Collection: Maturity Groups 000–IV (PI507670–PI 574486)* (US Department of Agriculture Technical Bulletin No. 1914, pp. 1–131). U.S. Department of Agriculture.

Hina, A., Cao, Y., Song, S., Li, S., Sharmin, R. A., & Elattar, M. A. (2020). High-resolution mapping in two RIL populations refines major "QTL hotspot" regions for seed size and shape in soybean (*Glycine max* L.). *International Journal of Molecular Sciences, 21*(3), 1040.

Hitz, W. D., Carlson, T. J., Kerr, P. S., & Sebastian, S. A. (2002). Biochemical and molecular characterization of a mutation that confers a decreased raffinose saccharide and phytic acid phenotype on soybean seeds. *Plant Physiology, 128*, 650–656.

Iowa State University Extension Service, Agricultural Marketing Resource Center. (2013). *Soybean Oil and Biodiesel Usage Projections and Balance Sheet*. Retrieved from: http://www.extension.iastate.edu/agdm/crops/.../biodieselbalancesheet.pdf .

Israel, D., Burton, J., & Kwanyuen, P. (2006). Genetic variability for phytic acid phosphorus and inorganic phosphorus in seeds of soybeans in maturity groups V, VI, and VII. *Crop Science, 46*, 67–71.

Israel, D., Kwanyuen, P., Walker, D., & Burton, J. (2007). Response of low seed phytic acid soybeans to increases in external phosphorus supply. *Crop Science, 47*, 2036–2041.

Kamal-Eldin, A., & Appelqvist, L. A. (1996). The chemistry and antioxidant properties of tocopherols and tocotrienols. *Lipids, 31*, 671–701.

Karr-Lilienthal, L. K., Grieshop, C. M., Spears, J. K., & Fahey Jr., G. C. (2005). Amino acid, carbohydrate, and fat composition of soybean meals prepared at 55 commercial U.S. soybean processing plants. *Journal of Agricultural and Food Chemistry, 53*, 2146–2150.

Kim, H. K., Kang, S. T., Cho, J. H., Choung, M. G., & Suh, D. Y. (2005). Quantitative trait loci associated with oligosaccharide and sucrose contents in soybean (*Glycine max* L.). *Journal of Plant Biology, 48*, 106–112.

Kim, J., Hong, S. B., Jung, W. S., Yu, C. Y., Ma, K. H., Gwag, J. G., & Chung, I. M. (2007). Comparison of isoflavone composition in seed, embryo, cotyledon, and seed coat of cooked-with-rice and vegetable soybean (*Glycine max* L.) varieties. *Food Chemistry, 102*, 738–744.

Kornegay, E. T. (1978). Feeding value and digestibility of soybean hulls for swine. *Journal of Animal Science, 47*(6), 1272–1280.

Krause, D., Easter, R., & Mackie, R. (1994). Fermentation of stachyose and raffinose by hind-gut bacteria of the weanling pig. *Letters in Applied Microbiology, 18*, 349–352.

Krishnan, H. B. (2001). Biochemistry and molecular biology of soybean seed storage proteins. *Journal of New Seeds, 2*, 1–25.

Kritchevsky, D., & Chen, S. C. (2005). Phytosterols—Health benefits and potential concerns: A review. *Nutrition Research, 25*, 413–428.

Kumar, R., Mahto, R. K., Kumar, R., Saini, M., & Talukdar, A. (2022). Development of seed shape software 'SEEDFIG' and study of seed size and seed shape traits in soybean (*Glycine max* [L.] Merr.). *The Pharma Innovation Journal, 11*(10), 1547–1550.

Kumar, R., Saini, M., Taku, M., Debbarma, P., Mahto, R. K., Ramlal, A., Sharma, D., Rajendran, A., Pandey, R., Gaikwad, K., & Lal, S. K. (2023a). Identification of quantitative trait loci (QTLs) and candidate genes for seed shape and 100-seed weight in soybean (*Glycine max* L. Merr.). *Frontiers in Plant Science, 13,* 1074245.

Kumar, R., Saini, M., Taku, M., Debbarma, P., Sharma, D., Pandey, R., Gaikwad, K., Lal, S. K., & Talukdar, A. (2023b). Genetic studies and mapping of quantitative trait loci (QTL) for yield and yield-related traits in vegetable and seed soybean (*Glycine max* L.)-derived populations. *Indian Journal of Genetics and Plant Breeding, 83*(3), 301–309.

Kumar, V., Rani, A., Rajpal, S., Srivastava, G., Ramesh, A., & Joshi, O. P. (2005). Phytic acid in Indian soybean: Genotypic variability and influence of growing location. *Journal of the Science of Food and Agriculture, 85*, 1523–1526.

Kwanyuen, P., & Burton, J. W. (2005). A simple and rapid procedure for phytate determination in soybeans and soy products. *Journal of the American Oil Chemists' Society, 82*, 81–85.

Lee, C. H., Oh, S. H., Yang, E. J., & Kim, Y. S. (2006). Effects of raw, cooked, and germinated small black soybean powders on dietary fiber content and gastrointestinal functions. *Food Science and Biotechnology, 15*, 635–638.

Lee, S., Van, K., Sung, M., Nelson, R., LaMantia, J., & McHale, L. K. (2019). Genome-wide association study of seed protein, oil, and amino acid contents in soybean from maturity groups I to IV. *Theoretical and Applied Genetics, 132*(6), 1639–1659.

Li, W., Zheng, D. H., Van, K. J., & Lee, S. H. (2008). QTL mapping for major agronomic traits across two years in soybean (*Glycine max* L. Merr.). *Journal of Crop Science and Biotechnology, 11*(3), 171–176.

Liener, I. E. (1994). Implications of antinutritional components in soybean foods. *Critical Reviews in Food Science and Nutrition, 34*, 31–67.

Liu, K. (1997). *Soybeans: Chemistry, Technology, and Utilization*. Chapman & Hall.

Liying, Z., Li, D., Qiao, S., Johnson, E. W., Li, B., Thacker, P. A., & Han, I. K. (2003). Effects of stachyose on performance, diarrhea incidence, and intestinal bacteria in weanling pigs. *Archives of Animal Nutrition, 57*, 1–10.

Maruyama, K., Ratnaparkhe, M. B., & Kole, C. (2016). *Genetics, Genomics, and Breeding of Soybean*. CRC Press.

Maruyama, N., Tecson-Mendoza, E. M., Maruyama, Y., Adachi, M., & Utsumi, S. (2005). Molecular design of soybean proteins for enhanced food quality. In K. Shetty, G. Paliyath, A. Pometto, & R. E. Levin (Eds.), *Food Biotechnology* (pp. 676–701). CRC Press.

Maughan, P., Maroof, M. A. S., & Buss, G. (2000). Identification of quantitative trait loci controlling sucrose content in soybean (*Glycine max*). *Molecular Breeding, 6*, 105–111.

Medic, J., Atkinson, C., & Hurburgh, C. R. (2014). Current knowledge in soybean composition. *Journal of the American Oil Chemists' Society, 91*, 363–384.

Middelbos, I. S., & Fahey, G. C., Jr. (2008). Soybean carbohydrates. In L. A. Johnson, P. J. White, & R. Galloway (Eds.), *Soybeans: Chemistry, Production, Processing, and Utilization* (pp. 269–296). AOCS Press.

Murphy, P. (2008). Soybean proteins. In L. A. Johnson, P. J. White, & R. Galloway (Eds.), *Soybeans: Chemistry, Production, Processing, and Utilization* (pp. 229–269). AOCS Press.

Murphy, P. A., Song, T., Buseman, G., Barua, K., Beecher, G. R., Trainer, D., & Holden, J. (1999). Isoflavones in retail and institutional soy foods. *Journal of Agricultural and Food Chemistry, 47*, 2697–2704.

Naeve, S. L., Orf, J. H., & O'Neill, T. (2006). *Quality of the United States Soybean Crop: 2006*. Report to the American Soybean Association and U.S. Soybean Export Council quality mission to Asia.

Nielsen, N. C., Bassuner, R., & Beaman, T. (1997). The biochemistry and cell biology of embryo storage proteins. In B. A. Larkins & I. K. Vasil (Eds.), *Cellular and Molecular Biology of Plant Seed Development* (Vol. 4, pp. 151–220). Springer.

Obendorf, R. L., Zimmerman, A. D., Ortiz, P. A., Taylor, A. G., & Schnebly, S. R. (2008). Imbibitional chilling sensitivity and soluble carbohydrate composition of low raffinose, low stachyose soybean seed. *Crop Science, 48*, 2396–2403.

Obendorf, R. L., Zimmerman, A. D., Zhang, Q., Castillo, A., Kosina, S. M., Bryant, E. G., Sensenig, E. M., Wu, J., & Schnebly, S. R. (2009). Accumulation of soluble carbohydrates during seed development and maturation of low-raffinose, low-stachyose soybean. *Crop Science, 49*, 329–341.

Oltmans, S. E., Fehr, W. R., Welke, G. A., & Cianzio, S. R. (2004). Inheritance of low-phytate phosphorus in soybean. *Crop Science, 44*(2), 433–435.

Oltmans, S. E., Fehr, W. R., Welke, G. A., Raboy, V., & Peterson, K. L. (2005). Agronomic and seed traits of soybean lines with low-phytate phosphorus. *Crop Science, 45*, 593–598.

Peregrine, E. K., Sprau, G. L., Cremeens, C. R., Handly, P., Kilen, T. C., Smith, J. R., Thomas, D. A., Sarins, J. D., & Nelson, R. L. (2008). *Evaluation of the USDA Soybean Germplasm Collection: Maturity Group V (FC 30265-PI 612614) and Maturity Groups VI–VIII (PI 416758-PI606432B)* (USDA Technical Bulletin No. 1920, pp. 1–367). U.S. Department of Agriculture.

Peregrine, E. K., Sprau, G. L., Cremeens, C. R., Nelson, R. L., Orf, J. H., & Thomas, D. A. (2008). *Evaluation of the USDA Soybean Germplasm Collection: Maturity Groups 000–IV (PI 578371-PI612761)* (USDA Technical Bulletin No. 1919, pp. 1–155). U.S. Department of Agriculture.

Pokorny, J., & Schmidt, S. (2011). Plant lipids and oils. In Z. Sikorski & A. Kolakowska (Eds.), *Chemical and Functional Properties of Food Lipids* (pp. 249–271). CRC Press.

Raboy, V., & Dickinson, D. B. (1993). Phytic acid levels in seeds of *Glycine max* and *G. soja* as influenced by phosphorus status. *Crop Science, 33*, 1300–1305.

Raboy, V., Dickinson, D. B., & Below, F. E. (1984). Variation in seed total phosphorus, phytic acid, zinc, calcium, magnesium, and protein among lines of *Glycine max* and *G. soja*. *Crop Science, 24*, 431–434.

Rajendran, A., Lal, S. K., Raju, D., & Ramlal, A. (2022). Associations of direct and indirect selection for pregermination anaerobic stress tolerance in soybean (*Glycine max*). *Plant Breeding, 141*(5), 634–643.

Rajendran, A., Lal, S. K., Raju, D., Mallikarjun, B. P., Ramlal, A., & Sharma, D. (2023). Waterlogging tolerance evaluation methods for soybean (*Glycine max* (L.) Merr.) at the pregermination stage. *Genetic Resources and Crop Evolution, 70*, 1–11.

Ramlal, A., Nautiyal, A., Baweja, P., Kumar, V., Mehta, S., Mahto, R. K., Tripathi, S., Shanmugam, A., Mallikarjuna, B. P., Raman, P., & Lal, S. K. (2022a). Angiotensin-converting enzyme inhibitory peptides and isoflavonoids from soybean (*Glycine max* (L.) Merr.). *Frontiers in Nutrition, 9,* 1068388.

Ramlal, A., Nautiyal, A., Baweja, P., Mahto, R. K., Mehta, S., Mallikarjuna, B. P., Vijayan, R., Saluja, S., Kumar, V., Dhiman, S. K., & Lal, S. K. (2022b). Harnessing heterosis and male sterility in soybean (*Glycine max* (L.) Merrill): A critical revisit. *Frontiers in Plant Science, 13,* 981768.

Redondo-Cuenca, A., Villanueva-Suárez, M. J., & Mateos-Aparicio, I. (2008). Soybean seeds and its by-product okara as sources of dietary fibre: Measurement by AOAC and Englyst methods. *Food Chemistry, 108*, 1099–1105.

Sakai, T., & Kogiso, M. (2008). Soy isoflavones and immunity. *Journal of Medical Investigation, 55*, 167–173.

Saldivar, X., Wang, Y. J., Chen, P., & Hou, A. (2011). Changes in chemical composition during soybean seed development. *Food Chemistry, 124*, 1369–1375.

Sattler, S. E., Cahoon, E. B., Coughlan, S. J., & DellaPenna, D. (2003). Characterization of tocopherol cyclase from higher plants and cyanobacteria: Evolutionary implications for tocopherol synthesis and function. *Plant Physiology, 132*, 2184–2195.

Setchell, K. (1998). Phytoestrogens: The biochemistry, physiology, and implications for human health of soy isoflavones. *American Journal of Clinical Nutrition, 68*, 1333S–1346S.

Sharma, S., Kaur, M., Goyal, R., & Gill, B. S. (2014). Physical characteristics and nutritional composition of some new soybean (*Glycine max* (L.) Merrill) genotypes. *Journal of Food Science and Technology, 51*, 551–557.

Sikorski, Z. Z., & Kolakowska, A. (2010). *Chemical and Functional Properties of Food Lipids* (pp. 249–271). CRC Press.

Skoneczka, J. A., Maroof, M. A. S., Shang, C., & Buss, G. R. (2009). Identification of candidate gene mutation associated with low stachyose phenotype in soybean line PI200508. *Crop Science, 49*, 247.

Smith, A. K. (1972). *Soybeans: Chemistry and Technology*. Avi Publishing Co.

Song, J., Liu, Z., Hong, H., Ma, Y., Tian, L., Li, X., & Qiu, L. J. (2016). Identification and validation of loci governing seed coat color by combining association mapping and bulk segregation analysis in soybean. *PLoS One, 11*(7), e0159064.

Soyatech Inc. (1994). *Soya Bluebook*. Bar Harbor, ME.

Stevenson, D. G., Doorenbos, R. K., Jane, J., & Inglett, G. E. (2006). Structures and functional properties of starch from seeds of three soybean (*Glycine max* (L.) Merr.) varieties. *Starch/Stärke, 58*(11), 509–519.

Taira, H. (1990). Quality of soybeans for processed foods in Japan. *Japan Agricultural Research Quarterly, 24*(3), 224–230.

Tao, Y., Mace, E. S., Tai, S., Cruickshank, A., Campbell, B. C., & Zhao, X. (2017). Whole genome analysis of candidate genes associated with seed size and weight in *Sorghum bicolor* reveals signatures of artificial selection and insights into parallel domestication in cereal crops. *Frontiers in Plant Science, 8*, 1237.

Tosh, S. M., & Yada, S. (2010). Dietary fibres in pulse seeds and fractions: Characterization, functional attributes, and applications. *Food Research International, 43*(2), 450–454.

USDA Global Market Analysis/Foreign Agricultural Service. (2022). *World Agricultural.*

Van Soest, P., & McQueen, R. (1973). The chemistry and estimation of fiber. *Proceedings of the Nutrition Society, 32*(2), 123–131.

Van, K., & McHale, L. K. (2017). Meta-analyses of QTLs associated with protein and oil contents and compositions in soybean (*Glycine max* (L.) Merr.) seed. *International Journal of Molecular Sciences, 18*(6), 1180.

Wang, T., Hammond, E. G., & Fehr, W. R. (1997). Phospholipid fatty acid composition and stereospecific distribution of soybeans with a wide range of fatty acid composition. *Journal of the American Oil Chemists' Society, 74*(12), 1587–1594.

Weaver, C. M., & Cheong, J. M. K. (2005). Soy isoflavones and bone health: The relationship is still unclear. *Journal of Nutrition, 135*(5), 1243–1247.

Wilcox, J. R., Premachandra, G. S., Young, K. A., & Raboy, V. (2000). Isolation of high seed inorganic P, low-phytate soybean mutants. *Crop Science, 40*(6), 1601–1605.

Wilson, L., Birmingham, V. A., Moon, D. P., & Snyder, H. E. (1978). Isolation and characterization of starch from mature soybeans. *Cereal Chemistry, 55*(5), 661–670.

Wilson, R. F. (2004). Seed composition. In *Soybeans: Improvement, Production, and Uses* (16th ed.). https://doi.org/10.2134/agronmonogr16.3ed.c.

Wolf, W. (1972). Soybean ultrastructure and its relationship to processing. In *Symposium: Seed Proteins* (pp. 231–241). AVI Publishing Co.

Wolf, W., & Briggs, D. (1985). Studies on the cold-insoluble fraction of the water-extractable soybean proteins. II. Factors influencing conformation changes in the 11S component. *Archives of Biochemistry and Biophysics, 76*, 377–393.

Xu, Y., Li, H. N., Li, G. J., Wang, X., Cheng, L. G., & Zhang, Y. M. (2011). Mapping quantitative trait loci for seed size traits in soybean (*Glycine max* L. Merr.). *Theoretical and Applied Genetics, 122*(3), 581–594.

Yadav, N. S. (1996). Genetic modification of soybean oil quality. In *Soybean: Genetics, Molecular Biology and Biotechnology* (pp. 165–188).

Yan, L., Hofmann, N., Li, S., Ferreira, M. E., Song, B., & Jiang, G. (2017). Identification of QTL with large effect on seed weight in a selective population of soybean with genome-wide association and fixation index analyses. *BMC Genomics, 18*(1), 1–11.

Yuan, F. J., Zhao, H. J., Ren, X. L., Zhu, S. L., Fu, X. J., & Shu, Q. Y. (2007). Generation and characterization of two novel low phytate mutations in soybean (*Glycine max* L. Merr.). *Theoretical and Applied Genetics, 115*(5), 945–957.

Zarkadas, C. G., Gagnon, C., Gleddie, S., Khanizadeh, S., Cober, E. R., & Guillemette, R. J. D. (2007). Assessment of the protein quality of fourteen soybean (*Glycine max* (L.) Merr.) cultivars using amino acid analysis and two-dimensional electrophoresis. *Food Research International, 40*(1), 129–146.

Zhang, H., Goettel, W., Song, Q., Jiang, H., Hu, Z., & Wang, M. L. (2020). Selection of *GmSWEET39* for oil and protein improvement in soybean. *PLoS Genetics, 16*(11), e1009114.

CHAPTER 3

Qualitative and Quantitative Traits of Soybeans

NISHA AGRAWAL, M. SHIVAKUMAR,* V. NATARAJ, GIRIRAJ KUMAWAT, PRIYA MEGHWAL, SATISH NAGAR, and MILIND B. RATNAPARKHE

Indian Institute of Soybean Research (ICAR–IISR), Indore, Madhya Pradesh, India

*Corresponding author

ABSTRACT

Soybean is an economically important oilseed crop grown worldwide due to the high value of its protein and oil content. Given its narrow genetic base, the study of soybean is essential as it serves as a beneficial crop for both human consumption and animal feed. Soybean has a wide range of available genetic resources, and its associated agronomic traits support numerous applications in breeding. Genome organization reveals that soybean is diploid and encodes qualitative and quantitative traits governed by multiple genes. Economically important traits are regulated by both qualitative and quantitative loci. Qualitative traits include agronomic characteristics such as leaf traits, flower color, stem pigmentation, pubescence, seed shape, seed traits, and pod traits. These traits support broad applications in genetic mapping and marker development for identifying genes that control key traits. Many quantitative trait loci (QTLs) governing important traits in soybean have been identified, though many remain to be explored. Candidate genes for quantitative traits have also been identified. Functional validation of many traits is still needed, and molecular markers tagging particular loci represent advanced areas of soybean research. Thus, soybean is a major model crop among legumes

Soybeans, Volume 2: Genome, Biotechnology, Stress, and Disease.
R. Ambika Rajendran, Ayyagari Ramlal, Sanjay Kumar Lal, and Godfree Chigeza (Eds.)

DOI: 10.1201/9781998511730-4

for studies in genomics, translational genomics, and proteomics. This chapter emphasizes genetic and molecular analysis of both qualitative and quantitative traits.

3.1 INTRODUCTION

Soybean (*Glycine max* (L.) Merr.) is an emerging crop among the legume family for oilseeds. Soybean is an important source of nutrients and food resources across the world. It is domesticated along with the wild-type soybean named *Glycine soja* Siebold and Zucc. Wild type soybean originated from East China 6000–9000 years ago, and then genetic variation adapted the crop to different areas. In Heilongjiang Province, located in the north of China, the soybean crop became one of the major crops at that time. Heritability and genetic variation are the sources of changes in the traits. Any type of change can be a qualitative type or a quantitative type due to modifications in the genome. When the desired character is controlled by one locus, a change in the locus brings phenotypic effects, resulting in the qualitative trait. Quantitative traits are controlled by many loci, and changes in either one or more bring the effect through epistatic interaction. The soybean crop poses an extremely narrow genetic base, as explained by Maranna et al. (2021) and Shivakumar et al. (2019). Delannay et al. (1983) and Gizlice et al. (1994) traced the gene pool of Northern American soybean, which has an 80% gene pool of northern and southern origins from introductions. The reduced potential genetic diversity makes crops more vulnerable to various stresses and less sustainable against crop loss. In several crops, single genetic male sterility causes a loss of function that is unfortunately linked to disease susceptibility (Ramlal et al., 2022).

3.2 QUALITATIVE TRAIT

Qualitative traits in genetics are traits that are governed by few genes or controlled by a single gene that result in distinct, easily distinguishable phenotypic variations. These traits are typically directed by alleles of a single gene that result in clear-cut, discrete phenotypes, such as the presence or absence of a trait, and different colors, shapes, or patterns. Qualitative traits in soybeans are provided in Table 3.1. It includes leaf shape, plant growth habit, pubescence/hair color, seed shape, leaf intensity of green color, leaf size of the lateral leaflet, pod color, Testa color, cotyledon/flower color, hairiness/pubescence density, hilum color, seed coat color, hypocotyl coloration, and anthocyanin pigment. Various studies were performed to analyze the characteristic changes that take place through genetic control. In this regard, Palmer et al. (2008) studied 45 different genotypes distributed in 5 groups for qualitative traits associated with soybean plants, including hair color, which showed great variations among them. Similarly, leaf shape, plant growth habit, hair color, and seed shape were distributed in four groups (Figure 3.1). However, for the pod color, leaf size of a lateral leaflet, leaf intensity of green color, and Testa color, the genotypes were distributed into three distinct groups, respectively. While only two groups were found with the color of flower and hairiness. The majority of genotypes exhibited white flower color, hairiness, and yellow-green colored Testa. When assessing diversity, it became evident that leaf shape, plant growth habit, seed shape, pod color, and leaf green color intensity displayed substantial morphological diversity compared to other qualitative traits. Furthermore, these qualitative traits have been extensively used for characterizing soybean genotypes. Bernard and Weiss (1973) studied the male-sterile lines for flower morphology and pollen viability and validated fertility restoration. The results found that for genetic studies, the cytoplasmic male sterile line is quite useful if a genotype will restore fertility in a hybrid combination with the sterile lines. Kao et al. (2021) analyzed 200 germplasm accessions and found significant variability among 14 qualitative traits, including plant type, flower color, stem color, hilum color, seed shape, seed coat color, hypocotyl coloration, number of branches, leaf color, leaflet size, leaflet shape, pubescence density, pubescence color, and pod set capacity, respectively. Kumawat et al. (2019) analyzed the genetic variability for qualitative and quantitative traits through interspecific hybridization (Figure 3.2). Some pigment-based analyses were performed, and Kovinich et al. (2010) suggested that in black seeds of soybean, anthocyanin pigments accumulate within the seed coat palisade cells (known as the epidermal layer). Before 2009, it was believed that only black soybean seed coats had major anthocyanin pigments, which are structurally simple and consist of 3-O-glucosides of cyanidin, delphinidin, and petunidin in most of the varieties (Choung et al., 2001). Lee et al. (2009) analyzed six additional anthocyanin components, including pelargonidin-3-O-glucoside, catechin-cyanidin-3-O-glucoside, and minor amounts of cyanidin-3-O-galactoside, delphinidin-3-O-galactoside, cyanidin, and peonidin-3-O-glucoside, respectively, from the black seed coats of soybean. This indicates that anthocyanin biosynthesis in the seed coat is more complex than previously reported, and genes coordinate the biosynthesis and accumulation of anthocyanins (Springob et al., 2003). Also, in soybean, Bernard and Weiss (1973) and Palmer and Kilen (1987) reported that six genetic loci (I, R, T, Wp, W1, and O) control seed coat color. Tuteja and Vodkin (2008) found that chalcone synthase, the branch-point enzyme that commits phenylpropanoid flux to flavonoid/anthocyanin biosynthesis, is encoded by a small gene family (CHS1–CHS9). Senda et al. (2004) and Tuteja et al. (2004) reported that the *CHS7* and *CHS8* genes play a major role in the seed coat through the RNAi silencing method. They also found that the *CHS7* and *CHS8* genes resulted in the inhibitor (I) locus reducing flux into the flavonoid pathway, causing the appearance of a yellow seed coat. Bernard and Weiss (1973) identified four alleles of this I locus (I, ii, ik, and i), each of which resulted in different spatial distributions of pigments in the seed coat of the crop.

Salas et al. (2006) studied the round-shaped soybean seeds, and the QTLs responsible for the seed geometry traits were mapped. They used 192 segregants from three different recombinant inbred lines (RILs) derived from Minsoy × Archer, Minsoy × Noir1, and Noir1 × Archer as mapping populations. Segregate populations were sown in two experimental locations with two replication trials: one in Los Andes, Chile, and another replication trial in East Lansing, MI, USA. Phenotypic traits were evaluated for seed length (SL), seed height (SH), seed width (SW), and seed volume (SV), and seed shape was determined using the formula of seed width × height × length. The heritability of seed shape was estimated through analysis of variance. A significant number of 19 QTLs with $\geq$ 3.7 LOD value were found on 10 linkage groups (LG) contributing to all the studied traits. Based on the populations and environments studied, only one QTL proved to be stable, while six were consistent in at least two populations across both environments. A single QTL accounted for 7.5% of the phenotypic variation in SH, 18.5% for SW, and four or fewer QTLs contributed to 30% of the genetic variation. They also observed strong correlations between all traits across all populations, ranging from 0.5 to 0.9. However, no significant correlation was found between SL and SW. In conclusion, the study revealed that seed shape is a genetically complex trait. Even when similar phenotypic parents were carefully selected, different genotypes, identified through molecular markers, can yield desirable transgressive segregants. In a similar study, Hina et al. (2020) focused on soybean seed size and shape to overall enhance yield and quality. They employed two related RIL populations (ZY and K3N) and conducted evaluations under various experimental conditions. The goal was to identify six seed-related traits for main and epistatic QTLs. Through composite interval mapping (CIM), they discovered a total of 88 QTLs, and with mixed-model-based composite interval mapping (MCIM), they found 48 QTLs. Around 15 QTLs were common to both methods, including two major and novel QTLs (qSW-1-1ZN and qSLT-20-1K3N). Additionally, they identified 51 and 27 QTLs for the first time using CIM and MCIM methods, respectively. Notably, they detected four major QTLs on chromosomes 06, 10, 13, and 20, which were considered hotspots in their research.

FIGURE 3.1 Qualitative traits (flower color, leaf type, growth habit, seed coat color, stem pigmentation, pod color, pod length, hilum color, seed shape, seed size) in soybean.

FIGURE 3.2 Genetic variability generated for qualitative and quantitative traits from inter-specific cross JS 335 × *G. soja* in soybean.

3.3 QUANTITATIVE TRAITS

Quantitative traits are influenced by multiple genes, contributing to significant variation in genotypes. Table 3.1 summarizes various quantitative traits, including 100-seed weight (100-SW), seed weight per plant (SW/plant), number of seeds per plant, number of effective pods, photoperiod (PP), number of nodes, plant height (PH), internode length, days to maturity (DTM), protein content, long juvenile trait (LJ), seedling shoot, and root architecture trait. Kao et al. (2021) examined 200 germplasm accessions and observed substantial variability in 15 quantitative traits, including PH, shelling percentage, seed architecture traits (including SL, width, thickness, immature seed thickness, immature SW, and immature SL in mm), leaflet length, leaflet width, stem length to the first pod, pod length, pod width, and 100 immature seed weight (100-SW). Similarly, quantitative traits like grain yield, biomass, 100-seed weight, number of branches, number of nodes, and number of pods were studied in different

TABLE 3.1 Details of the Qualitative and Quantitative Traits of Soybean

	Traits	Different Forms	References
Qualitative	Leaf shape	Oval, needle, ellipse, linear	Li et al. (2021a)
	Evolutionary type	Wild, semi-wild	Li et al. (2021a)
	Bloom	mud film, mud-free film, lustrous	Li et al. (2021a)
	Plant growth habit	Determinate, semi-determinate	Kao et al. (2021)
	Pubescence/hair color	Greyish white, Pale brown, brown	Kao et al. (2021)
	Seed shape	Round, oblate, oval, flat	Salas et al. (2006)
	Leaf intensity of green color	Green, dark green	Kao et al. (2021)
	Leaf size of lateral leaflet	Lanceolate, lanceolate to oblong, rhomboid, oval, elliptic	Kao et al. (2021)
	Pod color	Brown, tan, black	Guo (2023)
	Testa color	Yellow, green	Li et al. (2014)
	Cotyledon/flower color	Purple, white	Li et al. (2021a)
	Hairiness/ pubescence density	Absent, rare, sparse, medium, dense	Kao et al. (2021)
	Hilum color	Black, brown	Li et al. (2021a)
	Seed color	Black, brown, yellow, green, dichromatism	Li et al. (2021a)
	Seed coat color	Yellowish white, yellow, green pale brown, reddish brown	Kao et al. (2021)
	Hypocotyl (Anthocyanin) coloration	Green, purple	Bernard and Weiss (1973); Palmer and kilen (1987)
	Stem color	Light green, green, dark green	Kao et al. (2021)

TABLE 3.1 *(Continued)*

	Traits	Different Forms	References
Quantitative	100-seed weight (g)	H Satt541, B2 Satt304	Li (2006)
	Seed yield (g)	A1 Satt511	Li (2006)
	Seeds filling period	F Satt335	Li (2006)
	Seed hardness (%)	–	Keim et al. (1990)
	Number of seeds per plant (number)	–	Pedersen and Lauer (2004)
	Number of effective pods	–	Pedersen and Lauer (2004)
	Photoperiod (GDD)	–	Bernard (1971)
	Number of branches per plant (number)	–	He et al. (2014)
	Number of nodes (number)	*Glyma.06G094400, Glyma.06G147600, Glyma.19G160800.1,* and *Glyma.19G161100*	Li et al. (2021b)
	Plant height (cm)	L Satt652, Dt1 L	Lee et al. (1996); Li (2006);
	Lodging (%)	I Satt330, L Satt284, L Dt1	Lee et al. (1996); Li (2006)
	Internodes length (cm)	LGs A2, C2, E, F, G, I, and K	Josie et al. (2007)
	Days to maturity (days)	F Satt335, N Satt584, K	Li (2006)
	Protein content (%)	–	Tian et al. (2020)
	Long juvenile trait	–	Hartwig and Kiihl (1979); Tripathi et al. (2021)
	Seedling shoot (mm)	–	Prince et al. (2020)
	Root architecture trait (cm)	Satt315–I locus on chromosome 8	Manavalan et al. (2015)

mapping populations and crosses by various researchers across the country (Kumar et al., 2023; Maranna et al., 2021; Shivakumar et al., 2011, 2019) and revealed significant variability among the traits for improvement in soybean productivity.

3.3.1 SEED WEIGHT AND YIELD

Productivity-related agricultural traits like yield and seed weight are typically influenced by multiple genes, polygenic genes, or quantitative genes. Plant breeding and crop domestication have effectively increased the allelic combinations that boost yield. However, during the bottleneck events caused by domestication and modern breeding within adapted gene pools, many beneficial alleles have been overlooked or lost. Wild species of the crop have fewer or more allele combinations of genes controlling the agricultural traits (Tanksley and McCouch, 1997). It was found that segregating populations controlled 11–15 traits clustered in 3 intervals, of which 2 are associated with seed yield (Mansur et al., 1993). Mansur et al. (1993) studied these 15 traits, including those responsible for plant survival and growth, such as morphological traits, reproductive traits, and increasing the number of plants, including seed traits, seed yield, and carbon isotope discrimination ratios. The favorable alleles controlling the traits might be very beneficial if introduced into the elite cultivars. Further, PI 245331 contains the alleles favorable for 3 out of 9 traits, with 33.3% phenotypic variation in yield and yield component QTL, loci for seed yield, 100-SW, seed filling period, maturity (DTM), PH, and plant lodging (LDG) respectively. Several studies reported the positive alleles for yield from wild accessions in 81 different crop species (Tanksley et al., 1996; Xiao et al., 1998), and in soybean, higher yields must come from either more seeds per unit area or from a longer seed filling period from wild type soybean (Concibido et al., 2003; Kabelka et al., 2004).

Bárbaro et al. (2006) demonstrated that soybean grain yield is mainly influenced by the number of seeds per pod (NSPP) and 100-SW. According to Pedersen and Lauer (2004), the yield per unit area of the crop is determined by the number of seeds per unit area and ultimately depends on the number of plants per unit area, which results from the vigor and high germination of the variety under cultivation. While the NSPP depends on the number of ovules per placenta and the frequency of embryonic abortions. Carlson and Lersten (2004) suggested that ovaries contain one to four ovules, which is determined during the flowering stage and results in NSPP. Tischner et al. (2003) used the RIL populations derived from three cultivars' reciprocal crosses and found multiple significant QTL for average NSPP in soybean, accounting for 50% of the heritable variation. Moreover, they found three QTL (*Ms1*, *Ms6*, and *St5*) linked to male and female sterility, leaflet number (*Lf1*), and leaflet shape (*ln*), respectively. Among these genes associated with NSPP, the leaflet shape gene has a narrow leaflet and is tightly linked with the NSPP or has a major pleiotropic effect (Bernard and Weiss, 1973; Domingo, 1945; Takahashi, 1934). However, the evidence for genetic association between the two traits has not been resolved due to the presence of additional minor modifying genes for the leaflet shape in the same genetic background (Jeong et al., 2011; Weiss, 1970). Many researchers worked on this aspect and found the relationship between the leaflet shape in locus and seed yield using the isogenic lines (Mandl and Buss, 1981). You et al. (1995) analyzed the interaction of this trait using different cultivars. Moreover, Dinkins et al. (2002) studied using broad (*Ln/Ln*), heterozygous (*Ln/ln*), and narrow (*ln/ln*) leaflet types and found that the broad and narrow leaflet shape genotypes have similar seed yields, but the narrow leaflet shape plants produced smaller seed sizes compared to broad leaflet plants. Lee et al. (2001) concluded that the narrow leaflet-shaped plant produces more seeds compared to the broad-shaped plant. For higher seed production, Yu et al. (2008) developed multiple narrow leaflet-shaped soybean cultivars. Collectively, it was concluded that the two traits of leaflet shape and seed yield were either tightly linked or controlled by pleiotropic effects. Jeong et al. (2012), in another study, found that the narrow leaflet-shaped soybean varieties had more seeds per pod compared to broad leaflet varieties, governed by a single recessive gene responsible for leaflet shape. The homologous single gene of *Arabidopsis JAGGED* (*JAG*) responsible for lateral organ development and exerting a pleiotropic effect on fruit patterning was mapped to the 12.6-kb region (designated *Gm-JAGGED1*) in soybean. An amino acid (AA) substitution was found in this gene that encodes the EAR motif, resulting in broad (*Ln*) leaflet to narrow leaflet (*ln*). Carvalho et al. (2017) previously reported that the NSPP and 100-SW were negatively correlated. Li et al. (2018) believed that an increase in the number of four-seeded pods caused a decrease in seed size, ultimately leading to significant declines in yield gains. However, compared to other varieties of soybean, three-seeded pods have a greater number of pods and larger seeds (Tavares et al., 2013; Zhou et al., 2005). Recently, a RIL population of soybeans was analyzed for seed size and shape traits by Kumawat and Xu (2021), who identified a major locus, qSS2.

3.3.2 PLANT HEIGHT, LODGING AND MATURITY

Many researchers (Griffin and Palmer, 1987a, b, 1989; Kiang and Gorman, 1983; Kiang, 1987; Palmer and Kilen, 1987) worked on the study of QTL and inheritance of the traits in soybean. In this regard, Graef et al. (1988) studied eight isozyme loci and the associated quantitative traits responsible for the crosses between wild-type *G. soja* × *G. max*. Moreover, the association of phenotypes with the number of isozyme loci for different traits, including PH, lodging, plant type, and vining, was studied. Further, Graef et al. (1989) evaluated approximately 4000 BC2F4 lines at two locations in two replications for DTM, PH, LDG, plant type (PLT), and vining (VNG). Similarly, Lee et al. (1996) studied the traits in 111 lines derived from the cross PI 97100 × Coker 237 for the association of HT, LDG, and maturity. It was found that for HT and LDG, a Dt1 major locus was present on the LG L through combined and interval mapping, with the phenotypic variation of 67.7% and 56.4% for both traits. In addition, two more QTLs were identified for HT on LG H (KO07) and LG N (A516b). For plant maturity, two independent QTLs were identified on LG K between the R051 and N100 intervals, and between B032 and CpTI. Keim et al. (1990) found a genetic variation of 16–24% for the Restriction Fragment Length Polymorphism (RFLP) markers and QTLs associated with eight reproductive and morphological traits. Moreover, for DTM and PH-linked RFLP markers, R79 (31% variation) and G173 (53% variation) were found.

3.3.3 SEED HARDNESS

Seed germination in a crop depends upon the seed's hardness and dormancy, which is an undesirable trait in soybean. However, under waterlogging conditions, after sowing, seed hardness plays a considerable role during this period (Hou and Thseng, 1991). Kumawat et al. (2020) identified 14 germplasm accessions that are responsible for containing the functional alleles of GmHS1-1 for the seed hardness trait. Among these accessions, one was present in *G. soja* wild type, and the other 13 were *G. max* genotypes. Moreover, the hardness of the seed was studied by Keim et al. (1990) in soybean, who reported that seed germination rate, viability, and the quality of stored seeds are quantitative traits. Keim et al. (1990) identified 72 RFLP markers associated with seed hardness in a segregating population of a cross between *G. max* and *G. soja* and identified 5 independent variations in RFLP markers. They found a 71% variation in seed hardness between the RFLP markers.

3.3.4 SEED PROTEIN AND OIL

Hawbaker et al. (1993) studied the QTLs associated with 10 traits, including seed protein and seed weight, in three cultivars: BSR 101, Hodgson 78, and Jilin 3. They planted 63 BSR 101-derived lines, eight Hodgson 78 lines, and 42 Jilin 3 lines, along with control plants, in a randomized block design trial (RBD) for two years in two locations to assess phenotypic data. They found significant ($P < 0.05$) variations among these lines for all 10 quantitative traits. Hawbaker et al. (1993) observed 57.1% variation in BSR 101 lines, 87.5% in Hodgson 78 lines, and 76.2% in Jilin 3 lines for at least one trait. The significance for oil content was found to be 2.7%, and 25.7% for DTM. The protein percentage in soybean was quantitatively controlled and responsible for the quantity and quality of protein (Burton, 1985). He conducted soybean crossbreeding between adapted and exotic germplasms and found an average heritability of 0.83% for protein. Multiple studies have investigated isozyme loci associated with various traits using QTL analysis in soybean.

3.3.5 PHOTOPERIOD AND LONG JUVENILE TRAIT

Being a short-day crop, soybean readily flowers when the day length is shorter than its critical requirement. The photoperiodic trait, one of the quantitative traits harnessed by multiple genes, includes the photoperiod and maturity genes *E1* (Bernard, 1971), *E2* (Buzzell, 1971), *E3* and *E4* (Buzzell and Voldeng, 1980), *E5* (McBlain and Bernard, 1987), *E7* (Cober and Voldeng, 2001), *E8* (Cober et al., 2010), and *E10* (Samanfar et al., 2017), and these genes are present in many allelic forms. These genes are crucial for adapting soybean to different latitudes, making the plant either insensitive or sensitive to day length, leading to early or late flowering and maturity. Dominant gene forms result in late flowering and maturity, while recessive alleles cause early flowering and maturity. The short-day effect limits soybean growth in lower latitude regions, prompting the search for genotypes with an extended vegetative period under inductive conditions. Hartwig and Kiihl (1979) were the first to demonstrate the genetic control of the trait, that is, known as a long juvenile trait. E6, E9, and E11 (Bonato and Vello, 1999; Kong et al., 2014; Wang et al., 2019) studied the genes and reported that the recessive allele of these genes caused late flowering and maturity, adapted for the lower latitude for proper vegetative growth. Long juvenility trait and its inheritance have been reported in AGS 25 in India (Gupta et al., 2021a, b). Through QTL mapping, the association of *FT2a* allele and delayed flowering was reported by Tripathi et al. (2021). Harnessing long juvenility in breeding programs can lead to soybean varieties suitable for delayed monsoons and wider adaptability in nontraditional areas.

3.4 BREEDING APPROACH FOR THE QUALITATIVE AND QUANTITATIVE TRAIT

Transfer of qualitative and quantitative traits through conventional breeding and marker assisted breeding was successful in many crops, including soybean (Kumar et al., 2015; Maranna et al., 2016). For qualitative breeding, conventional methods are typically sufficient, as the transfer of traits can be visually observed. In contrast, for quantitative trait improvement, molecular breeding tools and techniques are invaluable. Molecular markers, in particular, have become a powerful tool for breeders. They aid in identifying new sources of variation and locating the genomic regions or loci that control inherited quantitative traits (QTLs). Linked markers associated with traits of interest have now become a useful tool for major crop species in the dissection of complex inherited QTLs like yield and seed weight traits (Li, 2006). Due to the genetic complexity and the effect of one or more alleles in controlling the quantitative trait, it is difficult to analyze in the segregating populations. Therefore, to resolve the problem of multiple alleles in segregating populations, advanced backcross populations have been proposed by Tanksley and Nelson (1996). Although backcrossing decreases the presence of donor-derived wild species alleles and allows many undesirable or lethal alleles to be readily exposed, it also facilitates the identification of favorable donor alleles. The linked favorable alleles can be easily transferred or introgressed into the elite germplasm pool with a lesser chance of acquiring undesirable or lethal alleles through recombination selection. Through the approaches of linkage mapping and backcross breeding, the alleles from the wild or exotic plant can be easily tagged in the genetic background of an elite cultivar (Li, 2006). Moreover, the transfer of alleles with their positions into the progeny can be performed in the elite cultivar. Time is also an important factor to be considered in backcross breeding from wild species. Tanksley et al. (1981) proposed the time-based selection of isozyme markers to decrease the number of backcross generations needed, thereby reducing the time required for introgression. Based on the marker loci linked to the trait, this approach may also be useful to increase the recovery of the cultivated phenotype and reduce the number of segregants to be evaluated for economically important traits in the field or trial locations. The use of genetic markers for the selection of quantitative traits has been a concern since Sax (1923), who reported a linkage between seed size and pigmentation in *Phaseolus vulgaris* L. Similarly, many researchers identified a linkage between molecular markers and genetic factors controlling quantitative traits (Suarez et al., 1991; Thompson and Thoday, 1979). Suarez (1989) reported that the traits were all quantitatively inherited, with the involvement of different gene actions. Suarez et al. (1991) justified the use of molecular markers based on the fact that the marker locus serves to identify the chromosomal region in which it is located and enables that region to be followed in inheritance or selection experiments. The QTL linked to the markers can then be selected indirectly, which reduces the need for costly performance trials necessary to evaluate the quantitative traits. To make the technique of selecting for QTL with marker genes more powerful and easily applicable, it would be desirable to have a large number of genes distributed throughout the genome with alleles interacting in a co-dominant manner. Each marker locus should produce no effect on the morphology of the organism and should be unequivocally identifiable, even if a large number of marker loci are segregating in the same cross. Griffin and Palmer (1987a) and Tanksley et al. (1982) fit the criteria of multiple loci inherited in quantitative traits. In this regard, many researchers have worked on analyzing quantitative traits in soybean through linked molecular markers for the photoperiodic trait, as mentioned in the above part about photoperiod and the long juvenile phase. Multiple loci govern the photoperiodic trait in soybean, enabling its adaptation to various latitudes. Different allelic forms of dominant and recessive genes control soybean's response to short and long days. The lipoxygenase trait, which affects the "beany" flavor in soybeans, also has various alleles governing this characteristic. Additionally, 100-SW in soybeans, a key factor in enhancing productivity, is influenced by allele-linked markers that facilitate the transfer of this trait for increased productivity. These traits are not effectively assessed through conventional breeding approaches. Marker-assisted breeding proves to be a valuable tool in addressing these complex genetic traits. Many researchers worked on them Hildebrand et al. (1980) reported that the acid phosphatase locus (*Ap*) was linked to the Kunitz-trypsin inhibitor locus (Ti) and found on chromosome 9 of soybean. However, Rennie et al. (1987) were not successful in their work related to finding the resistance loci of *Phytophthora megasperma* Drechs. to other isozyme loci. However, Griffin and Palmer (1987a) identified the inheritance of four aconitase and one endopeptidase isozyme, as well as the linkage between these and several other loci. They analyzed that the expression of allozymes and the monogenic nature of the five isozyme loci exhibited co-dominant inheritance. Additionally, they found that the ACO3 locus was located in LG 1, while all other isozyme variants were independently inherited.

3.5 CONCLUSION AND FUTURE PROSPECTS

Genetic studies are the foundation of crop improvement programs, encompassing both qualitative and quantitative traits. The limited genetic diversity in commercial crop cultivars can lead to genetic vulnerability and stagnant selection progress. To address this, researchers often introduce exotic or wild germplasm into adapted cultivars to boost genetic variability. One commonly used method of increasing genetic variability among adapted cultivars is the introgression of exotic or wild germplasm. Wild species have been used most frequently as sources of pest resistance, although they have also provided genes for improved yield in some crops. The main reasons for the use of unadopted or wild germplasm in introgression programs are: (i) to increase the genetic diversity of the crop as a safeguard against unpredictable biological and environmental hazards; (ii) as a source of genes for specific traits such as disease, pest, and stress resistance; and (iii) as a source of favorable alleles for yield. Enhancing our understanding of genomics and phenomics for both cultivated and wild types is crucial for further improving soybean yields and trait attributes. Integrating genetics, genomics, and phenomics for qualitative and quantitative traits will help us better understand soybean breeding programs. This ultimately results in the enhancement of crop yield and component traits through genome editing (GE) approaches.

KEYWORDS

- **genetic diversity**
- **genome editing**
- **germplasm introgression**
- **molecular markers**
- **qualitative and quantitative traits**
- **quantitative trait loci**
- **soybean genetics**

REFERENCES

Bárbaro, I. M., Mauro, A., Unêdatrevisoli, S. H., Arriel, N., & Costa, M. M. (2006). Path analysis and expected response in indirect selection for grain yield in soybean. *Crop Breeding and Applied Biotechnology, 6*(2), 151–159. https://doi.org/10.12702/1984-7033.v06n02a06.

Bernard, R. L. (1971). Two genes for time of flowering in soybeans. *Crop Science, 11*(2), 242–244. https://doi.org/10.2135/cropsci1971.0011183X001100020022x.

Bernard, R. L., & Weiss, M. G. (1973). Qualitative genetics. In B. E. Caldwell (Ed.), *Soybeans: Improvement, Production, and Uses* (1st ed.). American Society of Agronomy.

Bonato, E. R., & Vello, N. A. (1999). E6, a dominant gene conditioning early flowering and maturity in soybeans. *Genetics and Molecular Biology, 22*(2), 229–232. https://doi.org/10.1590/S1415-47571999000200016.

Burton, J. W. (1985). Breeding soybeans for improved protein quantity and quality. In *Proceedings of the 3rd World Soybean Research Conference* (pp. 361–367).

Buzzell, R. I. (1971). Inheritance of a soybean flowering response to fluorescent-daylength conditions. *Canadian Journal of Genetics and Cytology, 13*(4), 703. https://doi.org/10.1139/g71-100.

Buzzell, R. I., & Voldeng, H. D. (1980). Inheritance of insensitivity to long day length. *Soybean Genetics Newsletter, 7*, 26–29. https://lib.dr.iastate.edu/soybeangenetics/vol7/iss1/13 (accessed on 25 March 2025).

Carlson, J. B., & Lersten, N. R. (2004). Reproductive morphology. In H. R. Boerma & J. E. Specht (Eds.), *Soybeans: Improvement, Production, and Uses* (3rd ed., pp. 59–95). ASA, CSSA, and SSSA.

Carvalho, I. R., Nardino, M., Demari, G. H., Szareski, V. J., Follmann, D. N., Pelegrin, A. J., et al. (2017). Relations among phenotypic traits of soybean pods and growth habit. *African Journal of Agricultural Research, 6*(2), 450–458. https://doi.org/10.5897/AJAR2016.11660.

Choung, M., Baek, I. Y., Kang, S. T., et al. (2001). Isolation and determination of anthocyanins in seed coats of black soybean (*Glycine max* (L.) Merr.). *Journal of Agricultural and Food Chemistry, 49*(12), 5848–5851.

Cober, E. R., & Voldeng, H. D. (2001). Low R: FR light quality delays flowering of E7E7 soybean lines. *Crop Science, 41*(6), 1823–1826. https://doi.org/10.2135/cropsci2001.1823.

Cober, E., Molnar, S., Charette, M., & Voldeng, H. (2010). A new locus for early maturity in soybean. *Crop Science, 50*(2), 524–527. https://doi.org/10.2135/cropsci2009.04.0174.

Concibido, V., La Vallee, B., Mclaird, P., et al. (2003). Introgression of a quantitative trait locus for yield from *Glycine soja* into commercial soybean cultivars. *Theoretical and Applied Genetics, 106*(4), 575–582. https://doi.org/10.1007/s00122-002-1071-5.

Delannay, X., Rodgers, D. M., & Palmer, R. G. (1983). Relative genetic contributions among ancestral lines to North American soybean cultivars. *Crop Science, 23*(5), 944–949.

Dinkins, R. D., Keim, K. R., Farno, L., & Edwards, L. H. (2002). Expression of the narrow leaflet gene for yield and agronomic traits in soybean. *Journal of Heredity, 93*(5), 346–351.

Domingo, W. E. (1945). Inheritance of number of seeds per pod and leaflet shape in the soybean. *Journal of Agricultural Research, 70*, 251–268.

Gizlice, Z., Carter, T. E., Jr., & Burton, J. W. (1994). Genetic base for North American public soybean cultivars released between 1947 and 1988. *Crop Science, 34*(4), 1143–1151.

Graef, G. L. (1988). *Relation of Isozyme Genotypes to Quantitative Characters in Soybean* (Doctoral dissertation, Iowa State University, Ames, IA).

Graef, G. L., Fehr, W. R., & Cianzio, S. R. (1989). Relation of isozyme genotypes to quantitative characters in soybean. *Crop Breeding, Genetics, & Cytology, 29*(3), 683–688.

Griffin, J. D., & Palmer, R. G. (1987a). Inheritance and linkage studies with five isozyme loci in soybean. *Crop Science, 27*(5), 885–892.

Griffin, J. D., & Palmer, R. G. (1987b). Locating the *Spl* locus on soybean linkage group 1. *Journal of Heredity, 78*(2), 122–123.

Griffin, J. D., & Palmer, R. G. (1989). Genetic studies with two superoxide dismutase loci in soybean. *Crop Science, 28*(1), 122–127.

Guo, Y. (2023). Soybean pod coloration and yield loss. *Nature Food, 4*, 538. https://doi.org/10.1038/s43016-023-00808-8.

Gupta, S., Agrawal, N., Tripathi, R., Kumawat, G., Nataraj, V., Maranna, S., et al. (2021b). Long juvenility trait: A vehicle for commercial utilization of soybean (*Glycine max* (L.) Merr.) in lower latitudes. *Plant Breeding, 140*(4), 634–645. https://doi.org/10.1111/pbr.12926.

Gupta, S., Kumawat, G., Yadav, S., Tripathi, R., Agrawal, N., Maranna, S., et al. (2021a). Identification and characterization of a novel long juvenile resource AGS 25. *Genetic Resources and Crop Evolution, 68*(7), 1591–1602. https://doi.org/10.1007/s10722-020-01055-7.

Hartwig, E. E., & Kiihl, R. A. S. (1979). Identification and utilization of a delayed flowering character in soybean for short-day conditions. *Field Crops Research, 2*(2), 145–151. https://doi.org/10.1016/0378-4290(79)90017-0.

Hawbaker, M. S., Fehr, W. R., Mansur, L. M., Shoemaker, R. C., & Palmer, R. G. (1993). Genetic variation for quantitative traits in soybean lines derived from tissue culture. *Theoretical and Applied Genetics, 87*(1, 2), 49–53. https://doi.org/10.1007/BF00223743.

He, Q., Yang, H., Xiang, S., et al. (2014). QTL mapping for the number of branches and pods using wild chromosome segment substitution lines in soybean (*Glycine max* L. Merr.). *Plant Genetic Resources, 12*(S1), S172–S177.

Hildebrand, D. F., Orf, J. H., & Hymowitz, T. (1980). Inheritance of an acid phosphatase and its linkage with the Kunitz trypsin inhibitor in seed protein of soybeans. *Crop Science, 20*, 83–85.

Hina, A., Cao, Y., Song, S., Li, S., Sharmin, R. A., Elattar, M. A., Bhat, J. A., & Zhao, T. (2020). High-resolution mapping in two RIL populations refines major "QTL hotspot" regions for seed size and shape in soybean (*Glycine max* L.). *International Journal of Molecular Sciences, 21*(3), 1040. https://doi.org/10.3390/ijms21031040.

Hou, F. F., & Thseng, F. S. (1991). Studies on the flooding tolerance of soybean seed: Varietal differences. *Euphytica, 57*, 169–173. https://doi.org/10.1007/BF00023075.

Jeong, N., Moon, J. K., Kim, H. S., Kim, C. G., & Jeong, S. C. (2011). Fine genetic mapping of the genomic region controlling leaflet shape and number of seeds per pod in the soybean. *Theoretical and Applied Genetics, 122*, 865–874.

Jeong, N., Suh, S. J., Kim, M. H., Lee, S., Moon, J. K., Kim, H. S., & Jeong, S. C. (2012). *Ln* is a key regulator of leaflet shape and number of seeds per pod in soybean. *The Plant Cell, 24*(12), 4807–4818. https://doi.org/10.1105/tpc.112.104968.

Josie, J., Alcivar, A., Rainho, J., et al. (2007). Genomic regions containing QTL for plant height, internode length, and flower color in soybean (*Glycine max* L. Merr.). *Bios, 78*(4), 119–126.

Kabelka, E. A., Diers, B. W., Fehr, W. R., LeRoy, A. R., Baianu, I. C., You, T., Neece, D. J., & Nelson, R. L. (2004). Putative alleles for increased yield from soybean plant introductions. *Crop Breeding, Genetics & Cytology, 44*(3), 784–791.

Kao, C. F., He, S. S., Wang, C. S., Lai, Z. Y., Lin, D. G., & Chen, S. (2021). A modified Roger's distance algorithm for mixed quantitative–qualitative phenotypes to establish a core collection for Taiwanese vegetable soybeans. *Frontiers in Plant Science, 11*, 612106. https://doi.org/10.3389/fpls.2020.612106.

Keim, P., Diers, B. W., Olson, T. C., & Shoemaker, R. G. (1990). RFLP mapping in the soybean: Association between marker loci and variation in quantitative traits. *Genetics, 126*, 735–742.

Kiang, Y. T. (1987). Mapping three protein loci on a soybean chromosome. *Crop Science, 27*, 44–46.

Kiang, Y. T., & Gorman, M. B. (1983). Soybean. In S. D. Tanksley & T. J. Orton (Eds.), *Isozymes in Plant Genetics and Breeding. Part B* (pp. 295–328). Elsevier.

Kong, F., Nan, H., Cao, D., Li, Y., Wu, F., Wang, J., Lu, S., Yuan, X., Abe, J., Cober, E. R., & Liu, B. (2014). A new dominant gene *E9* conditions early flowering and maturity in soybean. *Crop Science, 54*, 2529–2535. https://doi.org/10.2135/cropsci2014.03.0228.

Kovinich, N., Arnason, J. T., Luca, V. D., & Miki, B. (2010). Coloring soybeans with anthocyanins? In D. Gang (Ed.), *The Biological Activity of Phytochemicals* (Vol. 41, pp. 47–57). Springer.

Kumar, R., Saini, M., Taku, M., Debbarma, P., Mahto, R. K., Ramlal, A., Sharma, D., Rajendran, A., Pandey, R., Gaikwad, K., Lal, S. K., & Talukdar, A. (2023). Identification of quantitative trait loci (QTLs) and candidate genes for seed shape and 100-seed weight in soybean (*Glycine max* L. Merr.). *Frontiers in Plant Science, 13*, 1074245. https://doi.org/10.3389/fpls.2022.1074245.

Kumar, V., Rani, A., Rawal, R., & Mourya, V. (2015). Marker-assisted accelerated introgression of null allele of Kunitz trypsin inhibitor in soybean. *Breeding Science, 65*, 447–452. https://doi.org/10.1270/jsbbs.65.447.

Kumawat, G., & Xu, D. (2021). A major and stable quantitative trait locus *qSS2* for seed size and shape traits in a soybean RIL population. *Frontiers in Genetics, 12*, 646102. https://doi.org/10.3389/fgene.2021.646102.

Kumawat, G., Maranna, S., Gupta, S., Tripathi, R., Agarwal, N., et al. (2020). Identification of novel genetic sources for agronomic and quality traits in soybean using multi-trait allele-specific genic marker assays. *Journal of Plant Biochemistry and Biotechnology, 30*(2). https://doi.org/10.1007/s13562-020-00580-x.

Kumawat, G., Yadav, A., Shivakumar, M., Gill, B. S., Patel, R. M., Gupta, S., Satpute, G. K., Chand, S., & Husain, S. M. (2019). Validation of QTLs for seed weight in a backcross population derived from an interspecific cross in soybean (*Glycine max* L. Merr.). *Journal of Oilseed Research, 36*(4), 210–216.

Lee, J. H., Kang, N. S., Shin, S. O., et al. (2009). Characterization of anthocyanins in the black soybean (*Glycine max* L.) by HPLC-DAD-ESI/MS analysis. *Food Chemistry, 112*, 226–231.

Lee, S. H., Bailey, M. A., Mian, M. A. R., Shipe, E. R., Ashley, D. A., Parrott, W. A., Hussey, R. S., & Boerma, H. R. (1996). Identification of quantitative trait loci for plant height, lodging, and maturity in a soybean population segregating for growth habit. *Theoretical and Applied Genetics, 92*, 516–523.

Lee, S. H., Park, K. Y., Lee, H. S., Park, E. H., & Boerma, H. R. (2001). Genetic mapping of QTLs conditioning soybean sprout yield and quality. *Theoretical and Applied Genetics, 103*, 702–709.

Li, D. (2006). *Soybean QTL for Yield and Yield Components Associated with Glycine Soja Alleles* (Doctoral dissertation, University of Kentucky). University of Kentucky Doctoral Dissertations. Retrieved from: https://uknowledge.uky.edu/gradschool_diss/331 (accessed on 25 March 2025).

Li, W. X., Wang, P., Zhao, H., Sun, X., Yang, T., Li, H., Hou, Y., Liu, C., Siyal, M., Raja, V. R., Hu, B., & Ning, H. (2021a). QTL for main stem node number and its response to plant densities in 144 soybean FW-RILs. *Frontiers in Plant Science, 12*, 666796. https://doi.org/10.3389/fpls.2021.666796.

Li, W., Peng, M., Wang, Z., Bi, Y., Liu, M., Wang, L., Di, S., Liu, J., Fan, C., Yang, G., & Yu, D. (2021b). The evaluation of agronomic traits of wild soybean accessions (*Glycine soja* Sieb. & Zucc.) in Heilongjiang Province, China. *Agronomy, 11*(3), 586. https://doi.org/10.3390/agronomy11030586.

Li, Y. H., Reif, J. C., Jackson, S. A., et al. (2014). Detecting SNPs underlying domestication-related traits in soybean. *BMC Plant Biology, 14*, 251. https://doi.org/10.1186/s12870-014-0251-1.

Li, Y. Y., Li, R. C., Cheng, C. G., Zhao, Y. Y., Liu, C. Y., Qi, Z. M., et al. (2018). Meta and overview analysis of QTL associated with pod and seed traits and candidate gene mining in soybean (*Glycine max* L. Merrill). *Journal of Agricultural Biotechnology, 26*, 1821–1833.

Manavalan, L. P., Prince, S. J., Musket, T. A., Chaky, J., Deshmukh, R., Vuong, T. D., et al. (2015). Identification of novel QTL governing root architectural traits in an interspecific soybean population. *PLoS ONE, 10*(3), e0120490.

Mandl, F. A., & Buss, G. R. (1981). Comparison of narrow and broad leaflet isolines of soybean. *Crop Science, 21*, 25–27.

Mansur, L. M., Lark, K. G., Kross, H., et al. (1993). Interval mapping of quantitative trait loci for reproductive, morphological, and seed traits of soybean (*Glycine max* L.). *Theoretical and Applied Genetics, 86*, 907–913. https://doi.org/10.1007/BF00211040.

Maranna, S., Nataraj, V., Kumawat, G., et al. (2021). Breeding for higher yield, early maturity, wider adaptability and waterlogging tolerance in soybean (*Glycine max* L.): A case study. *Scientific Reports, 11*, 22853. https://doi.org/10.1038/s41598-021-02064-x.

Maranna, S., Verma, K., Talukdar, A., Lal, S. K., Kumar, A., & Mukherjee, K. (2016). Introgression of null allele of Kunitz trypsin inhibitor through marker-assisted backcross breeding in soybean (*Glycine max* L. Merr.). *BMC Genetics.* https://doi.org/10.1186/s12863-016-0413-2.

McBlain, B. A., & Bernard, R. L. (1987). A new gene affecting the time of flowering and maturity in soybean. *Journal of Heredity, 78*, 160–162. https://doi.org/10.1093/oxfordjournals.jhered.a110349.

Palmer, R. G., & Kilen, T. C. (1987). Qualitative genetics and cytogenetics. In J. R. Wilcox (Ed.), *Soybeans: Improvement, Production, and Uses* (2nd ed., pp. 135–207). American Society of Agronomy.

Palmer, R. G., & Xu, M. (2008). Positioning three qualitative trait loci on soybean molecular linkage group E. *Journal of Heredity, 99*(6), 674–678. https://doi.org/10.1093/jhered/esn070.

Pedersen, P., & Lauer, J. G. (2004). Response of soybean yield components to management system and planting date. *Agronomy Journal, 96*(5). https://doi.org/10.2134/agronj2004.1372.

Prince, S. J., Vuong, T. D., Wu, X., Bai, Y., Lu, F., Kumpatla, S. P., Valliyodan, B., Shannon, J. G., & Nguyen, H. T. (2020). Mapping quantitative trait loci for soybean seedling shoot and root architecture traits in an interspecific genetic population. *Frontiers in Plant Science, 11*, 1284. https://doi.org/10.3389/fpls.2020.01284.

Ramlal, A., Nautiyal, A., Baweja, P., Kumar, M. R., Mehta, S., Pujari, M. B., Vijayan, R., Shukla, S., Kumar, V., Kumar, D. S., Lal, S. K., Raju, D., & Rajendran, A. (2022).

Harnessing heterosis and male sterility in soybean [*Glycine max* (L.) Merr.]: A critical revisit. *Frontiers in Plant Science, 13,* 981768. https://doi.org/10.3389/fpls.2022.981768.

Rennie, B. D., Beversdorf, W. D., & Buzzell, R. I. (1987). Assaying for linkage between Rps and isozyme loci in soybean. *Crop Science, 27*, 519–521.

Salas, P., Oyarzo-Llaipen, J. C., Wang, D., Chase, K., & Mansur, L. (2006). Genetic mapping of seed shape in three populations of recombinant inbred lines of soybean (*Glycine max* L. Merr.). *Theoretical and Applied Genetics, 113*(8), 1459–1466. https://doi.org/10.1007/s00122-006-0392-1.

Samanfar, B., Molnar, S. J., Charette, M., Schoenrock, A., Dehne, F., Golshani, A., et al. (2017). Mapping and identification of a potential candidate gene for a novel maturity locus, E10, in soybean. *Theoretical and Applied Genetics, 130*, 377–390. https://doi.org/10.1007/s00122-016-2819-7.

Sax, K. (1923). The association of size differences with seed coat pattern and pigmentation in *Phaseolus vulgaris*. *Genetics, 8*, 552–560.

Senda, M., Masuta, C., Ohnishi, S., et al. (2004). Patterning of virus-infected *Glycine max* seed coat is associated with suppression of endogenous silencing of chalcone synthase genes. *The Plant Cell, 16*, 807–818.

Shivakumar, M., Basavaraja, G. T., Salimath, P. M., Patil, P. V., et al. (2011). Identification of rust-resistant lines and their genetic variability and character association studies in soybean [*Glycine max* (L.) Merr.]. *The Indian Society of Genetics & Plant Breeding, 3*(14), 235–240.

Shivakumar, M., et al. (2019). NAM population—A novel genetic resource for soybean improvement: Development and characterization for yield and attributing traits. *Plant Genetic Resources: Characterization and Utilization, 17*(6), 545–553. https://doi.org/10.1017/S1479262119000352.

Springob, K., Nakajima, J., & Yamazaki, M. (2003). Recent advances in the biosynthesis and accumulation of anthocyanins. *Natural Product Reports, 20*, 288–303.

Suárez, J. C. (1989). Association between isozyme genotypes and quantitative traits in soybean (Publication No. 9245) [Doctoral dissertation, Iowa State University]. *Retrospective Theses and Dissertations*. https://lib.dr.iastate.edu/rtd/9245 (accessed on 25 March 2025).

Suárez, J. C., Graef, G. L., Fehr, W. R., & Cianzio, S. R. (1991). Association of isozyme genotype with agronomic and seed composition traits in soybean. *Euphytica, 52*, 137–146.

Takahashi, N. (1934). Linkage relation between the genes for the form of leaves and the number of seeds per pod of soybeans. *Japanese Journal of Genetics, 9*, 208–225.

Tanksley, S. D., & McCouch, S. R. (1997). Seed banks and molecular maps: Unlocking genetic potential from the wild. *Science, 277*, 1063.

Tanksley, S. D., Grandillo, S., Fulton, T. M., Zamir, D., Eshed, Y., Petiard, V., et al. (1996). Advanced backcross QTL analysis in a cross between an elite processing line of tomato and its wild relative *L. pimpinellifolium*. *Theoretical and Applied Genetics, 92*, 213–224.

Tanksley, S. D., Medina-Filho, H., & Rick, C. M. (1981). The effect of isozyme selection on metric characters in an interspecific backcross of tomato—Basis of an early screening procedure. *Theoretical and Applied Genetics, 60*, 291–296.

Tanksley, S. D., Medina-Filho, H., & Rick, C. M. (1982). Use of naturally-occurring enzyme variation to detect and map genes controlling quantitative traits in an interspecific backcross of tomato. *Heredity, 49*, 11–25.

Tanksley, S., & Nelson, J. (1996). Advanced backcross QTL analysis: A method for the simultaneous discovery and transfer of valuable QTLs from unadopted germplasm into elite breeding lines. *Theoretical and Applied Genetics, 92*, 191–203.

Tavares, L. C., Rufino, C., Brunes, A. P., Tunes, L. M., & Peske, S. T. (2013). Performance of soybean seeds under water stress: Yield and physiological quality of F1 generation. *Ciência Rural, 43*, 1357–1363. https://doi.org/10.1590/S0103-84782013000800003.

Thompson, J. N., Jr., & Thoday, J. M. (1979). *Quantitative Genetic Variation*. Academic Press.

Tian, X., Zhang, K., Liu, S., Sun, X., Li, X., Song, J., Qi, Z., Wang, Y., Fang, Y., Wang, J., Jiang, S., Yang, C., Tian, Z., Li, W. X., & Ning, H. (2020). Quantitative trait locus analysis of protein and oil content in response to planting density in soybean (*Glycine max* [L.] Merr.) seeds based on SNP linkage mapping. *Frontiers in Genetics, 11*, 563. https://doi.org/10.3389/fgene.2020.00563.

Tischner, T., Allphin, L., Chase, K., Orf, J. H., & Lark, K. G. (2003). Genetics of seed abortion and reproductive traits in soybean. *Crop Science, 43*, 464–473.

Tripathi, R., Agrawal, N., Kumawat, G., et al. (2021). QTL mapping for long juvenile trait in soybean accession AGS 25 identifies association between a functional allele of *FT2a* and delayed flowering. *Euphytica, 217*, 36. https://doi.org/10.1007/s10681-021-02775-2.

Tuteja, J. H., & Vodkin, L. O. (2008). Structural features of the endogenous *CHS* silencing and target loci in the soybean genome. *Crop Science, 48*, S49–S68.

Tuteja, J. H., Clough, S. J., Chan, W. C., et al. (2004). Tissue-specific gene silencing mediated by a naturally occurring chalcone synthase gene cluster in *Glycine max*. *The Plant Cell, 16*, 819–835.

Wang, F., Nan, H., Chen, L., Fang, C., Zhang, H., Su, T., Li, S., Cheng, Q., Dong, L., Liu, B., Kong, F., & Lu, S. (2019). A new dominant locus, *E11*, controls early flowering time and maturity in soybean. *Molecular Breeding, 39*, 70. https://doi.org/10.1007/s11032-019-0978-3.

Weiss, M. G. (1970). Genetic linkage in soybeans. Linkage group IV. *Crop Science, 10*, 368–370.

Xiao, J., Li, J., Grandillo, S., Ahn, S. N., Yuan, L., Tanksley, S. D., et al. (1998). Identification of trait-improving quantitative trait loci alleles from a wild rice relative, *Oryza rufipogon*. *Genetics, 150*, 899–909.

You, M. G., Liu, Y. B., Zhao, T. J., & Gai, J. Y. (1995). Effects of leaf shape on seed yield and its components in soybeans. *Soybean Genetics Newsletter, 22*, 66–70.

Yu, Y. H., Yu, H., Jeong, J., Park, H., Song, D., Kim, C., Kim, S., Lee, Y., Park, J., & Park, K. (2008). *A General Survey of Korean Legume Cultivars* (in Korean). National Institute of Crop Science.

Zhou, X. A., Wang, X. Z., Cai, S. P., Jun, W. X., Sha, A. H., Qiu, D. Z., et al. (2005). Relation of three-seed and four-seed pods with yield of RIL in soybeans. *Journal of Oil Crop Science, 27*, 22–25.

CHAPTER 4

Soybean Protein Quality and Its Improvement

BINGI PUJARI MALLIKARJUNA,[1,*] P. G. SURESHA,[2]
M. SHIVAKUMAR,[3] SANJAY KUMAR LAL,[4] K. GOPALAKRISHNA NAIDU,[5]
AKSHAY TALUKDAR,[4] AMBIKA RAJENDRAN,[4] and MANISHA SAINI[4]

[1]*ICAR–Indian Agricultural Research Institute, Regional Research Center, Dharwad, Karnataka, India*

[2]*DST–Agharkar Research Institute, Pune, Maharashtra, India*

[3]*Indian Institute of Soybean Research (ICAR–IISR), Indore, Madhya Pradesh, India*

[4]*Division of Genetics, Indian Council of Agricultural Research (ICAR)–Indian Agricultural Research Institute (IARI), Pusa Campus, New Delhi, India*

[5]*All India Coordinated Research Project (AICRP) on Soybean, University of Agricultural Sciences, Dharwad, Karnataka, India*

*Corresponding author

ABSTRACT

Soybean is an important food legume crop cultivated worldwide for its protein and oil content. It is the major source of vegetable oil and high-quality protein used for human consumption and livestock feed. Nutritionally, soybean protein is abundant in essential amino acids (AAs), vitamins, and flavonoids, and it has important health benefits. Breeding for high-protein soybean is highly desirable and is a major breeding objective worldwide. However, the quantitative nature of seed

---

Soybeans, Volume 2: Genome, Biotechnology, Stress, and Disease.
R. Ambika Rajendran, Ayyagari Ramlal, Sanjay Kumar Lal, and Godfree Chigeza (Eds.)

DOI: 10.1201/9781998511730-5

protein content, its negative correlation with seed yield and oil content, and its interaction with environmental effects make breeding progress challenging in developing soybean cultivars with higher yield, accompanied by high protein and oil. In this chapter, the effect of environmental factors on seed protein content and its quality, as well as the genetic improvement of soybean protein content through the utilization of the enormous genetic variation existing in cultivated and wild soybean genetic resources, is discussed. The genetic factors controlling seed protein content, AA content, and seed storage protein are presented. In addition, we summarize the integration of breeding approaches involving germplasm resources, genomics technologies, and high-throughput phenotyping for detecting genes related to seed protein content, along with genomics-assisted breeding methods such as MAS, MARS, GS, transgenic technology, and gene-editing tools.

4.1 INTRODUCTION

Soybean [*Glycine max* (L.) Merr.] is a vital source of edible oil for human consumption and protein meal for livestock consumption, which is also becoming an important component of the human diet. Its seed is rich in oil (ranging from 8.3% to 27.9%) and contains the highest protein concentration (ranging from 34.1% to 56.8%), which is higher than any major legume crop, thus making it a unique crop in the family Leguminosae (Li et al., 2019). During 2020, soybeans contributed 61% of the world's oilseed production, with 367.8 million metric tons harvested (Soystats, 2022). Although the crop is cultivated worldwide, the major countries producing soybeans are Brazil (139.5 million metric tons), the United States (114.7 million metric tons), Argentina (46.2 million metric tons), China (19.6 million metric tons), and India (10.5 million metric tons). Soybean originated in Southeast Asia and was first domesticated in China approximately 6000 to 9000 years ago from the wild soybean, *Glycine soja* Sieb. & Zucc. (Zhang et al., 2020). The crop was successfully introduced to North America for primary usage as vegetable oil. Soon, it was found that soybean meal (SM) is an important source of protein for animal feeds and human consumption. Soybean oil and protein meal are important products of soybeans and form one of the substantial sources of plant-derived edible oil and proteins used for human consumption and animal feed (Patil et al., 2017).

Proteins are vital components of human nutrition and health and are critical for the overall development of the human body. Adequate protein intake is required throughout the human life cycle, *viz.*, infancy, childhood, and adolescence. A major portion of a protein diet is derived from animal meats and plant-based proteins. Plant-based proteins are a major source of protein for the vegetarian population and are mainly found in legumes and nuts. Soybean is also considered an important source of protein for humans. It has been used as a basic food ingredient in traditional Asian foods (such as tofu, soymilk, natto, edamame, etc.) for thousands of years. Due to high seed protein content and the production of different varieties of meat analogues and milk substitutes, soybeans are gaining interest among the vegetarian population in Western countries (Qin et al., 2022b). In recent decades, vegetarian diets have become more desirable among the population due to ethical, ecological, and health-related reasons. In recent years, increasing attention has been paid to plant-based protein sources as more evidence demonstrates that individuals consuming plant-based proteins have lower health risks (such as cardiovascular disease (CVD) and other metabolic disorders) compared to those consuming animal-based foods. With the growing interest and adoption of vegetarian lifestyles, the high-quality protein content makes soybeans one of the most attractive food choices among vegetarians. As a result, the demand for soy foods has been increasing tremendously (from $1 billion in 2000 to $4.5 billion in 2013) (Patil et al., 2017). In addition, the poultry and swine industries depend mainly on SM for feeding. Because of its high protein concentration, SM has been a major source of protein for the livestock and poultry industries. Therefore, the sustainable production of soybean protein by developing superior varieties with high-protein content has become the major objective in soybean breeding and research. This chapter provides insight into soybean protein quality as well as approaches for the improvement of protein content and quality.

4.2 SOYBEAN PROTEIN

Soybean seed is rich in protein (about 40%) and also contains lipids, dietary fibers, unsaturated FAs (mono and polyunsaturated), carbohydrates (both soluble and insoluble), vitamins, minerals, phospholipids, and isoflavones. It is considered a complete vegetable protein because the essential amino acid (AA) profile of soybean seed is comparable to that of high-quality animal protein sources. Soybean protein also contains all the essential AAs required by the human body (Guo et al., 2022). Soybean protein complements well with other proteins and cereals that are deficient in some AAs that are present in soybean. Around 70–83% of soybean proteins have storage functions, and a higher intake of soy protein has been shown to lower the risk of prostate and breast cancer compared to other populations who depend on animal-based proteins. In addition, its consumption is good for kidneys, lowers blood pressure, reduces diabetes, and relieves depression (Singh et al., 2007). Also, there are claims that soy protein may lower total cholesterol concentrations and thereby reduce the risk of heart disease (Zhang et al., 2003). Thus, soybean protein can be a great substitute for animal-based proteins (Capriotti et al., 2014; Fukushima, 2004; Michelfelder, 2009). According to the protein digestibility-corrected AA scores (PDCAAS) adapted by the World Health Organization and the Food and Agriculture Organization, the protein quality of soybean is comparable to that of milk, egg, and beef proteins (Rizzo and Baroni, 2018).

On a dry weight basis, soybean protein contains approximately 40% crude protein and primarily consists of globulin (salt-soluble). Upon ultracentrifugation, soybean proteins can be separated into four major groups based on solubility, *viz.*, 2S, 7S, 11S, and 15S. The 2S fraction (20% of the soybean protein) consists mostly of albumins and low molecular weight proteins (such as 2S globulin and cytochrome C) and anti-nutritional factors such as lectins, P34, Kunitz trypsin inhibitor (KTI), and Bowman-Birk trypsin inhibitor (Hocine and Boye, 2007). Trypsin inhibitors, which affect digestibility, can be deactivated by preheating soy meal. Globulins are mainly present in the 7S, 11S, and 15S fractions (Liu, 2004). The 7S globulin (which mainly contains β-conglycinin, α-amylase, lipoxygenase, and hemagglutinin) and the 11S globulin (which consists of soybean glycinin) are the two major components (together accounting for 70% of soy protein) in the majority of soybean storage proteins. Hence, these two storage proteins have a greater impact on the nutritional quality of soybean and are targeted for genetic improvement. The proportion of glycinin to β-conglycinin protein is the key aspect reported to influence soybean protein quality and affects the functional properties of soy food products (Cai and Chang, 1999; Poysa et al., 2006). Compared to β-conglycinin, glycinin is rich in sulfur-containing AAs. Different glycinin subunits differ in the amount of methionine residues. The glycinin subunits G1, G2, and

G7 contain 6–7 methionine residues, whereas the subunits G3, G4, and G5 contain 5, 2, and 4 methionine residues per subunit, respectively (Nielsen et al., 1989). The 15S fraction is a polymer accounting for 10% of the soy protein (Sui et al., 2021). The nutritional value of soybean is limited by anti-nutritional factors such as lectins and lipoxygenase, which form about 1–2% of soybean proteins. The lipoxygenase enzyme is responsible for the oxidation of fats and imparting a beany flavor, which is not preferred by consumers (Jegadeesan and Yu, 2020). It is possible to reduce the beany flavor by heat inactivation of the lipoxygenase enzyme. However, the elimination of the lipoxygenase enzyme through breeding is the most economical and efficient method to reduce beany flavor.

4.3 GENETIC IMPROVEMENT OF SOYBEAN PROTEIN

Seed protein content of soybean is a quantitatively inherited trait governed by multiple genes, and its expression is affected by the interaction between genotype and environment. Seed protein content of soybean is reported to be negatively correlated with phenology (time to flowering and maturity) and yield-contributing traits such as plant height (PH), number of branches per plant, number of nodes per plant, and other quality traits such as oil content. Therefore, these genetic and environmental factors cause difficulty in breeding soybean cultivars with high seed yield, protein content, and oil (Patil et al., 2017). In order to break the undesirable negative correlation between desirable traits, the identification of molecular markers linked to genes/QTLs controlling protein, yield, and other quality traits is a prerequisite (Hwang et al., 2014). With the advancement in the development of genomic resources, soybean reference genome sequence, and whole-genome resequencing (WGRS) data have been made available (Song et al., 2013, 2015; Valliyodan et al., 2016). In addition, a number of quantitative trait loci (QTL) for soybean seed protein content have been identified (Table 4.1). The information on the QTLs and linked markers will be useful in the detection of genes underlying seed protein content and genomics-assisted breeding. Furthermore, novel genetic engineering methods such as transgenic and gene editing have immense potential in the functional characterization of genes involved in seed protein content and their application in the genetic improvement of seed protein content and quality. These approaches are discussed in detail in the following sections.

4.4 HARNESSING GENETIC VARIATION AND GERMPLASM RESOURCES FOR HIGH PROTEIN CONTENT

Existence of genetic variability is an essential prerequisite for crop improvement. Harnessing genetic diversity is the most economical way to develop superior cultivars with improved yield, enhanced resistance to pests and diseases, and improved seed protein and oil content. For seed protein content, there is abundant genetic variability among the diverse germplasm lines, including cultivated soybean. The database of the USDA soybean germplasm collection shows wide variation in seed protein content of cultivated soybean (31.7% to 57.9%) and wild soybean accessions (35.5%–56.9%) (https://npgsweb.ars-grin.gov/gringlobal/descriptors), suggesting that there is scope to genetically improve soybean protein content. Wild soybeans are a key reservoir for genes responsible for abiotic stress (drought and heat), biotic stress (disease and pest) resistance, and higher oil and protein contents (Lam et al., 2015). The wild soybean contains higher protein, as high as 55.7%, which is higher than that of cultivated soybean (Li, 1990). Li (1993) evaluated more than 5200 wild accessions of soybean collected in China and found an average seed protein content of 44.99%. Similarly, analysis of 6115 wild soybean lines revealed an average seed protein content of 45.36% (Wang et al., 1998). Wang et al. (2001) screened around 5929 wild soybean (*Glycine soja*) accessions collected from various parts of China and reported a substantial amount of genetic variability for soybean seed protein (29%–55.7%) with an average of 45.42%. They also observed higher protein content in wild type I than in semi-wild type II, which has a protein content higher than that of semi-wild type III. Similarly, Wee et al. (2018) evaluated 334 wild soybean (*G. soja*) accessions collected across several climatic zones of Japan and reported seed protein content that ranged from 48.6% to 57.0%. More than half of the accessions (59%) had protein contents of 54%, which is higher than the control *G. max* variety "Fukuyutaka" (which has a protein content of 44.6%). These results suggested that high protein *G. soja* accessions could be used to introgress the high seed protein content trait into popular soybean varieties. Xu et al. (1985) reported that the average seed protein content of wild soybean was higher (44.38%) than that of cultivated samples (42.15%). Kofsky et al. (2018) reported that due to selection for increased yield and oil content, the average seed protein content of *G. soja* was found to be higher than that of *G. max*.

These studies showed the presence of a considerable amount of genetic variability for seed protein in wild soybean germplasm resources, which could be used to increase protein content in elite cultivars.

Among the cultivated soybeans, a large variability for seed protein content has also been reported. Wang et al. (1998) evaluated a total of 21,050 accessions of soybean collected from different geographical locations in China and observed an average of 44.31% seed protein content. However, it was found that there was significant variation in the average protein content of soybean accessions from different provinces. In another study, Vollmann et al. (2000) found seed protein content in the range of 39.9% to 47.6% under high temperature and moderate rainfall conditions during the seed-filling stage. However, higher rainfall during seed filling and insufficient nitrogen fixation lead to a reduction in seed protein content (36.5%–34.7%). Grieshop and Fahey (2001) observed differences in crude protein content of soybeans collected from China (42.14%), Brazil (40.86%), and the United States (41.58%). Chung et al. (2003) reported seed protein content of 41.3% to 49.4% among the F_5–derived population from the cross between Asgrow A3733 (protein content = 41.8%–42.5%) and PI 437088A (protein content = 47.9%–48.7%). Similarly, Hyten et al. (2004) found protein concentration in the range of 39.97% to 45% among the F_6 recombinant inbred (RIL) population grown in six environments. Jun et al. (2008) evaluated an association mapping population consisting of 48 high-protein and 48 low-protein accessions and reported seed protein content of 31.7% to 57.4%. Shi et al. (2010) analyzed 105 food-grade soybean lines and observed seed protein content ranging from 37% to 50.1%. The germplasm from the United States had an average of 41.3% protein content, whereas the average seed protein content of germplasm from Japan and South Korea was 44.5%. Lee et al. (2021) observed significant variation (28.7% to 44.57%) in the protein content of soybean germplasm collected from various countries. They found the average seed protein content of soybean germplasm from South Korea (39.7%), North Korea (39.2%), China (39.0%), Japan (38.8%), the United States (38.0%), and Russia (37.2%). In addition, significant differences in average seed protein content of small (39.7%), medium (38.6%), and large-sized (39.3%) seeds were observed. These wild germplasm resources with high seed protein content will provide a crucial resource to elucidate and understand the genetic regulation of seed protein content in soybean. In addition, the utilization of germplasm resources in the functional analysis of genetic variation occurring in these germplasm lines is one of the major frontiers in the improvement of seed protein content (Patil et al., 2017).

4.5 EFFECT OF ENVIRONMENTAL FACTORS ON SEED PROTEIN CONTENT

The variability of soybean protein content and other seed composition traits is influenced by environmental factors such as temperature, solar radiation, altitude, and management practices (Assefa et al., 2019). High temperatures and reduced photoperiods may result in higher protein content due to increased translocation of nitrogen to the seed (Cure et al., 1982; Dornbos and Mullen, 1992; Gibson and Mullen, 1996). Wolf et al. (1982) reported a positive correlation between protein content and increased temperature. However, at higher temperatures, the AA content (except methionine) remained unaltered. In another study, Dornbos and Mullen (1992) found that soybean plants exposed to higher temperatures (35°C) during the seed filling stage contained around 4% higher protein content than plants exposed to lower temperatures (29°C). Similarly, Kumar et al. (2006) reported a positive linear relationship between seed protein content and temperature. Additionally, Ren et al. (2009) using proteomic studies observed significantly altered seed protein expression profiles under elevated temperatures due to upregulation of key enzyme involved in methionine biosynthetic pathway. Exposure to higher UV-B radiation is also reported to cause significant reduction in protein content (Reddy et al., 2016).

Soybean seed protein content is also reported to be affected by climatic and management factors. Goldflus et al. (2006) observed a significant difference in seed protein content and AAs (both essential and non-essential) under diverse growing conditions, including soil fertility, topography, and changes in climatic factors. Increased application of nitrogenous fertilizers is reported to cause increased seed protein concentration (Ham et al., 1975; Nakasathien et al., 2000; Rotundo and Westgate, 2009). In addition, maintenance of sufficient soil moisture during the reproductive phase is found to contribute to the accumulation of seed protein content (Wijewardana et al., 2019). Moisture deficit during the reproductive phase leads to a decline in seed protein concentration (Carrera et al., 2009). Furthermore, Kumar et al. (2006) reported that decreased precipitation will increase seed protein content. Therefore, various environmental and management factors may have a significant effect on seed protein content.

4.6 GENETIC ARCHITECTURE OF SOYBEAN PROTEIN CONTENT

Soybean seed protein content is reported as a quantitatively inherited trait with additive effects and is controlled by multiple genes affected by the environment and genotype × environment interactions (Patil et al., 2017). In addition, the interdependency of seed protein content with other seed composition traits results in a strong negative association of seed protein content with other component traits, and seed yield makes it challenging to improve seed protein content (Chung et al., 2003; Kim et al., 2016). Genetic linkage mapping is a useful approach for delineating the genetic and molecular basis of quantitative traits. A large number of QTL controlling seed protein content have been reported across every chromosome of the soybean genome (Guo et al., 2022). Diers et al. (1992) initially used RFLP markers to identify a major QTL for protein content on chromosome (Chr.) 20. This major QTL has been the focus of extensive studies and has consistently been mapped on Chr. 20 (Chung et al., 2003; Hwang et al., 2014; Jun et al., 2008; Nichols et al., 2006; Qi et al., 2016; Rodrigues et al., 2010; Sebolt et al., 2000; Specht et al., 2001; Warrington et al., 2015). This QTL is of particular interest because of its large additive effect found in different mapping populations (Csanádi et al., 2001; Fliege et al., 2022; Sebolt et al., 2000; Warrington et al., 2015) and across different environments (Brummer et al., 1997). Also, this high protein QTL has been tested in different backgrounds (Warrington et al., 2015). Nichols et al. (2006) narrowed down this QTL region to a 3-cM interval using BC_5F_5-derived NILs, which corresponds to the physical distance of 8.4 Mb, from 24.54 to 32.92 Mb on chr 20 (Bolon et al., 2010). Based on genome-wide association analysis (GWAS) with large genotypic and phenotypic data, this QTL region was reduced to a 3 Mb region located between 27.6 and 30.0 Mb on Chr 20 (Hwang et al., 2014) and to a <1 Mb window (Vaughn et al., 2014). Further refining of the candidate region to a narrow base pair region revealed only three plausible candidate genes (Bandillo et al., 2015). On resequencing 106 soybean lines, Valliyodan et al. (2016) identified a 2.4 Mb interval on Chr 20 containing three gene clusters associated with protein content, in which the candidate genes *Glyma20g19680*, *Glyma20g21030*,

TABLE 4.1 Information on QTL Mapping Studies for Seed Protein Content in Soybean

Cross	Mapping Population Studied	QTL Mapped on Linkage Group (Gm Composite2003)	% Phenotypic Variation Explained	Associated Loci	References
A81356022 × PI468916	F2:3	I, E, P, L, G	12–42%	K011_1, A407_1, A144_1, A688_1, SAC7_1, A242_1, A023_1, A245_2	Diers et al. (1992)
Noir 1 × Minsoy	F7-derived RIL	U7, U14	–	T155_1, Satt006, A329_2	Mansur et al. (1996)
M82806 × HHP	F2:4 F2:5 F2:6	A2, B1, C1, D1, E, F, G, H, I	–	A505_1, A109_1, A063_1, A398_1, A691_1, B174_1, K002_1, A816_1, A890_1, A235_1, A069_1, A144_1	Brummer et al. (1997)
Young × PI416937	F4-derived	C1.1, C1.2, E, J.1, N.1, P, GA1_25, GA1_27	–	gc197_1, EV3_1, A338_2, A463_1, A517_1, cr167_1, B166_1, A071_2, gc34_2, B142_1, A352_1, A199_3, cr274_1	Lee et al. (1996)
PI97100 × Coker237	F2-derived	E, H.1, K	–	A454_1, A566_2, A065_3, R051_2, Q043_1, A132_4	
Peking × Essex	F2:3	H, F	–	B072_1, B148_1	Qiu et al. (1999)
Minsoy × Archer	Recombinant inbred	C1, M	6–12%	R079_1, Satt578	Orf et al. (1999)
Noir 1 × Archer	Recombinant inbred	L	11%	Satt166	
Minsoy × Noir 1	Recombinant inbred	A1, C1	12–15%	T155_1, SOYGPATR	
G. max × *G. soja*	Backcross3	I	65%	Satt127	Sebolt et al. (2000)
Parker × *G. soja*	BC3F4-derived	I	65%	A144_1	
Minsoy × Noir 1	F7:11 RIL	A1, C1, K, M, O	–	B170_1, K001_1, Satt178, Satt567, Satt478	Specht et al. (2001)
Ma.Belle × Proto	F2	C2, D1a, K, M	4.7–7.1%	Satt077, Sct_028, Satt567, Satt196	Csanadi et al. (2001)
M91-212006 × SZG9652	F2:3 F2:4	A2	–	Ti	Vollmann et al. (2002)

TABLE 4.1 *(Continued)*

Cross	Mapping Population Studied	QTL Mapped on Linkage Group (Gm Composite2003)	% Phenotypic Variation Explained	Associated Loci	References
A3733 × PI437088A	F5-derived RILs	I	27.7	Satt239, Satt496	Chung et al. (2003)
Essex × Essex	F4:6	B1, L	–	Satt251, Satt373	Chapman et al. (2003)
Minsoy × Archer	RI	C1	15.4	L92_1, Satt578	Stombaugh et al. (2004)
N87-984-16 × TN93-99	F6	G	20.10%	Satt570	Panthee et al. (2005)
BSR 101 × LG82-8379	F5	G, A2, C1, C2, D1b, O, F, B2, N, H, M	3–16%	Satt409, Satt338, Satt363, Satt157, Satt358, Satt510, Satt168, Satt339, Satt142, Satt308	Kabelka et al. (2004)
Essex × Williams	F6,8 RIL	C2, F, K, M	–	Satt202, Satt277, Satt144, Satt335, Satt102, Satt539, Satt463, Satt540	Hyten et al. (2004)
Kefeng No. 1 × 1138-2	F2:7:10 RIL	B1, D1b, L	–	A118_1, Satt197, Satt560, A118_1, A520_1, A725_1	Gai et al. (2007)
RG10 × OX948	F6 and F7 RIL	A2, D2, I, B1, C2, G, I, B2, F	–	Satt187, Satt389, Satt571, Satt419, Satt270, Sat_123, Satt277, Satt394, Satt162, Satt063, Sat_090, Satt309, Satt569, Satt394	Reinprecht et al. (2006)
Charleston × Dongnong 594	F2:10 RIL	D1b, E, N, O	5.16–14.29%	Sat_135, Satt537, Satt151, Satt263, Satt117, Satt151, Satt257, Satt173, Satt581	Chen et al. (2007)
Jindou 23 × Huibuzhi	F13	B2, C2, G, H	–	Satt557, Satt100, Satt199, Satt012, Satt568, Satt442, Satt063, Satt070, Satt275, Satt038	Liang et al. (2010)
OAC Millennium × Heinong 38	F4:7 RIL	C2	–	Satt100	Palomeque et al. (2009)
Misuzudaizu × Moshidou Gong 503	F7:8 RIL	A2, C2, D2, E, I, G, L, K	3.4–20.9%	Satt239, Satt310, Satt384, Satt281, A104_2, Satt156, GM195, Q026, A378_1	Tajuddin et al. (2003)

TABLE 4.1 *(Continued)*

Cross	Mapping Population Studied	QTL Mapped on Linkage Group (Gm Composite2003)	% Phenotypic Variation Explained	Associated Loci	References
X3145-B-B-3–15 × AC Brant	F4:5 RIL	D1a, C2, I, E	–	Satt496, Satt575, Satt077, Satt520	Pandurangan et al. (2012)
Kefeng No. 1 × Nannong 1138-2	F2:7:10 RIL	B2	12.40%	Satt560	Zhang et al, (2004)
OAC Wallace × OAC Glencoe	F4:6 RIL	H, F, K, L, M	–	Satt317, Satt302, Satt510, Satt335, Satt001, Satt273, Satt182, Satt523, Satt323, Satt463	Eskandari et al. (2013)
ZDD09454 × Yudou12	F2:9	A1, A2, B1, C2, I, G, H, K, L, M	–	Satt211, Sat_213, GMAC7L, Satt336, Satt308, Sat_115, GMENOD2B, Sat_215, GMENOD2B, Satt475, Sat_156, Satt541, Satt469, Sat_094, Satt564, Sat_260, Satt143, Sat_195, AB002807, Satt270	Lu et al. (2012)
OAC Millenium × Heinong 38	F4:7 RIL	C2, I, K	–	Satt277, Satt100, Satt162, Satt273	Rossi et al. (2013)
HeFeng 47 × HeiNong 44 HeFeng 47 × HeiNong 37 DongNong 46 × KenJian 23	F5:6 RIL	A1, B1, B2, C1, C2, D1a, D2, F, G, I, K, L, M, N, O	–	Satt236, Satt449, Satt197, Satt070, Sat_189, Satt202, Satt070, Sat_337, Satt565, Satt565, Sat_404, Satt286, Satt202, Satt202, Satt316, Sat_414, Satt129, Satt468, Sat_414, Sat_069, Satt274, Satt157, Satt296, Satt157, Satt296, Satt226, Satt301, Satt311, Satt226, Satt543, Satt615, Satt669, Sat_365, Sat_074, Sct_033, Sat_262, Satt343, Sat_417, Satt193, Satt149, Sat_074, Satt343, Sat_262, Satt343, Sct_033, Sct_033, Satt334, Satt309, Sat_372, Sat_324, Satt367, Sat_087, Satt242, Sat_293, Satt102, Satt242, Sat_293, Satt264, Satt273, Satt481, Sat_286, Satt481, Sat_286, Satt201, Satt150, Sat_275, Satt022, Satt022, Sat_295, Satt530, Satt683, Satt549, Sat_091, Sat_341, Satt241, Satt241, Satt173, Satt259, Sat_282	Mao et al. (2013)

TABLE 4.1 *(Continued)*

Cross	Mapping Population Studied	QTL Mapped on Linkage Group (Gm Composite2003)	% Phenotypic Variation Explained	Associated Loci	References
SD02-4-59 × A02-381100	F:5 RIL	E1, C1, D1b, D2, I, K, N	–	Satt251, BARC-032333-08951, Satt338, BARCSOYSSR_04_1400, BARC-044521-08714, BARC-064861-18829, Satt274, Sat_289, BARCSOYSSR_17_0900, Sat_300, BARCSOYSSR_17_0646, BARCSOYSSR_17_0621, BARCSOYSSR_20_0049, Satt562, Satt587, Satt354, BARC-021887-04232, BARC-057049-14550, BARCSOYSSR_09_1123, BARCSOYSSR_09_1171, Sat_084, Satt584	Wang et al. (2014)
Benning (PI595645) × Danbaekkong (PI619083)	F5 RIL	B2, D2, E, I	–	BARC-018353-03589, BARC-056587-14511, Sat_273, BARC-027786-06670, BARC-025927-05161, Satt256, GSM0012, Satt354	Warrington et al. (2015)
Charleston × Dongnong594	F2:15-F2:18 RIL	D1a, D1b, K, O	–	Satt331, satt173, Satt358, Sat_001, Sat_119, Satt242, Satt203, Satt198, Sat_069, Sat_135	Qi et al. (2014)
Various × Various	Two subpopulations of 48 germplasm accessions each	A1, E, I, J, G, K, L, M, N, O	–	Satt385, Satt268, Satt564, Satt571, Satt405, Satt431, Satt242, Satt723, Satt551, Satt159, Satt653	Jun et al. (2008)
Hefeng 25 × Maple Arrow	F5:6 RIL	O	–	Satt550, Satt345	Li et al. (2009)
MD 96-5722 × Spencer	F5:7 RIL	B2	–	ss248275088, ss248293401	Akond et al. (2014)
Dongnong 46 × L-100	F5:8 RIL	B2, D2, I, G, H, K, M	–	Satt494, Satt710, Sat_122, Satt577, Satt543, Satt217, Satt163, Satt614	Soybase.org
Fiskeby III × Mandarin	F5:7 RIL	I	–	BARC-025987-05207	Hacisalihoglu et al. (2018)

Source: http://soybase.org/.

and *Glyma20g21080* were identified as candidate genes. Huang et al. (2018) detected 13 candidate genes related to soybean protein content and found that eight of them were highly expressed in mature soybean seeds. Recently, Fliege et al. (2022), using a fine mapping approach and cloning of the major seed protein QTL on Chr 20, identified the gene *Glyma.20G85100* controlling seed protein. Analysis of transgenic plants indicated that the high/low seed protein alleles of the cqSeed protein-003 QTL were caused by transposon insertion within the CCT domain protein encoded by the *Glyma.20G85100* gene.

In addition, a number of seed protein QTLs have been mapped on different linkage groups (LGs) (Table 4.1). To date, the Soybase database (http://soybase.org/) lists a total of 248 QTLs controlling seed protein content in soybean. The highest number of seed protein QTLs have been mapped on Chr. 20 (25 QTLs), followed by Chr. 6 (19 QTLs), Chr. 15 (18 QTLs), Chr. 9 (17 QTLs), and Chr. 18 (16 QTLs) (Guo et al., 2022). Out of these, two QTLs, namely cqPro-20 (on Chr. 20) and cqPro-15 (on Chr. 15), were recognized as confirmed QTLs based on a low error rate (< 0.01) and alleles of the same locus segregating in different test populations (Patil et al., 2017). QTLs for seed protein content have been identified on the LGs B1 and L (Chapman et al., 2003). Lu et al. (2013) and Mao et al. (2013) mapped QTLs for seed protein on Chr. 19. In another study, Kim et al. (2016) mapped QTL for high protein and low oil content from the line PI 407788A on Chr. 15. Based on high-density linkage mapping using an interspecific population derived from the cross *G. max* (Williams 82) × *G. soja* (PI 483460B), Patil et al. (2018) reported seed protein QTLs on Chrs. 6, 8, 13, 19, and 20. Also, the major QTL for protein and oil was mapped on Chr. 20 (qPro_20), suggesting a negative correlation between protein and oil. By comparing gene and sequence divergence between genomic regions containing a major seed protein QTL, Lestari et al. (2013) identified a total of 19 and 35 candidate genes on Chr. 20 and Chr. 10, respectively. Using chromosome segment substitution lines, a wild soybean allele responsible for high protein content was fine-mapped to a 329-kb region on Chr. 15 and predicted *Glyma.15g049200* as a candidate gene for high seed protein content (Yang et al., 2019). Recently, Qin et al. (2022) identified a haplotype block of 471 kb in size on Chr. 6 (from Chr6_18844283 to Chr6_19315351) associated with high seed protein content. This region contains seven candidate genes, including polynucleotidyl transferase (*Glyma.06G202900* and *Glyma.06G203100*), polygalacturonase activity (*Glyma.06G202600* and *Glyma.06G203000*), ATP synthase (*Glyma.06G203200*), and genes without annotation (*Glyma.06G202700* and *Glyma.06G202800*).

As far as the genetic control of variation in AAs is concerned, few studies have reported the identification of genomic regions (Panthee et al., 2006; Vaughn et al., 2014; Warrington et al., 2015). This may be due to the difficulty in accurate and high-throughput measurement of specific AAs, which requires expensive high-resolution nuclear magnetic resonance (HRNMR), high-performance liquid chromatography (HPLC), or gas chromatography (GC) platforms and wet chemistry methods (Ravindran et al., 2009). Nevertheless, about 113 genes controlling sulfur-containing AA enzymes have been reported in soybean (Guo et al., 2022). Panthee et al. (2006) used an F_6-derived recombinant population and identified seven QTLs controlling sulfur-containing AAs (cysteine and methionine) concentration. Fallen et al. (2013) reported one QTL each for cysteine (on Chr. 20) and methionine (on Chr. 13) using an $F_{5:9}$ RIL population. Recently, Malle et al. (2020) conducted genome-wide studies and reported nine QTLs for cysteine and methionine content.

The 7S and 11S globulins are the major storage proteins of soybean protein, and these play an important role in the functionality and nutritional quality of soybean seed protein. The expression of these two storage proteins is reported to be encoded by a multigene family. Genetic analysis indicated that the dominant locus *Scg-1* controls 7S globulin (Teraishi et al., 2001). A total of seven genes (designated as *Gy1–Gy7*) were found to encode 11S globulin in soybean. These genes have been mapped onto Chr. 3 (*Gy1*, *Gy2*, and *Gy6*), Chr. 10 (*Gy4*), Chr. 13 (*Gy5*), and Chr. 19 (*Gy3* and *Gy7*) (Beilinson et al., 2002; Diers et al., 1992; Nielsen et al., 1989). Several studies mapped QTLs for seed storage proteins, and the soybean database (www.soybase.org/) shows 10 and 7 QTLs for seed conglycinin and glycinin content, respectively. Using an F_6 derived RIL population of the cross N87-984-16 × TN93-99, Panthee et al. (2004) reported three QTLs (Satt461, Satt292, and Satt156) for glycinin on LG D2, I, and L, respectively. For conglycinin, two QTLs (Satt461 and Satt249) were detected on LG D2 and J. Further, Liu et al. (2009) detected a total of 28 QTLs for 7S and 11S globulin content in soybean. Association mapping by Jian et al. (2013) found that 14 SSR loci were associated with seed protein subunits. In another study, Ma et al. (2016) mapped 35 QTLs related to glycinin (11S) and β-conglycinin (7S) contents, the ratio of glycinin to β-conglycinin (RGC), and the sum of glycinin and β-conglycinin (SGC). Further, genetic mapping of 7S α' and 11S A-type storage protein subunits by Boehm Jr et al. (2018) identified QTLs on Chromosome 3 (11S A1), Chromosome 10 (7S α' and 11S A4), and Chromosome 13 (11S A3). The identification of genomic regions and linked markers will help to discover genes underlying glycinin and β-conglycinin contents in soybean seeds and further application in molecular breeding for protein quality improvement in soybean.

4.7 INTEGRATION OF MULTIPLE BREEDING APPROACHES FOR IMPROVING SOYBEAN PROTEIN CONTENT

4.7.1 UTILIZATION OF GERMPLASM RESOURCES WITH HIGH PROTEIN CONTENT

Large germplasm accessions are a valuable source of genes for broadening genetic variability and developing superior varieties with improved yield, resistance to biotic and abiotic stresses, quality, and protein content, especially for crops with narrow genetic diversity such as soybean. Crop wild species are a very important source of novel genetic diversity that has otherwise been selected out or lost in the course of time as a result of domestication and breeding activities intended to develop uniform crop varieties (as in the case of protein content). Several breeding programs have demonstrated successful development of high-protein soybean varieties. For instance, Lai et al. (2005) developed a new soybean line, Longpin 8807, which has a high protein content (48.29%). Similarly, Carter et al. (2010) developed the high-protein line N6202, a F_4-derived selection from the cross N6201 × N95-7390. Interestingly, this line has an unusual combination of high protein (45.7%) and favorable yield (>90%) over the control variety NC-Roy. In another study, Du and Yu (2010) made a cross between cultivated soybean (N23674) and wild line (BB52), having protein contents of 42.49% and 40.09%, respectively. They identified a superior line with a higher protein content of 49.13% and enhanced AA concentration among $F_{4:5}$ progenies. Similarly, Eickholt et al. (2019) made an interspecific cross between *G. max* cultivar N7103 and *G. soja* wild soybean PI366122 and identified seven breeding lines that exhibited higher seed protein content compared to N7103 (43.7%). One breeding line (USDA-MxS-CS1-7) had elevated seed protein (47.7%) compared with N7103. They also identified 15 lines with higher cysteine content and two lines with elevated levels of methionine and cysteine AAs compared to the domesticated parent. Details on the list of soybean germplasm/varieties having higher seed protein content are presented elsewhere (Carter et al., 2010; Guo et al., 2022). These studies showed that wild germplasm is a valuable genetic resource for novel traits, including agronomic and seed composition traits. In addition, elite cultivars/breeding lines and landraces can be used to develop mapping populations and reference sets, and they can also be included in training populations for genomic selection (GS) approaches (Varshney et al., 2013). When the objective is to improve protein content through conventional breeding, one should take into consideration the level of protein content in parental genotypes, heritability, and gene action of the protein content. It is reported that if the female parent has a higher protein content than the male parent, the protein content in the hybrid progenies is affected by the male parent (Wang et al., 2013). In addition, breeders have to consider the correlation of seed protein content with other agronomically important traits for developing varieties with higher yield and protein content.

4.7.2 HIGH THROUGHPUT PHENOTYPING OF PROTEIN AND AMINO ACIDS

In recent decades, the development of next-generation sequencing technologies (NGS) and genomic resources has led to a considerable rise in the throughput and accuracy of gene discovery, which in turn has enabled great precision and efficiency in molecular breeding programs (Varshney et al., 2021). In the post-NGS era, high throughput and precise phenotyping challenges still remain a major stumbling block to genetically dissecting agronomically important traits, thereby hindering breeding progress. Rapid and enhanced phenotyping capabilities for various important traits will be crucial to realize the full potential of genomics-assisted breeding methods. In soybean, breeding for high seed protein will greatly benefit from precise and efficient analysis of seed protein and AAs. In soybean, near-infrared spectroscopy is a non-destructive method that can be efficiently used to measure the seed protein content of a large number of samples. This is a very popular method used by breeders for protein content analysis (Choung et al., 2001). For the analysis of specific AAs,

HPLC or GC platforms, HR-NMR, and wet chemistry methods are being used (Ravindran et al., 2009). In addition, high-throughput non-destructive phenotyping approaches such as high-resolution Fourier transform near-infrared Spectroscopy (FT-NIR) and Raman spectroscopy will precisely analyze seed protein content (Lee et al., 2013). FT-NIR spectroscopy facilitates rapid and accurate analysis of soybean seed protein and specific AA content. These non-destructive high-throughput phenotyping (HTP) approaches are desirable for analyzing seed protein content without sacrificing the seed sample. With the help of accurate and HTP, genetic variation underlying the phenotypic effects can be identified, which in turn helps to identify superior genotypes having desirable gene combinations using molecular breeding approaches.

4.8 MUTATION BREEDING APPROACH

Induced mutagenesis has been a very useful method for generating novel variability in crops with narrow genetic variability. In soybean, induced mutagenesis has been used initially to develop morphological variations, including plant architecture, phenology, resistance to biotic stress, seed coat color, etc. Furthermore, soybean mutants with altered seed quality traits such as seed protein content, KTi activity, and lipoxygenase enzymes were developed (Guo et al., 2005; Khan and Tyagi, 2013; Shi et al., 2015; Yu et al., 1995; Zhang et al., 2018). Induced mutagenesis is an effective means to improve soybean protein content. Qiu and Gao (1988) reported that EMS produced higher frequencies of mutants with high seed protein content and high heritability among M_2 and M_3 lines. Similarly, treating soybean seeds with 0.4% EMS followed by early generation selection has been recommended for improving seed protein content (Wang et al., 1989). Additionally, Yu et al. (1995) identified three high protein content mutant lines in the M_4 generation of the EMS-induced population. Treatment with a higher dose of gamma irradiation has also been reported to increase protein content in soybean (Qing et al., 1996). Induced mutagenesis with physical (γ-ray and thermal neutron) and chemical (EMS and sodium azide) mutagens has been successfully used to create genetic variation and breed high-protein soybean cultivars such as Heinong 41 (45.23%) and 90–3527 (47.53%) (Wang et al., 2000). Similarly, Guo et al. (2005) isolated several mutants with high protein content (up to 47.60%) by treating soybean seeds with γ-ray, thermal neutron, EMS, and sodium azide.

In addition, mutagenesis followed by hybridization has been used to breed high protein content soybean cultivars such as Qihuang 34 and Kexin 3. Recently, Wei et al. (2019) identified EMS-derived mutants (M_7 generation) with an average protein content higher (48.17%) than that of the wild type parent "Zp661" (41.19%). These mutagens are valuable tools for creating genetic variation for the identification of desirable mutants, and also for gene discovery and reverse genetics studies with TILLING or NGS technology. With the advancements in NGS technologies, it is now possible to efficiently screen large mutant populations to identify genome variations, including copy number variations (CNV) and structural variations (SVs) (Bolon et al., 2014).

4.9 GENOMICS ASSISTED BREEDING

The integration of genomic resources with breeding techniques is crucial for increasing genetic gain in a short time for developing superior cultivars. With the rapid advancement in genomics technologies, a number of high-quality genome assemblies, WGRS, and pangenomes are accessible in soybean, highlighting significant SVs in the genome. This information, along with the soybean genomic resources, can be applied for developing improved cultivars. The genomic breeding approaches, such as marker-assisted selection (MAS), marker-assisted back-crossing (MABC), Marker-Assisted Recurrent Selection (MARS), haplotype breeding, genome editing (GE) technology, and GS, can be used to design improved crop varieties (Varshney et al., 2021). In soybean, the application of genomics-assisted breeding for improving seed protein content has enabled the identification of QTLs and molecular markers (satt182, satt419, and satt239) linked to protein content (Chung et al., 2013; Eskandari et al., 2013; Reinprecht et al., 2006; Tajuddin et al., 2003), which will make selection for high protein content more efficient. For the major QTLs, it is possible to develop NILs by backcrossing, and these NILs form valuable genetic resources for understanding gene functions or the effect of gene/QTL on phenotypic expression. Song et al. (2014) used the MABC approach to develop a NIL (cgy-2–NIL) containing the mutant cgy-2 allele, associated with the absence of the allergenic α-subunit of β-conglycinin. Since seed protein content is reported to be a quantitative trait, pyramiding desirable QTLs into a single background through MABC is a challenging approach. In such conditions, other breeding approaches such as MARS can be effectively employed. In earlier studies, Brim and Burton (1979) and Wilcox (1998) effectively utilized recurrent selection for increasing seed protein content in soybean. Genetic improvement of complex traits, whose phenotypic variation is controlled by a large number of small-effect QTLs, requires efficient breeding approaches for the identification and subsequent utilization of trait variation. With the development of increased throughput and reduced costs of sequencing technologies, greater accessibility to whole-genome sequence information has led to the application of novel breeding approaches like GS, which uses genome-wide marker information to improve selection efficiency and the rate of genetic gain for soybean protein content. Using 50K SNP data on a large number of diverse soybean accessions and vast historical phenotypic data, Jarquin et al. (2016) developed genomic prediction models for yield and quality traits such as protein and oil. The resulting genomic prediction model explained predictive ability up to 0.92 for protein and oil and 0.72 for yield. Duhnen et al. (2017) suggested that the accuracy of predicting protein content can be improved through the inclusion of epistatic effects in the GS model. Further, Stewart-Brown et al. (2019) observed that increasing the size of the training set would improve the predictive ability of the model up to 0.81 for protein content. In addition, the accuracy of prediction for protein content was found to increase significantly with the size of the training population and the number of markers used in the model (Beche et al., 2021). Recently, Sun et al. (2022) found that improved prediction accuracy can be achieved by the use of intragenic markers and the inclusion of progeny lines in the training population established with germplasm resources. Another promising GAB approach is haplotype-based breeding, which involves the identification of superior haplotype blocks and a set of haplotype-informed markers that enable the selection of desired haplotype blocks to attain ideal genetic combinations for parental selection and improved performance of progenies (Varshney et al., 2021).

4.10 MODERN BIOTECHNOLOGICAL TOOLS FOR BREEDING HIGH PROTEIN SOYBEAN

The traditional breeding methods used for crop improvement with respect to protein content are laborious, low in selection efficiency, and time-consuming. To improve seed protein content, more effective breeding approaches are necessary. The recent decades have witnessed enormous progress in genome sequencing technologies, and the publication of reference genomes and transcriptomes provides access to detailed genomic information, leading to the accurate identification of genes controlling any trait (Lambirth et al., 2015). In addition, the development of efficient germplasm-independent plant transformation methods and regeneration technology enables the application of genome engineering (such as transgenics and gene editing) approaches in soybean (Chilcoat et al., 2017; Wu et al., 2021). Use of these technologies is highly advantageous to improve the selection efficiency of seed protein and AA content in soybeans. In soybeans, transgenic technology has been used to enhance AAs such as methionine (up to 12–20%) and cysteine (up to 15–35%) by transforming the 15 kDa *δ-zein* gene (Dinkins et al., 2001). Similarly, Li et al. (2005) transformed the 27 kDa *δ-zein* gene and obtained transgenic soybean plants that had higher methionine (up to 15.49–18.57%) and cysteine (up to 26.97–29.33%) content. In another study, El-Shemy et al. (2007) used a modified *Gy1* (A1aB1b) proglycinin gene with synthetic DNA encoding four continuous methionine (*V3-1*) to connect the *hpt* gene and the modified green fluorescent protein sGFP (*S65T*) gene. The resultant plasmid was transformed into soybean by particle bombardment, resulting in transgenic soybean expressing *hpt* and *V3-1* genes with higher accumulation of glycinin.

The increasing volume of genomics data, along with bioinformatics resources, has fostered researchers to use modern breeding approaches like CRISPR/Cas9 for crop improvement (Gao et al., 2022; Varshney et al., 2021). Until now, the CRISPR/Cas9 method has been used in soybean to successfully modify key traits such as plant architecture, flowering time, and seed oil profile by mutating the genes *GmSPL9*, *GmFT2a*, and *FAD2-2*, respectively (Siddique, 2022). With respect to seed storage proteins, Li et al. (2019) used the CRISPR/Cas9 system for editing soybean storage protein genes by testing sgRNAs to target nine different major storage protein genes. DNA mutations were detected in three storage protein genes in soybean hairy roots, with a mutation frequency of 3.8% to 43.7%. These studies provide an important resource for soybean researchers to breed soybean varieties with enhanced protein content.

4.11 CONCLUSION AND FUTURE PROSPECTS

Soybean is a unique legume crop containing oil (~20%) and protein (~40%). It is rich in protein, containing nutritionally vital essential AAs except for sulfur-containing cysteine and methionine. Breeding for high seed protein content is one of the major breeding objectives around the world. Although there is considerable genetic variability among the cultivated and wild species of the genus *Glycine*, little success has been achieved in the improvement of protein content and quality. This is largely due to the complex nature of seed protein content, and its negative association with yield and oil content makes it difficult to develop cultivars with high yield and high protein content. However, the past few decades have witnessed the development of large-scale soybean genomic resources and tools that have been applied to understand and delineate the genetic architecture of seed protein content. For seed protein content, more than 240 QTLs have been reported for soybean protein content, and the major QTL on Chr. 20 is the region of interest, which exhibited distinct haplotypes across the different populations. With the availability of well-annotated reference genomes, a wealth of genomic data serves as a useful resource for genomics-assisted breeding approaches such as GWAS and GS. In addition, developments in HTP technologies for measuring protein content, seed storage proteins, and AA content will help to integrate with genomic data for understanding and elucidating the genes underlying protein content on a whole-genome scale. More importantly, the gene regulatory networks involved in soybean protein content and their interaction with important traits such as oil content and seed yield-contributing traits will be understood. The integration of conventional breeding and genomics with modern biotechnological tools such as transgenic technology and gene editing will facilitate the development of high-yielding soybeans with improved protein content and quality.

KEYWORDS

- **essential amino acids**
- **gene editing**
- **genetic improvement**
- **genomics-assisted breeding**
- **high-throughput phenotyping**
- **quantitative trait loci**
- **soybean protein content**

REFERENCES

Akond, M., Liu, S., Boney, M., Kantartzi, S. K., Meksem, K., Bellaloui, N., Lightfoot, D. A., & Kassem, M. A. (2021). Identification of quantitative trait loci (QTL) underlying protein, oil, and five major fatty acids' contents in soybean. *American Journal of Plant Sciences, 5*, 158–167.

Assefa, Y., Purcell, L. C., Salmeron, M., Naeve, S., Casteel, S. N., Kovács, P., Archontoulis, S., Licht, M., Below, F., Kandel, H., Lindsey, L. E., Gaska, J., Conley, S., Shapiro, C., Orlowski, J. M., Golden, B. R., Kaur, G., Singh, M., Thelen, K., Laurenz, R., Davidson, D., & Ciampitti, I. A. (2019). Assessing variation in US soybean seed composition (protein and oil). *Frontiers in Plant Science, 10*, 298.

Bandillo, N., Jarquin, D., Song, Q., Nelson, R., Cregan, P., Specht, J., & Lorenz, A. (2015). A population structure and genome-wide association analysis on the USDA soybean germplasm collection. *The Plant Genome, 8*, 1–13.

Beche, E., Gillman, J. D., Song, Q., Nelson, R., Beissinger, T., Decker, J., Shannon, G., & Scaboo, A. M. (2021). Genomic prediction using training population design in interspecific soybean populations. *Molecular Breeding, 41*, 1–15.

Beilinson, V., Chen, Z., Shoemaker, R. C., Fischer, R. L., Goldberg, R. B., & Nielsen, N. C. (2002). Genomic organization of glycinin genes in soybean. *Theoretical and Applied Genetics, 104*, 1132–1140.

Boehm, J. D. J., Nguyen, V., Tashiro, R. M., Anderson, D., Shi, C., Wu, X., Woodrow, L., Yu, K., Cui, Y., & Li, Z. (2018). Genetic mapping and validation of the loci controlling 7S α' and 11S A-type storage protein subunits in soybean (*Glycine max* L. Merr.). *Theoretical and Applied Genetics, 131*(3), 659–671.

Bolon, Y. T., Joseph, B., Cannon, S. B., Graham, M. A., Diers, B. W., Farmer, A. D., May, G. D., Muehlbauer, G. J., Specht, J. E., Tu, Z. J., Weeks, N., Xu, W. W., Shoemaker, R. C., & Vance, C. P. (2010). Complementary genetic and genomic approaches help characterize the linkage group I seed protein QTL in soybean. *BMC Plant Biology, 10*, 41.

Brim, C., & Burton, J. (1979). Recurrent selection in soybeans. II. Selection for increased percent protein in seeds. *Crop Science, 19*, 494–498.

Brummer, E. C., Graef, G. L., Orf, J., Wilcox, J. R., & Shoemaker, R. C. (1997). Mapping QTL for seed protein and oil content in eight soybean populations. *Crop Science, 37*(2), 370.

Cai, T., & Chang, K. C. (1999). Processing effect on soybean storage proteins and their relationship with tofu quality. *Journal of Agricultural and Food Chemistry, 47*, 720–727.

Capriotti, A. L., Caruso, G., Cavaliere, C., Samperi, R., Stampachiacchiere, S., Chiozzi, R. Z., & Laganà, A. (2014). Protein profile of mature soybean seeds and prepared soybean milk. *Journal of Agricultural and Food Chemistry, 62*, 9893–9899.

Carrera, C., Martínez, M. J., Dardanelli, J., & Balzarini, M. (2009). Water deficit effect on the relationship between temperature during the seed fill period and soybean seed oil and protein concentrations. *Crop Science, 49*(3), 990–998.

Carter, T. E., Jr., Rzewnicki, P. E., Burton, J. W., Villagarcia, M. R., Bowman, D. T., Taliercio, E., & Kwanyuen, P. (2010). Registration of N6202 soybean germplasm with high protein, favorable yield potential, large seed, and diverse pedigree. *Journal of Plant Registrations, 4*, 73–79.

Chapman, A., Pantalone, V. R., Ustun, A., Allen, F. L., Landau-Ellis, D., Trigiano, R. N., & Gresshoff, P. M. (2003). Quantitative trait loci for agronomic and seed quality traits in an F2 and F4:6 soybean population. *Euphytica, 129*, 387–393.

Chathurika, W., Firas, A., Alsajri, J., Trenton, I. L., Jason, K., Bobby, G., Brien, H. W., Wei, G., & Raja Reddy, K. (2019). Physiological assessment of water deficit in soybean using midday leaf water potential and spectral features. *Journal of Plant Interactions, 14*(1), 533–543.

Chen, Q. S., Zhang, Z. C., Liu, C. Y., Xin, D. W., Qiu, H. M., Shan, D. P., Shan, C. Y., & Hu, G. H. (2007). QTL analysis of major agronomic traits in soybean. *Agricultural Sciences in China, 6*, 399–405.

Chilcoat, D., Liu, Z. B., & Sander, J. (2017). Use of CRISPR/Cas9 for crop improvement in maize and soybean. In D. P. Weeks & B. Yang (Eds.), *Progress in Molecular Biology and Translational Science* (Vol. *149*, pp. 27–46). Academic Press.

Choung, M. G., Baek, I. Y., Kang, S. T., Han, W. Y., Shin, D. C., Moon, H. P., & Kang, K. H. (2001). Non-destructive method for selection of soybean lines contained high protein and oil by near infrared reflectance spectroscopy. *Korean Journal of Crop Science, 46*, 401–406.

Chung, J., Babka, H., Graef, G., Staswick, P., Lee, D., Cregan, P., Shoemaker, R., & Specht, J. (2003). The seed protein, oil, and yield QTL on soybean linkage group. *Crop Science, 43*, 1053–1067.

Csanadi, G., Vollmann, J. G., Stift, G., & Lelley, T. (2001). Seed quality QTLs identified in a molecular map of early maturing soybean. *Theoretical and Applied Genetics, 103*, 912–919.

Cure, J. D., Patterson, R. P., Raper, J. C. D., & Jackson, W. A. (1982). Assimilate distribution in soybeans as affected by photoperiod during seed development. *Crop Science, 22*, 1245–1250.

Diers, B., Keim, P., Fehr, W., & Shoemaker, R. (1992). RFLP analysis of soybean seed protein and oil content. *Theoretical and Applied Genetics, 83*, 608–612.

Dinkins, R. D., Reddy, M. S. S., Meurer, C. A., Yan, B., Trick, H., Thibaud-Nissen, F. O., Finer, J. J., Parrott, W. A., & Collins, G. B. (2001). Increased sulfur amino acids in soybean plants overexpressing the maize 15 kDa zein protein. *In Vitro Cellular & Developmental Biology–Plant, 37*, 742–747.

Dornbos, D. L., & Mullen, R. E. (1992). Soybean seed protein and oil contents and fatty acid composition adjustments by drought and temperature. *Journal of the American Oil Chemists' Society, 69*, 228–231.

Du, L. L., & Yu, B. J. (2010). Analysis of salt tolerance, agronomic traits, and seed quality of *Glycine max*, saltborn *Glycine soja*, and their hybrids. *Chinese Journal of Oil Crop Sciences, 32*, 77–82.

Duhnen, A., Gras, A., Teyssedre, S., Romestant, M., Claustres, B., Dayde, J., & Mangin, B. (2017). Genomic selection for yield and seed protein content in soybean: A study of breeding program data and assessment of prediction accuracy. *Crop Science, 57*, 1325–1337.

Eickholt, D., Carter, T. E., Taliercio, E., Dickey, D., Dean, L. O., Delheimer, J., & Li, Z. (2019). Registration of USDA-Max 9 Soja Core Set-1, recovering 99% of wild soybean genome from PI 366122 in 17 agronomic interspecific germplasm lines. *Journal of Plant Registrations, 13*, 217–236.

El-Shemy, H. A., Khalafalla, M. M., Fujita, K., & Ishimoto, M. (2007). Improvement of protein quality in transgenic soybean plants. *Biologia Plantarum, 51*, 277–284.

Eskandari, M., Cober, E. R., & Rajcan, I. (2013). Genetic control of soybean seed oil II: QTL and genes that increase oil concentration without decreasing protein or with increased seed yield. *Theoretical and Applied Genetics, 126*(6), 1677.

Fallen, B. D., Hatcher, C. N., Allen, F. L., Kopsell, D. A., Saxton, A. M., Chen, P., Kantartzi, S. K., Cregan, P. B., Hyten, D. L., & Pantalone, V. R. (2013). Soybean seed amino acid content QTL detected using the universal soy linkage panel 1.0 with 1,536 SNPs. *Journal of Plant Genomics Science, 1*, 68–79.

Fliege, C. E., Ward, R. A., Vogel, P., Nguyen, H., Quach, T., Guo, M., Viana, J. P. G., Dos Santos, L. B., Specht, J. E., Clemente, T. E., Hudson, M. E., & Diers, B. W. (2022). Fine mapping and cloning of the major seed protein quantitative trait loci on soybean chromosome 20. *The Plant Journal, 110*(1), 114–128.

Fukushima, D. (2004). Soy proteins. In R. Y. Yada (Ed.), *Proteins in Food Processing*. Woodhead Publishing.

Gai, J., Wang, Y., Wu, X., & Chen, S. (2007). A comparative study on segregation analysis and QTL mapping of quantitative traits in plants—with a case in soybean. *Frontiers of Agriculture in China, 1*, 1–7.

Gibson, L. R., & Mullen, R. E. (1996). Soybean seed composition under high day and night growth temperatures. *Journal of the American Oil Chemists' Society, 73*, 733–737.

Goldflus, F., Ceccantini, M., & Santos, W. (2006). Amino acid content of soybean samples collected in different Brazilian states, harvest 2003/2004. *Brazilian Journal of Poultry Science, 8*, 105–111.

Grieshop, C. M., & Fahey, G. C. (2001). Comparison of quality characteristics of soybeans from Brazil, China, and the United States. *Journal of Agricultural and Food Chemistry, 49*, 2669–2673.

Guo, B., Sun, L., Jiang, S., Ren, H., Sun, R., Wei, Z., Hong, H., Luan, X., Wang, L., Wang, X., Xu, D., Li, W., Guo, C., & Qiu, L. J. (2022). Soybean genetic resources contributing to sustainable protein production. *Theoretical and Applied Genetics, 135*, 4095–4121.

Guo, Y. H., Wang, P. Y., Xu, D. C., Meng, L. F., Li, F. M., Dong, D. J., Lin, W. G., Wu, J. J., Zhang, L., Zhong, P., & Han, L. (2005). Study on mutagenesis and improvement of soy protein content. *Soybean Bulletin, 6*, 12–13.

Hacisalihoglu, G., Burton, A. L., Gustin, J. L., Eker, S., Asikli, S., Heybet, E. H., Ozturk, L., Cakmak, I., Yazici, A., Burkey, K. O., Orf, J., & Settles, A. M. (2018). Quantitative trait loci associated with soybean seed weight and composition under different phosphorus levels. *Journal of Integrative Plant Biology, 3*, 232–241.

Ham, G. E., Liener, I. E., Evans, S. D., Frazier, R. D., & Nelson, W. W. (1975). Yield and composition of soybean seed as affected by N and S fertilization. *Agronomy Journal, 67*, 293–297.

Hocine, L., & Boye, J. I. (2007). Allergenicity of soybean: New developments in identification of allergenic proteins, cross-reactivities, and hypo-allergenization technologies. *Critical Reviews in Food Science and Nutrition, 47*, 127–143.

Hwang, E. Y. Q., Song, G., Jia, J. E., Specht, D. L., Hyten, J., Costa, P. B., & Cregan, P. B. (2014). A genome-wide association study of seed protein and oil content in soybean. *BMC Genomics, 15*, 1.

Hyten, D. L., Pantalone, V. R., Sams, C. E., Saxton, A. M., Landau-Ellis, D., Stefaniak, T. R., & Schmidt, M. E. (2004). Seed quality QTL in a prominent soybean population. *Theoretical and Applied Genetics, 109*(3), 552–561.

Jarquin, D., Specht, J., & Lorenz, A. (2016). Prospects of genomic prediction in the USDA soybean germplasm collection: Historical data creates robust models for enhancing selection of accessions. *G3: Genes, Genomes, Genetics, 6*, 2329–2341.

Jegadeesan, S., & Yu, K. (2020). Food-grade soybean breeding: Current status and future directions. In M. Hasanuzzaman (Ed.), *Legume Crops–Prospects, Production and Uses*. IntechOpen.

Jian, S., Wen, Z., Li, H., Yuan, D., Li, J., Zhang, H., Ye, Y., & Lu, W. (2013). Identification of QTLs for glycinin (11S) and β-conglycinin (7S) fractions of seed storage protein in soybean by association mapping. *Acta Agronomica Sinica, 38*, 820–828.

Jun, T. K., Van, M., Kim, S., Lee, & Walker, D. (2008). Association analysis using SSR markers to find QTL for seed protein content in soybean. *Euphytica, 162*, 179–191.

Kabelka, E. A., Diers, B. W., Fehr, W. R., Leroy, A. R., Baianu, I. C., You, T., Neece, D. J., & Nelson, R. L. (2004). Putative alleles for increased yield from soybean plant introductions. *Crop Science, 44*, 784–791.

Khan, M. H., & Tyagi, S. D. (2013). A review on induced mutagenesis in soybean. *Journal of Cereals and Oilseeds, 4*(2), 19–25.

Kim, M., Schultz, S., Nelson, R. L., & Diers, B. W. (2016). Identification and fine mapping of a soybean seed protein QTL from PI 407788A on chromosome 15. *Crop Science, 56*, 219–225.

Kofsky, J., Zhang, H., & Song, B. H. (2018). The untapped genetic reservoir: The past, current, and future applications of the wild soybean (*Glycine soja*). *Frontiers in Plant Science, 9*, 949.

Kumar, V., Rani, A., Solanki, S., & Hussain, S. M. (2006). Influence of growing environment on the biochemical composition and physical characteristics of soybean seed. *Journal of Food Composition and Analysis, 19*, 188–195.

Lai, Y. C., Lin, H., Fang, W. C., Yao, Z. C., Qi, N., Wang, Q. X., Yang, X. F., & Li, H. (2005). Research on the excellent resource of wild soybean screen appraise and utilization in Heilongjiang. *Chinese Agricultural Science Bulletin, 21*, 379–382.

Lam, H. M., Xu, X., Liu, X., Chen, W., Yang, G., Wong, F. L., Li, M. W., He, W., Qin, N., Wang, B., Li, J., Jian, M., Wang, J., Shao, G., Wang, J., SSM, S., & Zhang, G. (2015). Resequencing of 31 wild and cultivated soybean genomes identifies patterns of genetic diversity and selection. *Nature Genetics, 42*, 1053–1059.

Lambirth, K. C., Whaley, A. M., Blakley, I. C., et al. (2015). A comparison of transgenic and wild-type soybean seeds: Analysis of transcriptome profiles using RNA-Seq. *BMC Biotechnology, 15*, 89.

Lee, H., Cho, B. K., Kim, M. S., Lee, W. H., Tewari, J., Bae, H., Sohn, S. I., & Chi, H. Y. (2013). Prediction of crude protein and oil content of soybeans using Raman spectroscopy. *Sensors and Actuators B: Chemical, 185*, 694–700.

Lee, J. S., Kim, H. S., & Hwang, T. Y. (2021). Variation in protein and isoflavone contents of collected domestic and foreign soybean (*Glycine max* (L.) Merrill) germplasms in Korea. *Agriculture, 11*, 735.

Lee, S. H., Bailey, M. A., Mian, M. A. R., Carter, T. E., Shipe, E. R., Ashley, D. A., Parrott, W. A., Hussey, R. S., & Boerma, H. R. (1996). RFLP loci associated with soybean seed protein and oil content across populations and locations. *Theoretical and Applied Genetics, 93*, 649–657.

Lestari, P., Van, K., Lee, J., Kang, Y. J., & Lee, S. H. (2013). Gene divergence of homeologous regions associated with a major seed protein content QTL in soybean. *Frontiers in Plant Science, 4*, 176.

Li, D. M., Sun, M. M., Han, Y. P., Teng, W. L., & Li, W. B. (2010). Identification of QTL underlying soluble pigment content in soybean stems related to resistance to soybean white mold (*Sclerotinia sclerotiorum*). *Euphytica, 172*, 49–57.

Li, F. (1993). Studies on the ecological and geographical distribution of the Chinese resources of wild soybean (*Glycine soja*). *Scientia Agricultura Sinica, 26*, 47–55.

Li, F. S. (1990). *Catalogue of Chinese Wild Soybean Resources.* China Agricultural Press.

Li, S., Xu, H., Yang, J., & Zhao, T. (2019). Dissecting the genetic architecture of seed protein and oil content in soybean from the Yangtze and Huaihe River Valleys using multi-locus genome-wide association studies. *International Journal of Molecular Sciences, 20*, 3041.

Li, Z. W., Meyer, S., Essig, J. S., Liu, Y., Schapaugh, M. A., Muthukrishnan, S., Hainline, B. E., & Trick, H. N. (2005). High-level expression of maize γ-zein protein in transgenic soybean (*Glycine max*). *Molecular Breeding, 16*, 11–20.

Liang, H. Z., Yu, Y. L., Wang, S. F., Lian, Y., Wang, T. F., Wei, Y. L., Gong, P. T., Liu, X. Y., Fang, X. J., & Zhang, M. C. (2010). QTL mapping of isoflavone, oil, and protein contents in soybean (*Glycine max* L. Merr.). *Agricultural Sciences in China, 9*, 1108–1116.

Liu, K. (2004). Soybeans as a powerhouse of nutrients and phytochemicals. In K. Liu (Ed.), *Soybeans as Functional Foods and Ingredients*. AOCS Publishing.

Liu, S. H., Zhou, R. B., Yu, D. Y., & Chen, S. Y., & Gai, J. Y. (2009). QTL mapping of protein-related traits in soybean (*Glycine max* (L.) Merr.). *Acta Agronomica Sinica, 35*, 2139–2149.

Lu, W., Wen, Z., Li, H., Yuan, D., Li, J., Zhang, H., Huang, Z., Cui, S., & Du, W. (2013). Identification of the quantitative trait loci (QTL) underlying water-soluble protein content in soybean. *Theoretical and Applied Genetics, 126*(2), 425–433.

Ma, Y., Kan, G., Zhang, X., Wang, Y., Zhang, W., Du, H., & Yu, D. (2016). Quantitative trait loci (QTL) mapping for glycinin and beta-conglycinin contents in soybean (*Glycine max* L. Merr.). *Journal of Agricultural and Food Chemistry, 64*, 3473–3483.

Malle, S., Eskandari, M., Morrison, M., & Belzile, F. (2020). Genome-wide association identifies several QTLs controlling cysteine and methionine content in soybean seed including some promising candidate genes. *Scientific Reports, 10*, 1–14.

Mansur, L. M., Orf, J. H., Chase, K., Jarvik, T., Cregan, P. B., & Lark, K. G. (1996). Genetic mapping of agronomic traits using recombinant inbred lines of soybean. *Crop Science, 36*, 1327–1336.

Mao, T., Jiang, Z., Han, Y., Teng, W., Zhao, X., Li, W., & Morris, B. (2013). Identification of quantitative trait loci underlying seed protein and oil contents of soybean across multi-genetic backgrounds and environments. *Plant Breeding, 132*, 630–641.

Michelfelder, A. J. (2009). Soy, a complete source of protein. *American Family Physician, 79*(1), 43–47.

Nakasathien, S., Israel, D. W., Wilson, R. F., & Kwanyuen, P. (2000). Regulation of seed protein concentration in soybean by supra-optimal nitrogen supply. *Crop Science, 40*(5), 1277–1284.

Nichols, D., Glover, K., Carlson, S., Specht, J., & Diers, B. (2006). Fine mapping of a seed protein QTL on soybean linkage group I and its correlated effects on agronomic traits. *Crop Science, 46*, 834–839.

Nielsen, N. C., Dickinson, C. D., Cho, T. J., Thanh, V. H., & Scallon, B. J., & Fischer, R. L. (1989). Characterization of the glycinin gene family in soybean. *The Plant Cell, 1*, 313–328.

Orf, J. H., Chase, K., Jarvik, T., Mansur, L. M., Cregan, P. B., Adler, F. R., & Lark, K. G. (1999). Genetics of soybean agronomic traits: I. Comparison of three related recombinant inbred populations. *Crop Science, 39*, 1642–1651.

Palomeque, L., Liu, L. J., Li, W., Hedges, B., Cober, E. R., & Rajcan, I. (2009). QTL in mega-environments, II. Agronomic trait QTL co-localized with seed yield QTL detected in a population derived from a cross of high-yielding adapted × high-yielding exotic soybean lines. *Theoretical and Applied Genetics, 119*, 429–436.

Pandurangan, S., Pajak, A., Molnar, S. J., Cober, E. R., Dhaubhadel, S., Hernández-Sebastià, C., Kaiser, W. M., Nelson, R. L., Huber, S. C., & Marsolais, F. (2012). Relationship between asparagine metabolism and protein concentration in soybean seed. *Journal of Experimental Botany, 63*(8), 3173–3184.

Panthee, D. R., Kwanyuen, P., Sams, C. E., West, D. R., Saxton, A. M., & Pantalone, V. R. (2004). Quantitative trait loci for β-conglycinin (7S) and glycinin (11S) fractions of soybean storage protein. *Journal of the American Oil Chemists' Society, 81*, 1005–1012.

Panthee, D. R., Pantalone, V. R., Sams, C. E., Saxton, A. M., West, D. R., Orf, J. H., & Killam, A. S. (2006). Quantitative trait loci controlling sulfur-containing amino acids, methionine, and cysteine, in soybean seeds. *Theoretical and Applied Genetics, 112*, 546–553.

Panthee, D. R., Pantalone, V. R., West, D. R., Saxton, A. M., & Sams, C. E. (2005). Quantitative trait loci for seed protein and oil concentration, and seed size in soybean. *Crop Science, 45*(5), 2015–2022.

Patil, G., Mian, R., Vuong, T., Pantalone, V., Song, Q., Chen, P., Shannon, G. J., Carter, T. C., & Nguyen, H. T. (2017). Molecular mapping and genomics of soybean seed protein: A review and perspective for the future. *Theoretical and Applied Genetics, 130*, 1975–1991.

Patil, G., Vuong, T. D., Kale, S., Valliyodan, B., Deshmukh, R., Zhu, C., Wu, X., Bai, Y., Yungbluth, D., Lu, F., Kumpatla, S., Grover Shannon, J., Varshney, R. K., & Nguyen, H. T. (2018). Dissecting genomic hotspots underlying seed protein, oil, and sucrose content in an interspecific mapping population of soybean using high-density linkage mapping. *Plant Biotechnology Journal, 16*, 1939–1953.

Poysa, V., Woodrow, L., & Yu, K. (2006). Effect of soy protein subunit composition on tofu quality. *Food Research International, 39*, 309–317.

Qi, Z., Han, X., Hou, M., Xin, D., Wang, Z., Zhu, R., Hu, Z., Jiang, H., Li, C., Liu, C., Hu, G., & Chen, Q. (2014). QTL analysis of soybean oil content under 17 environments. *Canadian Journal of Plant Science, 94*, 245–261.

Qi, Z., Pan, J., Han, X., Qi, H., Xin, D., Li, W., Mao, X., Wang, Z., Jiang, H., & Liu, C. (2016). Identification of major QTLs and epistatic interactions for seed protein concentration in soybean under multiple environments based on a high-density map. *Molecular Breeding, 36*, 1–16.

Qin, J., Wang, F., Zhao, Q., Shi, A., Zhao, T., Song, Q., Ravelombola, W., An, H., Yan, L., Yang, C., & Zhang, M. (2022a). Identification of candidate genes and genomic selection for seed protein in soybean breeding pipeline. *Frontiers in Plant Science, 16*, 882732.

Qin, P., & Wang, T., & Luo, Y. (2022b). A review on plant-based proteins from soybean: Health benefits and soy product development. *Journal of Agricultural and Food Research, 7*, 100265.

Qing, Y. Z., Zhang, F. H., Fan-Zhan, B., & Huang, X. L. (1996). Studies on the radiation effect of 60Co gamma-rays on the soybean plant. *J. Henan Agric. Sci., 6*, 15–17.

Qiu, B. X., Arelli, P. R., & Sleper, D. A. (1999). RFLP markers associated with soybean cyst nematode resistance and seed composition in a 'Peking' × 'Essex' population. *Theor. Appl. Genet., 98*, 356–364.

Qiu, G. J., & Gao, S. (1988). Studies on physically and chemically induced soybean mutations of high protein and oil content and their genetic pattern. *Improvement of Grain Legume Production Using Induced Mutations.* Proceedings of a workshop, Pullman, Washington, USA, 1–5 July (pp. 421–433).

Ravindran, V., Morel, P. C. H., Rutherfurd, S. M., & Thomas, D. V. (2009). Endogenous flow of amino acids in the avian ileum as influenced by increasing dietary peptide concentrations. *Br. J. Nutr., 101*, 822–828.

Reddy, K. R., Patro, H., Lokhande, S., Bellaloui, N., & Gao, W. (2016). Ultraviolet-B radiation alters soybean growth and seed quality. *Food Nutr. Sci., 7*, 55–66.

Reinprecht, Y., Poysa, V. W., Yu, K., Rajcan, I., Ablett, G. R., & Pauls, K. P. (2006). Seed and agronomic QTL in low linolenic acid, lipoxygenase-free soybean (*Glycine max* (L.) Merrill) germplasm. *Genome, 49*(12), 1510–1527.

Ren, C., Bilyeu, K. D., & Beuselinck, P. (2009). Composition, vigor, and proteome of mature soybean seeds developed under high temperature. *Crop Sci., 49*, 1010 1022.

Rizzo, G., & Baroni, L. (2018). Soy; Soy foods and their role in vegetarian diets. *Nutrients, 10*, 43.

Rodrigues, J. I. S., Miranda, F. D., Borges, L. L., Silva, M. F., Good-God, P. I. V., Piovesan, N. D., Barros, E. G., Cruz, C. D., & Moreira, M. A. (2010). QTL mapping for protein and oil content in soybeans in a Brazilian germplasm. *Pesquisa Agropecuária Brasileira, 45*, 472–480.

Rossi, M. E., Orf, J. H., Liu, L. J., Dong, Z., & Rajcan, I. (2013). Genetic basis of soybean adaptation to North American vs. Asian mega-environments in two independent populations from Canadian × Chinese crosses. *Theor. Appl. Genet., 126*, 1809–1823.

Rotundo, J. L., & Westgate, M. E. (2009). Meta-analysis of environmental effects on soybean seed composition. *Field Crops Res., 110*, 147–156.

Sebolt, A., Shoemaker, R., & Diers, B. (2000). Analysis of a quantitative trait locus allele from wild soybean that increases seed protein concentration in soybean. *Crop Sci., 40*, 1438–1444.

Shi, A., Chen, P., Zhang, B., & Hou, A. (2010). Genetic diversity and association analysis of protein and oil content in food-grade soybeans from Asia and the United States. *Plant Breed., 129*, 250–256.

Shi, Z., Liu, S., Noe, J., Arelli, P., Meksem, K., & Li, Z. (2015). SNP identification and marker assay development for high-throughput selection of soybean cyst nematode resistance. *BMC Genom., 16*, 314.

Siddique, S. (2022). Role of CRISPR/Cas9 in soybean (*Glycine max* L.) quality improvement. In T. Ohyama, Y. Takahashi, N. Ohtake, T. Sato, & S. Tanabata (Eds.), *Soybean–Recent Advances in Research and Applications*. IntechOpen.

Singh, P., Kumar, R., Sabapathy, S., & Bawa, A. (2007). Functional and edible uses of soy protein products. *Compr. Rev. Food Sci.*, *7*(1), 14–28.

Song, B., Shen, L., Wei, X., Guo, B., Tuo, Y., Tian, F., Han, Z., Wang, X., Li, W., & Liu, S. (2014). Marker-assisted backcrossing of a null allele of the α-subunit of soybean (*Glycine max*) β-conglycinin with a Chinese soybean cultivar. The development of improved lines. *Plant Breed.*, *133*, 638–648.

Song, Q., Hyten, D. L., Jia, G., Quigley, C. V., Fickus, E. W., Nelson, R. L., & Cregan, P. B. (2015). Fingerprinting soybean germplasm and its utility in genomic research. *G3 Genes Genom. Genet.*, *5*, 1999–2006.

Song, Q., Hyten, D., Jia, G., Quigley, C., Fickus, E., & Nelson, R. (2013). Development and evaluation of SoySNP50K: A high-density genotyping array for soybean. *PLoS One*, *8*, e54985.

Soystat. (2022). *SoyStats 2022*. Retrieved from: http://soystats.com/ (accessed on 25 March 2025).

Specht, J. E., Chase, K., Macrander, M., Graef, G. L., Chung, J., Markwell, J. P., Germann, M., Orf, J. H., & Lark, K. G. (2001). Soybean response to water. *Crop Sci.*, *41*(2), 493–509.

Stewart-Brown, B. B., Song, Q. J., Vaughn, J. N., & Li, Z. L. (2019). Genomic selection for yield and seed composition traits within an applied soybean breeding program. *G3 Genes Genom. Genet.*, *9*, 2253–2265.

Stombaugh, S. K., Orf, J. H., Jung, H. G., Chase, K., Lark, K. G., & Somers, D. A. (2004). Quantitative trait loci associated with cell wall polysaccharides in soybean seed. *Crop Sci.*, *44*, 2101–2106.

Sui, X., Zhang, T., & Jiang, L. (2021). Soy protein: Molecular structure revisited and recent advances in processing technologies. *Annu. Rev. Food Sci. Technol.*, *12*, 119–147.

Sun, R., Sun, B., Tian, Y., Su, S., Zhang, Y., Zhang, W., Wang, J., Yu, P., Guo, B., & Li, H. (2022). The practical soybean breeding pipeline by developing high throughput functional array ZDX1. *Theor. Appl. Genet.*, *135*, 1413–1427.

Tajuddin, T., Watanabe, S., Yamanaka, N., & Harada, K. (2003). Analysis of quantitative trait loci for protein and lipid contents in soybean seeds using recombinant inbred lines. *Breed. Sci.*, *53*, 133–140.

Teraishi, M., Takahashi, M., Hajika, M., Matsunaga, R., Uematsu, Y., & Ishimoto, M. (2001). Suppression of soybean β-conglycinin genes by a dominant gene: *Scg-1*. *Theor. Appl. Genet.*, *103*, 1266–1272.

Valliyodan, B., Qiu, D., Patil, G., Zeng, P., Huang, J., Dai, L., Chen, C., Li, Y., Joshi, T., & Song, L. (2016). Landscape of genomic diversity and trait discovery in soybean. *Sci. Rep.*, *6*, 23598.

Varshney, R. K., Bohra, A., Yu, J., Graner, A., Zhang, Q., & Sorrels, M. E. (2021). Designing future crops: Genomics-assisted breeding comes of age. *Trends Plant Sci.*, *26*(6), 631–649.

Varshney, R. K., Mohan, S. M., Gaur, P. M., Gangarao, N., Pandey, M. K., Bohra, A., Sawargaonkar, S. L., Chitikineni, A., Kimurto, P. K., & Janila, P. (2013). Achievements and prospects of genomics-assisted breeding in three legume crops of the semi-arid tropics. *Biotechnology. Adv.*, *31*, 1120–1134.

Vaughn, J. N., Nelson, R. L., Song, Q., Cregan, P. B., & Li, Z. (2014). The genetic architecture of seed composition in soybean is refined by genome-wide association scans across multiple populations. *G3 Genes Genom. Genet.*, *4*, 2283–2294.

Vollmann, J., Winkler, J., Fritz, C. N., Grausgruber, H., & Ruckenbauer, P. (2000). Spatial field variations in soybean (*Glycine max* [L.] Merr.) performance trials affect agronomic characters and seed composition. *Eur. J. Agron.*, *12*(1), 13–22.

Wang, K., Li, F., & Cheema, A. A. (2001). Studies on the distribution of wild soybean (*Glycine soja*) in China. *Pak. J. Biol. Sci.*, *4*(2), 149–155.

Wang, L. L., Liu, C. Y., Jiang, Z. F., Yao, B. C., Shen, Y. R., Ma, Z. Z., Xin, D. W., Han, X., Tao, L. T., Hu, G. H., & Chen, Q. S. (2014). Analysis of QTL underlying protein content of soybean in multi-environments. *Chin. J. Oil Crop Sci.*, *36*, 443–449.

Wang, P. Y., Wang, L. Z., & Zhang, J. Z. (1989). Induced protein content mutation in soybean. *Soybean Genet. Newsl.*, *16*, 38–40.

Wang, P. Y., Xu, D. C., Guo, Y. H., Meng, L. F., & Zhao, X. N. (2000). Induced mutation for soybean quality. *Acta Agric. Nucl. Sin.*, *14*, 21–23.

Wang, W. Z., Liu, X. Y., Cao, Y. S., & Zhang, M. (1998). Study on protein content of soybean germplasm resources in China. *Crop Var. Res.*, *1*, 35, 36.

Wang, X. F., Ma, W., & Fu, J. (2013). Correlation between protein content in filial generation and yield of soybean. *Soybean Sci.*, *32*, 573–575.

Warrington, C., Abdel-Haleem, H., Hyten, D., Cregan, P., Orf, J., Killam, A., Bajjalieh, N., Li, Z., & Boerma, H. (2015). QTL for seed protein and amino acids in the Benning × Danbaekkong soybean population. *Theor. Appl. Genet.*, *128*, 839–850.

Wee, C. D., Hashiguchi, M., Ishigaki, G., Muguerza, M., Oba, C., Abe, J., Harada, K., & Akashi, R. (2018). Evaluation of seed components of wild soybean (*Glycine soja*) collected in Japan using near-infrared reflectance spectroscopy. *Plant Genet. Resour.*, *16*, 94–102.

Wei, Z. Y., Li, Z. F., Liu, Z. X., & Qiu, L. Q. (2019). The selection effect and germplasm innovation for high protein content on EMS-mutated Zp661 progenies. *J. Plant Genet. Resour.*, *20*, 1579–1587.

Wijewardana, C., Reddy, K. R., Krutz, L. J., Gao, W., & Bellaloui, N. (2019). Drought stress has transgenerational effects on soybean seed germination and seedling vigor. *PLoS One*, *14*(9), e0214977.

Wilcox, J. R. (1998). Increasing seed protein in soybean with eight cycles of recurrent selection. *Crop Sci.*, *38*(6), 1536.

Wolf, R. B., Cavins, J. F., Kleiman, R., & Black, L. T. (1982). Effect of temperature on soybean seed constituents: Oil, protein, moisture, fatty acids, amino acids, and sugars. *J. Am. Oil Chem. Soc.*, *59*, 230–232.

Wu, Y. C., Guo, B. F., Gu, Y. Z., Luan, X. Y., Qiu, H. M., Liu, X. L., Li, H. Y., & Qiu, L. J. (2021). Mapping of a new quantitative locus qPRO-19-1 associated with seed crude protein content in soybean (*Glycine max* L.). *Journal of Plant Genetic Resources, 22*, 139–148.

Xu, B. (1985). The protein resources of *G. max* in China. *Soybean Science, 3*(4), 327–331.

Yang, H., Wang, W., He, Q., Xiang, S., Tian, D., Zhao, T., & Gai, J. (2019). Identifying a wild allele conferring small seed size, high protein content, and low oil content using chromosome segment substitution lines in soybean. *Theoretical and Applied Genetics, 132*, 2793–2807.

Yu, X. P., Du, L. E., & Wei, Y. C. (1995). EMS-induced soybean mutation can screen germplasm resources with high protein or high fat content. *Crop Variety Resources, 1*, 24–26.

Zhang, H. Y., Goettel, W., Song, Q. J., Jiang, H., Hu, Z. B., Wang, M. L., & An, Y. C. (2020). Selection of *GmSWEET39* for oil and protein improvement in soybean. *PLOS Genetics, 16*, e1009114.

Zhang, M. J., Li, Z. F., Yu, L. L., Wang, J., & Qiu, L. J. (2018). Identification and screening of protein subunit variation germplasm from both mutants and natural population in soybean. *Crops, 34*, 44–50.

Zhang, W. K., Wang, Y. J., Luo, G. Z., Zhang, J. S., He, C. Y., Wu, X. L., Gai, J. Y., & Chen, S. Y. (2004). QTL mapping of ten agronomic traits on the soybean (*Glycine max* L. Merr.) genetic map and their association with EST markers. *Theoretical and Applied Genetics, 108*(6), 1131–1139.

Zhang, X., Shu, X., Gao, Y., Yang, G., Li, Q., Li, H., & Jin, F. (2003). Soy food consumption is associated with lower risk of coronary heart disease in Chinese women. *Journal of Nutrition, 133*, 2874–2878.

CHAPTER 5

Soybean Databases: A Means to Explore Soybean Information

ASHLESHA MANTA,[1] DIVYA SHARMA,[1] AAKASH SINGH,[1] BYONKESH NONGTHONGBAM,[1] FATIMA NAZISH KHAN,[2] AYYAGARI RAMLAL,[3,4] NISHA SOGAN,[5] SOURAV SINGH DEO,[5] and APARNA NAUTIYAL[5,*]

[1]*Department of Botany, University of Delhi, New Delhi, Delhi, India*

[2]*Department of Biotechnology, Jamia Millia Islamia, New Delhi, India*

[3]*Division of Genetics, ICAR–Indian Agricultural Research Institute (IARI), Pusa Campus, New Delhi, Delhi, India*

[4]*School of Biological Sciences, Universiti Sains Malaysia (USM), Georgetown, Penang, Malaysia*

[5]*Department of Botany, Deshbandhu College, University of Delhi, New Delhi, India*

*Corresponding author

ABSTRACT

Soybean (*Glycine max* (L.) Merr.) is an extremely important food and economic crop. Apart from being a pivotal oil-yielding crop, it also serves as a promising model organism among legumes. Thus, there is a need to decipher its genetic intricacies for various crop improvement programs, ranging from traditional breeding methods to modern biotechnological approaches. This requires the use of various genetic and genomic resources, which are stored in databases. Biological databases are electronic interfaces that store biological data and information available for scientific study.

Soybeans, Volume 2: Genome, Biotechnology, Stress, and Disease.
R. Ambika Rajendran, Ayyagari Ramlal, Sanjay Kumar Lal, and Godfree Chigeza (Eds.)

DOI: 10.1201/9781998511730-6

The contents of these databases can be searched, compared, retrieved, and analyzed. Multiple soybean databases serve as repositories comprising extensive data on soybean, such as genome, transcriptome, and protein data; genetic and physical maps; ESTs; microarray; and cDNA data. These databases and tools offer platforms for molecular analysis techniques such as functional gene annotation, genotype comparisons and visualizations, and analysis of genomic features. For example, SoyBase is a comprehensive and easily accessible web repository that includes genetic, genomic, and associated data for soybean. Several such databases consolidate relevant data in one place, making access easier and aiding in crop improvement. This chapter provides an overview of important soybean databases and their utility as essential bioinformatic tools in soybean research.

5.1 INTRODUCTION

Soybean (*Glycine max* (L.) Merr.), also known as soja bean or soya bean, is a predominantly self-pollinated annual legume in the pea family (Fabaceae) (Rajendran and Lal, 2020; Ramlal et al., 2022). It is a leading oil crop on a global scale, accounting for about 18% of global vegetable oil production. Soybean meal (SM) is fed to livestock as an important source of protein because of its high crude protein content (44.1%) and has been known to increase milk and meat production. Apart from its importance as an oil crop and animal feed, soybean is also consumed by humans because of its high nutritional value, as several soy products like tofu, soymilk, miso, etc. It has been reported to prevent several diseases, including cancer, because of the presence of isoflavones (Ali et al., 2020; Ramlal et al., 2023a-c; Rizzo and Baroni, 2018; Tidke et al., 2015). Owing to the growing significance of soybean as a global crop, it is frequently subjected to crop improvement. This ranges from traditional breeding methods to several modern biotechnological approaches and requires the use of various genetic and genomic resources that are stored in databases. These databases contain several computational assets such as tools, packages, and web resources that are publicly accessible and have led to scientific advancement (Sheikh et al., 2022; Sreenivasaiah and Kim, 2010).

Biological databases are electronic interfaces that store biological information compiled from scientific experiments, published literature, high-throughput experimentation technology, and computational analysis.

They include data from several studies such as genomics, proteomics, metabolomics, microarray gene expression, and phylogenetic studies (Altman, 2004). Data is directly transmitted to biological databases for indexing, organizing, and improvement. They assist users in locating relevant biological data by making it available in a computer-readable format. The primary goals of biological databases are not only to store, organize, and share data in a structured and searchable manner in order to facilitate human data retrieval and visualization, but also to make data mining tools easily accessible, thereby saving time and resources (Helmy et al., 2016; Zou et al., 2015).

The following are the main characteristics of biological databases (Ramlal et al., 2021):

1. **Data Heterogeneity:** Databases contain a wide range of data types. For example, they store sequence-based information, DNA and RNA from numerous organisms (humans, mice, etc.), as well as high-dimensional data obtained from microarray tests, and so on. They also include graphics (regulatory networks, genomic maps) and forms (docking sites of molecules, etc.).

2. **High Data Volume:** There is an abundance of biological data to support extensive inquiries in a variety of domains and orientations.

3. **Data Sharing:** Data saved in these can be accessed from any location (laboratory or workstation) by any scientific community.

4. **Dynamic:** Every day, new data is generated in laboratories, which might occasionally contradict existing data, so databases must be updated when new data arrives.

5. **Large-Scale Data Integration:** Data shared and collected from around the world contain a wide range of information, such as structural or sequence-based information, and so on.

Thus, biological databases can be said to combine massive volumes of omics data, acting as critical resources and progressively becoming vital for scientists ranging from wet-lab biologists to in silico bioinformaticians (Zou et al., 2015). Soybean databases serve as storehouses for a large amount of data on soybean biological information, as well as technologies that provide a platform for studying this data. The information stored in these databases can be used for the genetic enhancement of soybean for various crop improvement programs (Rajendran and Lal, 2020; Rajendran, 2022).

This chapter will comprise a brief overview of some soybean databases along with the tools available for data analysis. It also highlights the merits and demerits associated with soybean databases while elucidating the significance of these databases and their implications in biological research, including gene ontology enrichment analysis, QTL mapping, and the identification of candidate genes for agronomic traits, among others. Furthermore, this chapter will cover the barriers and impediments to establishing, using, and implementing these databases, such as limited data, proprietary issues, and lack of funding, etc. Lastly, it focuses on future perspectives regarding how these databases can be improved and how they can be further implemented for crop improvement.

5.2 SOYBEAN DATABASES

Soybean databases can be defined as a collection of biological data on soybeans that can be searched, retrieved, analyzed, and cross-referenced. These databases are updated regularly by adding relevant genetic and genomic information on soybeans that are continuously becoming available because of biotechnological advancements. Several of these databases have tools that allow users to explore and perform biological data analyses. Soybean databases can be classified based on data types like genome and genetic data, functional networks and co-expression data, protein-related data, expressed sequence tags (ESTs), manual and image phenotype data, and so forth (Petereit et al., 2022). This section will comprise a brief detail on some relevant soybean databases as well as the tools associated with data analysis.

5.2.1 SOYBASE

SoyBase (https://soybase.org), a U.S. Department of Agriculture-Agricultural Research Service (USDA-ARS) database, houses genetic and genomic data on soybeans that is freely available for users. Ever since its development in 1990, it has continued incorporating data on soybeans ranging from genome sequence data of multiple cultivars/species to resequencing and chip-based variant information for thousands of accessions (Brown et al., 2020). SoyBase also holds the most recent physical, genetic, and genomic sequence maps integrated with quantitative trait locus (QTL) information collected from several years of QTL mapping. It also has a number of freely accessible data analysis tools. Furthermore, it provides links to external soybean resources, thus making it a complete collection of all soybean information.

5.2.1.1 FEATURES AND TOOLS OF SOY BASE

1. **Genetic Map:** It displays genetic and physical maps using the comparative map viewer, Cmap (Grant et al., 2010; Stein et al., 2022). All the genetic and sequence maps have been incorporated with linkage groups (LGs), genetic and molecular markers, as well as over 3100 QTLs describing more than 90 traits. All the map features, like QTLs, gene markers, and LGs, have external links, allowing users access to all the related textual data in the database.
2. **Genome Browser:** It contains the genomic sequence of the reference genome, cultivar Williams 82. SoyBase utilizes the GMOD genome browser GBrowse for sequence viewing (Donlin, 2009; Grant et al., 2017). The browser also houses a tool that can be used to search for or locate SNPs in chromosomes that are present in the SoySNP50K iSelect BeadChip, which is an Illumina Infinium BeadChip containing about 50,000 SNPs. The browser holds a separate database called SoyTEdb, which houses information related to transposable elements in the soybean genome (Du et al., 2010a).
3. **Data:** It comprises *G. max* and *G. soya* Data Store, which houses genetic and genomic data related to these two species. It also has tools that enable users to download genetic data and sequences. The SoyNAM database within this section holds genotype and phenotype data on 40 cultivars that have been developed by Nested Association Mapping of several genes controlling important agronomic traits (Brown et al., 2020; Song et al., 2017).
4. **Expression:** The RNA-Seq Atlas holds gene expression profiles of 14 tissues. This allows users to gain a better understanding of differential gene expression in various tissues (Söllner et al., 2017; Xanthopoulou et al., 2021). It also has a section that displays gene expression samples categorized by experiments. The Expression Explorer tool can be used to explore the tissue and developmental

expression patterns of a gene model. The tissue and developmental patterns are presented in both pictorial and tabular form.

5. **Mutants:** The fast neutron (FN) InDels section holds data on FN generated mutant populations (Bolon et al., 2011, 2014). There are mutant browsers that enable users to browse mutants by sample name, trait, image, or phenotype. There is also a gene and locus list of FN deleted genes, which allows for quick identification of mutants. It also allows a BLAST search against genomic sequences of all of the FN-induced indels. Thus, it can be used to determine if a gene of interest is covered by one of the indels.
6. **Projects:** It contains data on several important projects on soybeans. They are listed along with a brief description of each project, along with links to all the relevant literature associated with a particular project.
7. **Tools:** It houses several freely accessible data analysis tools. For example, the Soybean Pangenome Search Page enables users to perform a BLAST search against the SoyBase Soybean Cultivar Genomes Collection sequences. It also permits pan-genome analysis by comparing the sequence of interest with the sequences in the SoyBase Soybean Cultivar Genomes Collection (Bayer et al., 2021; Torkamaneh et al., 2021). Another important tool is the Soybean Metabolic Pathways tool, which houses all relevant information on the metabolic pathways of soybean, such as the compounds, genes/proteins/RNAs, and reactions involved in the pathways (Karp et al., 2010; Zhang et al., 2010). It also has several computational tools that allow data analyses, such as annotating pathway information and identifying pathways in common, etc.
8. **Community:** It contains additional data on soybeans, ranging from links to external data sources to information about ongoing projects in soybean research. It also holds information on soybean workshops, the SoyBase committee, as well as job availability at SoyBase.

Furthermore, the Help and Tutorials section in SoyBase acts as a guide on how to explore and use the database. It has links to various YouTube videos that demonstrate how to use the numerous data analysis tools present in the database. It also has a guide on how to upload data to SoyBase.

5.2.2 SOYKB

SoyKB or Soybean Knowledge Base (https://soykb.org) was developed as a comprehensive multi-omic web resource for soybean. It houses data on soybean genomics, transcriptomics, proteomics, and metabolomics, as well as molecular breeding data, such as QTL, traits, and germplasm information (Joshi et al., 2013). It also has several data analysis tools, like the Affymetrix probe ID mapper, sequence similarity tools, metabolite pathway viewer, chromosome visualizer, etc. Since its commencement in 2010, it has provided an interface to integrate multi-omics data with molecular breeding (Joshi et al., 2016).

5.2.2.1 FEATURES AND TOOLS OF SOYKB

1. **Search:** It contains information about six entities, which are genes, miRNA/sRNAs, metabolites, SNPs, PI lines, and traits. It enables a user to browse through any related information on a particular entity. For example, in the case of the gene card, users can look for information on a gene simply by entering the gene ID or its chromosome coordinates, and all the associated data on the gene, ranging from gene name and family to its functional annotation, are displayed on the browser.
2. **Browse:** It enables users to browse through several datasets like gene families, homologous genes, phosphorylation data, FN mutant data, and differential expression data. The Genome Browser within this section allows users to visualize gene models, supporting evidence, SNPs, and other experimental data (Joshi et al., 2012). The *in silico* breeding program has several data analysis tools that assist in the molecular breeding of soybean by incorporating QTL data with genomic variations and germplasm information. The NGS Resequencing Browser contains various genetic data generated from next-generation sequencing (NGS) technology and allows for the identification of SNPs and indels against the soybean reference genome. Another important browser, SoyTD, allows for the identification, classification, and extensive characterization of transporter proteins by using various genetic and genomic data (Deshmukh et al., 2020).

3. **Tools:** It contains several data analysis tools, some of which are multiple sequence alignment (MSA), chromosome visualizer, mutant finder, and the phylogeny tool, etc. One of the most significant tools is the Pathway Viewer, which can be used to explore the different genes and metabolites expressed in a particular pathway. Another tool, the protein BioViewer, shows primary protein sequences, predicted secondary structures, phosphorylation positions, amino acid (AA) characteristics, functional domain positions, and trans-membrane helix positions (Joshi et al., 2013; Zdobnov and Apweiler, 2001).
4. **Data Files:** It permits users to download different types of data on genes, such as chromosome coordinates, sequences, domain information, and experimental data. Users can also upload several types of data, including gene family, functional annotations, microarray experimental data, transcriptomics experimental data, proteomics experimental data, and metabolomics experimental data. It has a feature named "Extract Genome Sequence," which allows users to extract soybean gene sequences from any desired position. The Public Data Files feature has links to several external literature sources and associated data on NGS and SNPs.
5. **Analytics:** It contains the Pegasus Genomic Variation (PGen) Workflow, which is a tool for studying the genomic variation of soybean using next-generation resequencing data with large-scale inputs (Deelman et al., 2005; Joshi et al., 2016). It also allows customization for other plant species.
6. **Information:** It houses information on the complete architecture of SoyKB, data sources with links to correlated literature, soybean nutritional facts, links to external soybean databases, as well as workshop information on SoyKB. Furthermore, there are links to a video and a user manual that assist users on how to operate SoyKB.

5.2.3 SFGD

Soybean Functional Genomics Database or SFGD (http://bioinformatics.cau.edu.cn/SFGD) is a biological database that provides a platform for soybean functional genomics by integrating genome and transcriptome data, such as coding genes, full-length cDNA and miRNA sequences, microarray expression experiments, deep sequencing experiments, and single nucleotide polymorphisms (SNPs) (Yu et al., 2014). Furthermore, it also contains information on the acyl-lipid pathways of soybean by integrating plant metabolic network (PMN) data and *Arabidopsis* lipid-related genes, as well as various tools for the analysis of soybean omics data.

5.2.3.1 FEATURES AND TOOLS OF SFGD

1. **GBrowse:** It functions as a genome viewer. It also has various transcriptional data, such as soybean protein-coding genes, full-length cDNAs, probe set consensus sequences, miRNA precursors, microarray expression experiments, and deep sequencing data integrated into it, along with links to another webpage that displays associated information on each keyword (Yu et al., 2014).
2. **Gene Co-Expression Network:** It displays a soybean gene co-expression regulatory network from 14 microarray expression experiments (Obayashi and Kinoshita, 2009; Yu et al., 2014). The network contains more than 23,000 genes represented by probe sets. The Cytoscape Web software displays the network with each gene being presented with its associated co-expression genes, along with a table showing their annotations (Lopes et al., 2010).
3. **Acyl-Lipid Metabolic Pathways:** It houses information on 15 soybean acyl lipid metabolism pathways by integrating data from PMN and predictions using *Arabidopsis* lipid-related genes. It enables users to study the different metabolic pathways by displaying related information such as metabolism alias names, mass-to-charge ratios, and chemical formulae (Yu et al., 2014). It also shows the number of genes and enzymes involved in a particular pathway.
4. **Motif:** It displays all the motifs present and the frequencies of their occurrence in a particular FASTA sequence of a promoter entered by the user. Users can also submit a list of genes, and all the significant motifs in the promoter of those genes will be displayed.

5. **Expression Pattern:** It enables users to study the expression patterns of genes under specific conditions. It contains two tools named "Expression Pattern Search" and "Tissue/Time Specific Search." The Expression Pattern Search tool provides information on whether a gene is up-regulated or down-regulated based on the user's selection, while the Tissue/Time Specific Search only shows those probe sets that are highly expressed in a particular tissue or time as selected by the user. There is also an external link for "agriGO," which can be used for gene ontology enrichment analysis (Du et al., 2010b; Yu et al., 2014).

Furthermore, the General Search page allows users to look for information based on one or a list of gene names, flcDNA IDs, probe set IDs, or specific keywords. The BLAST option enables users to make BLAST searches against soybean cDNA, flcDNA, probe set consensus, and protein sequences. The Download option also permits users to download all coding genes' sequences and their annotations (Yu et al., 2014).

Apart from these three databases, there are numerous other soybean databases housing different types of data and analysis tools. A list of several more soybean databases is provided in Table 5.1.

TABLE 5.1 Available Soybean Databases and Their Roles

Database	Functions	Website	References
SoyFN	Exploring functional networks of soybean genes and miRNAs.	https://nclab.hit.edu.cn/SoyFN	Xu et al. (2014)
Plant-Impute DB	Genotype imputation by integrating genotype data and whole-genome resequencing data.	https://gonglab.hzau.edu.cn/Plant_imputeDB/	Gao et al. (2020)
SoyDB	Knowledge database of soybean transcription factors.	http://casp.rnet.missouri.edu/soydb/	Wang et al. (2010)
SoyProDB – Soybean Seed Protein Database	Identification of soybean seed proteins.	http://bioinformatics.towson.edu/Soybean_Seed_Proteins_2D_Gel_DB/Home.aspx	Tavakolan et al. (2013)
Rsoy – RIKEN Soybean	Soybean full-length cDNA database.	http://spectra.psc.riken.jp/menta.cgi/rsoy/index	Umezawa et al. (2008)

TABLE 5.1 *(Continued)*

Database	Functions	Website	References
SoyXpress	Soybean transcriptome database.	http://soyxpress.agrenv.mcgill.ca	Cheng and Strömvik (2008)
SoyGD	Genome browser housing soybean genomic and physical map data.	http://soybeangenome.siu.edu	Shultz et al. (2006)
SGMD	Provides an integrated view of soybean interactions with soybean cyst nematode; contains genomic, EST, and microarray data.	http://psi081.ba.ars.usda.gov/SGMD/default.htm	Alkharouf et al. (2003)
SoyVCF 2 Genomes	Identifies the closest accession in soybean germplasm collections.	http://pgl.gnu.ac.kr/soy_vcf2genome/	Ha et al. (2019)
SoyNet	Database of co-functional networks for soybean.	http://www.inetbio.org/soynet	Kim et al. (2017)
SMDB	Explores gene differences in plant genotypes, including genetically modified and disease-resistant crops.	http://www.bioinfoindia.org/smdb/	Kumar et al. (2014)
Soybean Research and Information Network (SRIN)	Showcases soybean research and its importance, houses information on agronomics, diseases, pests, and provides resources.	https://soybeanresearchinfo.com/	Supported by the North Central Soybean Research Program (NCSRP) and United Soybean Board (USB)
National Soybean Checkoff Research Database	Provides information about all soybean research projects funded by the Soybean Checkoff from 2008 to present in the U.S.	https://www.soybeanresearchdata.com/	Supported by the United Soybean Board (USB)
SoyCSN	Network analysis and prediction of tissue-specific transcriptomic data in soybean.	http://soykb.org/SoyCSN	Wang et al. (2019)

5.3 MERITS AND DEMERITS

5.3.1 MERITS

1. Biological databases provide a wide range of opportunities and give access to their stored data (Ramlal et al., 2021), which makes it possible for researchers to more thoroughly examine published data and use the genetic diversity of soybeans for crop improvement (Petereit et al., 2022).
2. The accessibility of soybean genomic information and related phenotype and marker databases lays the groundwork for next-generation breeding approaches, including Genome-Wide Association Study and Genomic Prediction (Shi et al., 2021), genome-wide comparison for selection signatures (Cortés et al., 2021), and machine learning (Schrider and Kern, 2018).
3. Soybean databases also make it possible to use genomic-assisted backcrossing rather than marker-assisted backcrossing, which will enhance soybean breeding (Bhat and Yu, 2021; Petereit et al., 2022).
4. Soybean databases aid in data integration to enhance gene annotation and find links between allelic variations and phenotypic traits.
5. In response to newer findings in soybean genome research, soybean databases are regularly updated (Torkamaneh et al., 2021).
6. These databases allow users to access tools to study and analyze genetic maps, mutant soybean lines, and gene families, and compare variable gene expression patterns (Petereit et al., 2022).
7. Soybean databases can also be employed for exploration, such as conserved regions (Torkamaneh et al., 2021), gene identification, and QTL mapping (Karikari et al., 2020), as well as estimation of disease resistance (Rolling et al., 2020).
8. Soybean databases, that is, SoyBase, keep records of important soybean projects, including Soybean Nested Association Mapping (SoyNam) (Song et al., 2017), Haplotype Map of soybean (Torkamaneh et al., 2020), and Screening of an EMS mutant soybean population (Espina et al., 2018).

9. Orthologous Gene Identification: Genomic and genetic information from many legume crops is gathered by the legume information service (LIS, legumeinfo.org). Links to the LIS system gene family and phylogeny predictions are available on every page of a soybean gene profile at SoyBase, allowing users to find orthologous genes in other legume species (Berendzen et al., 2021; Brown et al., 2020).
10. Soybean Pedigree: From 1940 to the present, pedigree information on soybeans is collected for test strains and releases, including every soybean strain in the plant variety protection (PVP) database. Over 15,000 soybean lines and commercial cultivars' parent data are listed in the dataset. This approach helps identify phylogenetic relatives and relationships among soybean lineages (Brown et al., 2020).

5.3.2 *DEMERITS*

1. Integrative soybean research has become more challenging because of the growth in genomic and phenotypic data for the crop. Datasets are frequently dispersed across several databases, making it difficult for researchers to retrieve all the pertinent information that could be analyzed (Petereit et al., 2022).
2. Numerous published databases have substandard metadata, which hinders in-depth research (Petereit et al., 2022).
3. There is a significant amount of data that is duplicated across datasets for many reasons, such as repetitive submission of the same or overlapping sequences by the authors, revision of annotations, deletion of EST data, and substandard database management that fails to recognize the redundancy. Because of this, some primary databases become large and challenging to use for data retrieval (Xiong, 2006).
4. Often, the same gene sequence is found using various names, leading to duplicate entries and data confusion. Alternatively, unrelated genes with similar names can also be listed in the repositories (Xiong, 2006).

5. Some errors, however, are merely the result of omissions or typing errors. These errors in annotations can be harmful because the majority of newly discovered sequences are functionally annotated based on their similarity to previously annotated sequences in databases. Consequently, a flawed annotation can simply be applied to all the similar gene sequences across the entire database (Klimke et al., 2011; Xiong, 2006).

5.4 SIGNIFICANCE AND ITS IMPLICATIONS

Several soybean databases, like SoyBase and SoyGD and so forth, integrate the exponentially growing genetic and genomic data, making it easily accessible to the soybean research community. In addition, they also contain the expanding genomic and phenotypic datasets, which could be employed to display the relationship between agronomic traits (phenotypic trait diversity) and allelic variants (genomic diversity), which is important for applied crop improvement programs (Petereit et al., 2022). Such databases are a pool of various information that could be subjected to a plethora of biological research.

5.4.1 SIGNIFICANCE

1. They provide professionally curated data, encompassing genomic sequence variants, epigenetic maps, regulatory RNA data, functional networks and co-expression data, pan-genome gene visualization, and manual and image phenotype (Petereit et al., 2022).
2. Analysis tools for searching sequence similarity, contig sequence annotation, soybean ontologies, unigene sequences, and so forth, are available in soybean databases such as SoyBase (Grant et al., 2009).
3. They house common reference soybean genomes like the "Williams 82" genomic sequence and pan-genome sequence, which could be generally used for the improvement of agronomic traits (Brown et al., 2020; Grant et al., 2009).

5.4.2 IMPLICATIONS

Soybean databases are updated continuously and can be employed for many biological research studies (Figure 5.1) as follows.

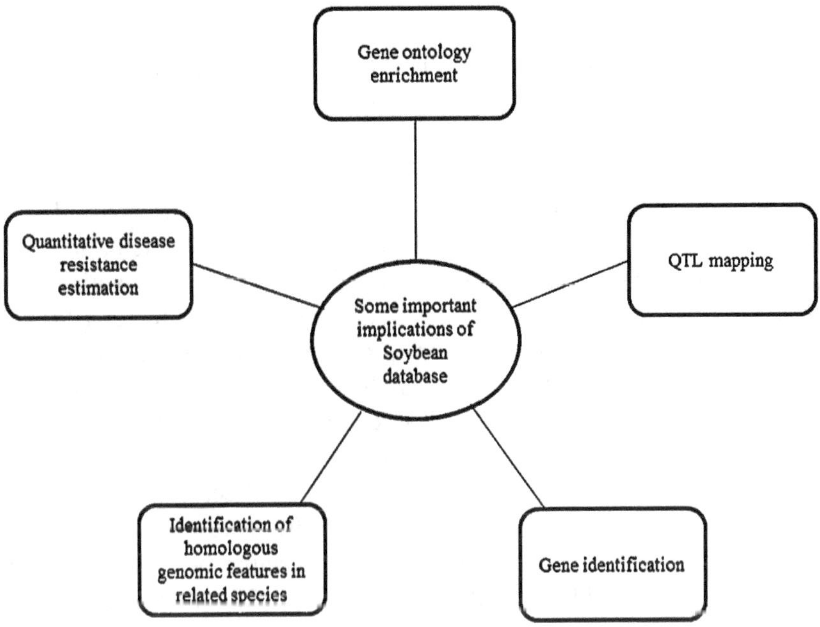

FIGURE 5.1 Some important implications of soybean database in biological research.

5.4.2.1 GENE ONTOLOGY ENRICHMENT ANALYSES

Due to inadequate genetic variability of the common reference genome "Wm82," pangenomes, which are species-representative genomes, were constructed. Recently, *PanSoy* was built using the GmHapMap dataset (Torkamaneh et al., 2021), annotated using Phytozome (https://phytozome.jgi.doe.gov/; Goodstein et al., 2012), and furnished with the EUPAN pipeline (Hu et al., 2017) using Wm82 as a reference. In addition, it holds 1086 Mb (~11%) with 3621 protein-coding genes (Torkamaneh et al., 2021), including 1659 novel genes. The gene ontology enrichment analysis can be performed for *PanSoy* genes using the GO term enrichment tool supported by SoyBase (Grant et al., 2009) based on the Wm82 genome. This analysis shows the role of the genes that are responsible for many biological processes. Using this analysis, it was recently found that the variant genes (~7%) in the PanSoy genome significantly involve defense responses, plant development, and signaling, and also show high evolutionary rates, which reveal a parallel scenario of their role in the agronomic traits of cultivated soybean (Torkamaneh et al., 2021).

5.4.2.2 QUANTITATIVE TRAIT LOCI MAPPING AND IDENTIFICATION OF CANDIDATE GENE FOR AGRONOMIC TRAITS

More than 1000 quantitative trait loci/nucleotides (QTL/QTNs), controlling 90 agronomic traits, have been reported in soybean. Around 300 QTLs for soybean seed weight, 241 QTLs for protein content, and 315 QTLs for oil content, distributed over 20 chromosomes, are available on SoyBase (www.soybase.org) via linkage mapping. Recently, 231 QTLs/QTNs for soybean seed weight were predicted by using a genome-wide association study (GWAS), and 36 candidate genes were identified via a co-expression network approach (Karikari et al., 2020; Qi et al., 2020). By the integrated approach of differential expression and co-expression analysis, 124 candidate genes that significantly affect soybean seed oil content have been reported (Yang et al., 2019). These identified QTLs and candidate genes can be employed for molecular breeding efforts to improve soybean yield and quality (Chen et al., 2022; Ikram et al., 2020).

5.4.2.3 QUANTITATIVE DISEASE RESISTANCE ESTIMATION

Quantitative disease resistance (QDR) is a race-unspecific genetic resistance or immunity that functions via the interconnected expression of multiple genes (Kou and Wang, 2010). Recently, a GWAS was performed in *Phytophthora sojae* Kaufm. & Gerd. introduced soybean plants to identify genetic markers that are linked to QDR loci (QDRL), using genotype data that is available on the SoySNP5OKiSelect Bead Chip (Song et al., 2013), downloaded from SoyBase. In addition, 48 genetic markers present on chromosome 13 were identified, representing 44 QDRL that depict the complex action of QDR to *P. sojae* (Rolling et al., 2019). This QDR estimation helps us to identify QDR alleles and their role in genomic selection (GS).

5.4.2.4 IDENTIFICATION OF HOMOLOGOUS GENOMIC FEATURES IN RELATED SPECIES

Many QTLs and potential candidate genes that are related to agronomic traits have been reported in soybean databases. This information could be used for marker-assisted selection and subjected to the map-based identification of polymorphisms in related species. Moreover, the identification of markers and polymorphisms in different accessions is a prerequisite for pacing crop improvement programs. In soybean, there is a negative correlation between the two important agronomic traits, that is, seed protein and oil content, making the improvement of both traits difficult through conventional breeding (Hwang et al., 2014). However, the molecular markers associated with QTLs controlling soybean seed protein and oil yield could be identified and used to solve the problem of improving both traits (seed protein and oil content).

5.5 CHALLENGES AND BARRIERS

Soybean databases provide a comprehensive collection of highly usable biological data that may be applied to a range of crop improvement programs. Despite this, there are a number of barriers and impediments to establishing, using, and implementing these databases, which are as follows.

1. **Limited Databases on Soybean Proteomics:** Proteomic-based techniques allow for the comprehensive identification of proteins, their isoforms, and their prevalence in each tissue, as well as the characterization of each protein's biochemical and cellular functions, and the analysis of protein regulation and its relationship to other regulatory networks, making soybean proteomics research important for crop improvement (Chandramouli and Qian, 2009; Wu et al., 2010). Due to recent biotechnological advancements, there is a huge amount of data generated on soybean multi-omics, including high-throughput proteomic data (Joshi et al., 2012). However, there have been minimal attempts to compile and integrate this data, resulting in limited databases housing information on soybean proteomics.

2. **Proprietary Issues:** The most basic barrier to using databases is that many of them are inaccessible to the majority of the public. Some are proprietary databases compiled by private companies, while others are collections owned by academic researchers or university departments that have never been made public. Even if the information in a database is publicly available, it is impractical if it's not available online (Pool and Esnayra, 2000; Ramlal et al., 2021).

3. **Disparate Terminology:** The subject of bioinformatics is one that is constantly evolving. Since the terminologies in the field of bioinformatics are continually changing, it is difficult to openly exchange ideas and acquire access to resources (Ramlal et al., 2021).

4. **Interoperability:** Lack of interoperability is another typical problem with using databases. Working across multiple databases requires dealing with differences in how the various collections organize their data. Thus, resolving complicated queries to vast collections of databases requires a high level of user expertise and, in the absence of it, may restrict access (Pool and Esnayra, 2000).

5. **Lack of Funding:** Databases necessitate continuous access not only for validation but also for re-analysis, testing new hypotheses, and developing new technologies/platforms. Finding financial support for maintenance and development to best serve the scientific community is a common challenge for most data resources and biological repositories (Chandras et al., 2009; Ellis and Kalumbi, 1999). Some publicly accessible soybean databases have been decommissioned owing to a lack of maintenance.

6. **Redundancy of Data:** Due to a number of factors, including repeated submission of identical or overlapping sequences by the same or different authors, annotation revision, an excessive amount of EST data deposition, and inadequate database administration, the databases contain a significant amount of information duplication (Xiong, 2006). As a result, some soybean databases have become excessively large and unwieldy for information retrieval.

7. **Maintaining the Integrity of Databases:** Databases can hold billions of bytes of information; therefore, errors are unavoidable, and these flaws might cause unexpected problems in data analysis. Errors in databases can occur for a variety of reasons, including mistakes made when data is entered into a database or when data is transferred into the database from journal articles and is reproduced inaccurately. It is also possible that inaccuracies in the sources were integrated into the database (Goldberg et al., 2010; Pool and Esnayra, 2000). After being introduced into a database, errors can easily spread to multiple systems, disrupting the integrity of several databases.

5.6 FUTURE PERSPECTIVES

All soybean databases, such as Soybase, SoyGD, and SoyKB, and so forth, are continuously updated with more datasets and tools, which majorly expand the scope of the soybean breeding program. For instance, new tools such as Gene Expression Explorer (for accessing extensive expressive data), ortholog identification in other legume species (genetic and genomic data from other legumes), GCViT diversity browser (exploring natural variation among accessions), and pan-genome viewer (displaying pan-genome sets) have been recently included in SoyBase (Brown et al., 2020). The genomic databases and tools stored could be employed to dissect the genetic basis for many agronomic traits, nullifying the complications in the soybean breeding program, specifically for seed composition traits. Many recent approaches to breeding have been performed successfully, witnessing the potent propensities for next-generation breeding techniques shortly.

No doubt, it is fundamental to reveal the source of environmental adaptation in crop breeding programs. Genome-environment association (GEA) has been recently used as a new striking approach to study stress tolerance (drought tolerance, heat tolerance, etc.) in other legumes (Kalve et al., 2022; López-Hernández and Cortés, 2019). The extensive availability of soybean marker datasets enables GEA to be a promising approach to study soybean environmental adaptation in the near future. Besides, the availability of genomic datasets, along with associated phenotypic data, sows a seed for the generation of soybean breeding approaches like genomic prediction (Shi et al., 2021), machine learning (Schrider et al., 2018), genome-wide scans of selection signatures (Cortes et al., 2020), and speed breeding (Varshney et al., 2021).

The well-curated soybean datasets available are poised to replace marker-assisted backcrossing with genome-assisted backcrossing, which will speed up the future soybean breeding program. Recent techniques like long-read sequencing have been used to construct high-quality reference genomes and pangenomes, covering a larger number of accessions. Moreover, it is believed that this next-level sequencing could be effectively used to study high-fidelity variation mapping across populations by generating larger pangenomes. As shown by Zhuang, Wang, Li, Cannon, Grimwood, Hu, Fan, Landis, Schmutz, Jackson, Zhang, Doyle, Zhang, and Ma, pangenomes are about to expand into the higher-level taxon, and it may positively influence the advances in functional genomics, leveraging the potential data quantity and quality for soybean crop improvement. All the available data stored in such databases will enable new techniques to solve the problems in soybean breeding programs and also meet the growing demands for soybean products on a larger scale.

5.7 CONCLUSION

Soybean is a major food and oil-yielding crop with high economic value. Improvement in soybean yield and quality requires traditional biotechnological approaches. In this regard, soybean databases are fundamental and serve a significant role in contemporary soybean research, establishing themselves as one of the primary elements in the soybean plant breeder's toolkit. Data retrieval and knowledge extraction are the two main objectives of a soybean database. With the aid of the refined tools provided by several different soybean databases, it is possible to search, collect, correlate, and examine the enormous amount of soybean data, which includes genome, transcriptome, proteomic data, genetic and physical maps, ESTs, microarray, and cDNA information. In order to face the problems brought on by a rising global population and consumption, new methods of breeding soybean cultivars might be developed using the valuable data stored inside these soybean databases. This chapter attempts to briefly describe some of the major soybean databases available while also highlighting the challenges associated with them. It also tries to shed light on their significance and how they have been implemented in research.

KEYWORDS

- **bioinformatics tools**
- **crop improvement**
- **expressed sequence tags**
- **functional gene annotation**
- **genetic and physical maps**
- **genomic resources**
- **soybean databases**

REFERENCES

Ali, W., Ahmad, M., İftakhar, F., Qureshi, M., & Ceyhan, A. (2020). Nutritive potentials of soybean and its significance for human health and animal production: A review. *Eurasian Journal of Food Science and Technology, 4*(1), 41. Retrieved from: https://dergipark.org.tr/en/pub/ejfst/issue/56793/790298 (accessed on 25 March 2025).

Alkharouf, N. W. (2004). SGMD: The Soybean Genomics and Microarray Database. *Nucleic Acids Research, 32*(90001), 398D–400. https://doi.org/10.1093/nar/gkh126.

Altman, R. B. (2004). Editorial: Building successful biological databases. *Briefings in Bioinformatics, 5*(1), 4, 5. https://doi.org/10.1093/bib/5.1.4.

Bayer, P. E., Valliyodan, B., Hu, H., Marsh, J. I., Yuan, Y., Vuong, T. D., et al. (2021). Sequencing the USDA core soybean collection reveals gene loss during domestication and breeding. *The Plant Genome, 15*(1), e20109. https://doi.org/10.1002/tpg2.20109.

Berendzen, J., Brown, A. V., Cameron, C. T., Campbell, J. D., Cleary, A. M., Dash, S., et al. (2021). The Legume Information System and associated online genomic resources. *Legume Science, 3*(3), e74. https://doi.org/10.1002/leg3.74.

Bhat, J. A., & Yu, D. (2021). High-throughput NGS-based genotyping and phenotyping: Role in genomics-assisted breeding for soybean improvement. *Legume Science, 3*(3), e81. https://doi.org/10.1002/leg3.81.

Bolon, Y. T., Haun, W. J., Xu, W. W., Grant, D., Stacey, M. G., Nelson, R. T., et al. (2011). Phenotypic and genomic analyses of a fast neutron mutant population resource in soybean. *Plant Physiology, 156*(1), 240–253. https://doi.org/10.1104/pp.110.170811.

Bolon, Y. T., Stec, A. O., Michno, J. M., Roessler, J., Bhaskar, P. B., Ries, L., et al. (2014). Genome resilience and prevalence of segmental duplications following fast neutron irradiation of soybean. *Genetics, 198*(3), 967–981. https://doi.org/10.1534/genetics.114.170340.

Brown, A. V., Conners, S. I., Huang, W., Wilkey, A. P., Grant, D., Weeks, N. T., et al. (2020). A new decade and new data at SoyBase, the USDA-ARS soybean genetics and genomics database. *Nucleic Acids Research, 49*(D1), D1496–D1501. https://doi.org/10.1093/nar/gkaa1107.

Chandramouli, K., & Qian, P. Y. (2009). Proteomics: Challenges, techniques, and possibilities to overcome biological sample complexity. *Human Genomics and Proteomics, 2009*(1), 239204. https://doi.org/10.4061/2009/239204.

Chandras, C., Weaver, T., Zouberakis, M., Smedley, D., Schughart, K., Rosenthal, N., et al. (2009). Models for financial sustainability of biological databases and resources. *Database, 2009*(0), bap017. https://doi.org/10.1093/database/bap017.

Chen, Q., Liu, B., Ai, L., Yan, L., Lin, J., Shi, X., et al. (2022). QTL and candidate genes for heterophylly in soybean based on two populations of recombinant inbred lines. *Frontiers in Plant Science, 13,* 961619. https://doi.org/10.3389/fpls.2022.961619.

Cheng, K., & Strömvik, M. V. (2008). SoyXpress: A database for exploring the soybean transcriptome. *BMC Genomics, 9*(1), 368. https://doi.org/10.1186/1471-2164-9-368.

Cortés, A. J., López-Hernández, F., & Osorio-Rodriguez, D. (2020). Predicting thermal adaptation by looking into populations' genomic past. *Frontiers in Genetics, 11,* 564515. https://doi.org/10.3389/fgene.2020.564515.

Deelman, E., Singh, G., Su, M. H., Blythe, J., Gil, Y., Kesselman, C., et al. (2005). Pegasus: A framework for mapping complex scientific workflows onto distributed systems. *Scientific Programming, 13*(3), 219–237. https://doi.org/10.1155/2005/128026.

Deshmukh, R., Rana, N., Liu, Y., Zeng, S., Agarwal, G., Sonah, H., et al. (2020). Soybean transporter database: A comprehensive database for identification and exploration of natural variants in soybean transporter genes. *Physiologia Plantarum, 171*(4), 756–770. https://doi.org/10.1111/ppl.13287.

Donlin, M. J. (2009). Using the Generic Genome Browser (GBrowse). *Current Protocols in Bioinformatics, 28*(1). https://doi.org/10.1002/0471250953.bi0909s28.

Du, J., Grant, D., Tian, Z., Nelson, R. T., Zhu, L., Shoemaker, R. C., et al. (2010a). SoyTEdb: A comprehensive database of transposable elements in the soybean genome. *BMC Genomics, 11*(1). https://doi.org/10.1186/1471-2164-11-113.

Du, Z., Zhou, X., Ling, Y., Zhang, Z., & Su, Z. (2010b). agriGO: A GO analysis toolkit for the agricultural community. *Nucleic Acids Research, 38*(suppl_2), W64–W70. https://doi.org/10.1093/nar/gkq310.

Ellis, L. B., & Kalumbi, D. (1999). Financing a future for public biological data. *Bioinformatics, 15*(9), 717–722. https://doi.org/10.1093/bioinformatics/15.9.717.

Espina, M. J., Ahmed, C. M. S., Bernardini, A., Adeleke, E., Yadegari, Z., Arelli, P., et al. (2018). Development and phenotypic screening of an ethyl methane sulfonate mutant population in soybean. *Frontiers in Plant Science, 9,* 394. https://doi.org/10.3389/fpls.2018.00394.

Gao, Y., Yang, Z., Yang, W., Yang, Y., Gong, J., Yang, Q. Y., et al. (2020). Plant-ImputeDB: An integrated multiple plant reference panel database for genotype imputation. *Nucleic Acids Research, 49*(D1), D1480–D1488. https://doi.org/10.1093/nar/gkaa953.

Goldberg, S. I., Niemierko, A., Shubina, M., & Turchin, A. (2010). "Summary Page": A novel tool that reduces omitted data in research databases. *BMC Medical Research Methodology, 10*(1), 91. https://doi.org/10.1186/1471-2288-10-91.

Goodstein, D. M., Shu, S., Howson, R., Neupane, R., Hayes, R. D., Fazo, J., et al. (2011). Phytozome: A comparative platform for green plant genomics. *Nucleic Acids Research, 40*(D1), D1178–D1186. https://doi.org/10.1093/nar/gkr944.

Grant, D., & Nelson, R. T. (2017). SoyBase: A comprehensive database for soybean genetic and genomic data. In *The Soybean Genome* (pp. 193–211). https://doi.org/10.1007/978-3-319-64198-0_13.

Grant, D., Nelson, R. T., Cannon, S. B., & Shoemaker, R. C. (2009). SoyBase, the USDA-ARS soybean genetics and genomics database. *Nucleic Acids Research, 38*(suppl_1), D843–D846. https://doi.org/10.1093/nar/gkp798.

Ha, J., Jeon, H. H., Woo, D. U., Lee, Y., Park, H., Lee, J., et al. (2019). Soybean-VCF2Genomes: A database to identify the closest accession in soybean germplasm collection. *BMC Bioinformatics, 20*(S13), 384. https://doi.org/10.1186/s12859-019-2859-5.

Helmy, M., Crits-Christoph, A., & Bader, G. D. (2016). Ten simple rules for developing public biological databases. *PLOS Computational Biology, 12*(11), e1005128. https://doi.org/10.1371/journal.pcbi.1005128.

Hu, Z., Sun, C., Lu, K., Chu, X., Zhao, Y., Lu, J., et al. (2017). EUPAN enables pan-genome studies of a large number of eukaryotic genomes. *Bioinformatics, 33*(15), 2408–2409. https://doi.org/10.1093/bioinformatics/btx170.

Hwang, E. Y., Song, Q., Jia, G., Specht, J. E., Hyten, D. L., Costa, J., et al. (2014). A genome-wide association study of seed protein and oil content in soybean. *BMC Genomics, 15*(1). https://doi.org/10.1186/1471-2164-15-1.

Ikram, M., Han, X., Zuo, J. F., Song, J., Han, C. Y., Zhang, Y. W., et al. (2020). Identification of QTNs and their candidate genes for 100-seed weight in soybean (*Glycine max* L.) using multi-locus genome-wide association studies. *Genes, 11*(7), 714. https://doi.org/10.3390/genes11070714.

Joshi, T., Fitzpatrick, M. R., Chen, S., Liu, Y., Zhang, H., Endacott, R. Z., et al. (2013). Soybean Knowledge Base (SoyKB): A web resource for integration of soybean translational genomics and molecular breeding. *Nucleic Acids Research, 42*(D1), D1245–D1252. https://doi.org/10.1093/nar/gkt905.

Joshi, T., Patil, K., Fitzpatrick, M. R., Franklin, L. D., Yao, Q., Cook, J. R., et al. (2012). Soybean Knowledge Base (SoyKB): A web resource for soybean translational genomics. *BMC Genomics, 13*(S1). https://doi.org/10.1186/1471-2164-13-s1-s15.

Joshi, T., Wang, J., Zhang, H., Chen, S., Zeng, S., Xu, B., et al. (2016). The evolution of Soybean Knowledge Base (SoyKB). *Methods in Molecular Biology,* 149–159. https://doi.org/10.1007/978-1-4939-6658-5_7.

Kalve, S., Gali, K. K., & Tar'an, B. (2022). Genome-wide association analysis of stress tolerance indices in an interspecific population of chickpea. *Frontiers in Plant Science, 13,* 933277. https://doi.org/10.3389/fpls.2022.933277.

Karikari, B., Wang, Z., Zhou, Y., Yan, W., Feng, J., & Zhao, T. (2020). Identification of quantitative trait nucleotides and candidate genes for soybean seed weight by multiple models of genome-wide association study. *BMC Plant Biology, 20*(1), 404. https://doi.org/10.1186/s12870-020-02604-z.

Karp, P. D., Paley, S. M., Krummenacker, M., Latendresse, M., Dale, J. M., Lee, T. J., et al. (2009). Pathway Tools version 13.0: Integrated software for pathway/genome informatics and systems biology. *Briefings in Bioinformatics, 11*(1), 40–79. https://doi.org/10.1093/bib/bbp043.

Kim, E., Hwang, S., & Lee, I. (2016). SoyNet: A database of co-functional networks for soybean (*Glycine max*). *Nucleic Acids Research, 45*(D1), D1082–D1089. https://doi.org/10.1093/nar/gkw704.

Klimke, W., O'Donovan, C., White, O., Brister, J. R., Clark, K., Fedorov, B., et al. (2011). Solving the problem: Genome annotation standards before the data deluge. *Standards in Genomic Sciences, 5*(1), 168–193. https://doi.org/10.4056/sigs.2084864.

Kou, Y., & Wang, S. (2010). Broad-spectrum and durability: Understanding of quantitative disease resistance. *Current Opinion in Plant Biology, 13*(2), 181–185. https://doi.org/10.1016/j.pbi.2009.12.010.

Kumar, A., Pratap, A., Udar, U., Chauhan, R. S., & Singh, T. R. (2014). SMDB: Soybean Marker Database. *International Journal of Computational Biology, 3*(2), 44. https://doi.org/10.34040/ijcb.3.2.2014.27.

Lopes, C. T., Franz, M., Kazi, F., Donaldson, S. L., Morris, Q., & Bader, G. D. (2010). Cytoscape Web: An interactive web-based network browser. *Bioinformatics, 26*(18), 2347–2348. https://doi.org/10.1093/bioinformatics/btq430.

López-Hernández, F., & Cortés, A. J. (2019). Last-generation genome–environment associations reveal the genetic basis of heat tolerance in common bean (*Phaseolus vulgaris* L.). *Frontiers in Genetics, 10,* 954. https://doi.org/10.3389/fgene.2019.00954.

Nelson, R., Wiesner-Hanks, T., Wisser, R., & Balint-Kurti, P. (2017). Navigating complexity to breed disease-resistant crops. *Nature Reviews Genetics, 19*(1), 21–33. https://doi.org/10.1038/nrg.2017.82.

Obayashi, T., & Kinoshita, K. (2009). Rank of correlation coefficient as a comparable measure for biological significance of gene coexpression. *DNA Research, 16*(5), 249–260. https://doi.org/10.1093/dnares/dsp016.

Petereit, J., Marsh, J. I., Bayer, P. E., Danilevicz, M. F., Thomas, W. J. W., Batley, J., et al. (2022). Genetic and genomic resources for soybean breeding research. *Plants, 11*(9), 1181. https://doi.org/10.3390/plants11091181.

Pool, R., & Esnayra, J. (2000). *Bioinformatics: Converting Data to Knowledge.* National Academies Press. https://doi.org/10.17226/9990.

Qi, Z., Song, J., Zhang, K., Liu, S., Tian, X., Wang, Y., et al. (2020). Identification of QTNs controlling 100-seed weight in soybean using multilocus genome-wide association studies. *Frontiers in Genetics, 11,* 689. https://doi.org/10.3389/fgene.2020.00689.

Rajendran, A., & Lal, S. K. (2020). Assessing the need of pre-germination anaerobic stress-tolerant varieties in Indian soybean (*Glycine max* (L.) Merrill). *National Academy Science Letters, 43*(7), 593–597. https://doi.org/10.1007/s40009-020-00937-9.

Rajendran, A., Lal, S. K., & Raju, D. (2022). Associations of direct and indirect selection for pregermination anaerobic stress tolerance in soybean (*Glycine max*). *Plant Breeding, 141*(5), 634–643. https://doi.org/10.1111/pbr.13048.

Ramlal, A., Ahmad, S., Kumar, L., Khan, F. N., & Chongtham, R. (2021). From molecules to patients: The clinical applications of biological databases and electronic health records. In R. Khalid & N. Dey (Eds.), *Translational Bioinformatics in Healthcare and Medicine* (pp. 107–125). Elsevier. https://doi.org/10.1016/b978-0-323-89824-9.00009-4.

Ramlal, A., Bhat, I., Nautiyal, A., Baweja, P., Mehta, S., Kumar, V., Tripathi, S., Mahto, R. K., Saini, M., Mallikarjun, B. P., Saluja, S., Lal, S. K., Subramaniam, S., & Fawzy, I. M. (2023a). *In silico* analysis of angiotensin-converting enzyme (ACE) inhibitory compounds obtained from soybean (*Glycine max* (L.) Merr.). *Frontiers in Physiology, 14*, 1172684. https://doi.org/10.3389/fphys.2023.1172684.

Ramlal, A., Lal, S. K., & Vidyasagar, S. (2023b). Editorial: Advances in breeding for waterlogging tolerance in plants. *Frontiers in Plant Science, 14*, 1284730. https://doi.org/10.3389/fpls.2023.1284730.

Ramlal, A., Nautiyal, A., Baweja, P., Kumar Mahto, R., Mehta, S., Pujari Mallikarunja, B., et al. (2022). Harnessing heterosis and male sterility in soybean (*Glycine max* (L.) Merrill): A critical revisit. *Frontiers in Plant Science, 13,* 981768. https://doi.org/10.3389/fpls.2022.981768.

Ramlal, A., Nautiyal, A., Lal, S. K., & Chigeza, G. (2023c). Editorial: A wonder legume, soybean: Prospects for improvements. *Frontiers in Plant Science, 14*, 1294185. https://doi.org/10.3389/fpls.2023.1294185.

Rizzo, G., & Baroni, L. (2018). Soy, soy foods and their role in vegetarian diets. *Nutrients, 10*(1), 43. https://doi.org/10.3390/nu10010043.

Rolling, W., Lake, R., Dorrance, A. E., & McHale, L. K. (2020). Genome-wide association analyses of quantitative disease resistance in diverse sets of soybeans (*Glycine max* (L.) Merr.) plant introductions. *PLoS ONE, 15*(3), e0227710. https://doi.org/10.1371/journal.pone.0227710.

Schrider, D. R., & Kern, A. D. (2018). Supervised machine learning for population genetics: A new paradigm. *Trends in Genetics, 34*(4), 301–312. https://doi.org/10.1016/j.tig.2017.12.005.

Sheikh, K., Ramlal, A., & Khan, F. N. (2022). Computational resources for oncology research: A comprehensive analysis. In *Computational Intelligence in Oncology* (pp. 65–92). Springer. https://doi.org/10.1007/978-981-16-9221-5_4.

Shi, A., Gepts, P., Song, Q., Xiong, H., Michaels, T. E., & Chen, S. (2021). Genome-wide association study and genomic prediction for soybean cyst nematode resistance in USDA common bean (*Phaseolus vulgaris*) core collection. *Frontiers in Plant Science, 12*, 624156. https://doi.org/10.3389/fpls.2021.624156.

Shultz, J. L., Kurunam, D., Shopinski, K., Iqbal, M. J., Kazi, S., Zobrist, K., et al. (2006). The soybean genome database (SoyGD): A browser for display of duplicated, polyploid, regions and sequence-tagged sites on the integrated physical and genetic maps of *Glycine max*. *Nucleic Acids Research, 34*(90001), D758–D765. https://doi.org/10.1093/nar/gkj050.

Söllner, J. F., Leparc, G., Hildebrandt, T., Klein, H., Thomas, L., Stupka, E., et al. (2017). An RNA-Seq atlas of gene expression in mouse and rat normal tissues. *Scientific Data, 4*(1), 170185. https://doi.org/10.1038/sdata.2017.185.

Song, Q., Yan, L., Quigley, C., Jordan, B. D., Fickus, E., Schroeder, S., et al. (2017). Genetic characterization of the soybean nested association mapping population. *The Plant Genome, 10*(2). https://doi.org/10.3835/plantgenome2016.10.0109.

Sreenivasaiah, P. K., & Kim, D. H. (2010). Current trends and new challenges of databases and web applications for systems-driven biological research. *Frontiers in Physiology, 1,* 147. https://doi.org/10.3389/fphys.2010.00147.

Stein, L. D., Mungall, C., Shu, S., Caudy, M., Mangone, M., Day, A., et al. (2002). The Generic Genome Browser: A building block for a model organism system database. *Genome Research, 12*(10), 1599–1610. https://doi.org/10.1101/gr.403602.

Tavakolan, M., Alkharouf, N. W., Khan, F. H., & Natarajan, S. (2013). SoyProDB: A database for the identification of soybean seed proteins. *Bioinformation, 9*(3), 165–167. https://doi.org/10.6026/97320630009165.

Tidke, S. A., Ramakrishn, D., Kiran, S., Kosturkova, G., & Ravishanka, G. A. (2015). Nutraceutical potential of soybean: Review. *Asian Journal of Clinical Nutrition, 7*(2), 22–32. https://doi.org/10.3923/ajcn.2015.22.32.

Torkamaneh, D., Laroche, J., Valliyodan, B., O'Donoughue, L., Cober, E., Rajcan, I., et al. (2020). Soybean (*Glycine max*) haplotype map (GmHapMap): A universal resource for soybean translational and functional genomics. *Plant Biotechnology Journal, 19*(2), 324–334. https://doi.org/10.1111/pbi.13466.

Torkamaneh, D., Lemay, M., & Belzile, F. (2021). The pan-genome of the cultivated soybean (PanSoy) reveals an extraordinarily conserved gene content. *Plant Biotechnology Journal, 19*(9), 1852–1862. https://doi.org/10.1111/pbi.13600.

Umezawa, T., Sakurai, T., Totoki, Y., Toyoda, A., Seki, M., Ishiwata, A., et al. (2008). Sequencing and analysis of approximately 40,000 soybean cDNA clones from a full-length-enriched cDNA library. *DNA Research, 15*(6), 333–346. https://doi.org/10.1093/dnares/dsn024.

Varshney, R. K., Bohra, A., Roorkiwal, M., Barmukh, R., Cowling, W. A., Chitikineni, A., et al. (2021). Fast-forward breeding for a food-secure world. *Trends in Genetics, 37*(12), 1124–1136. https://doi.org/10.1016/j.tig.2021.08.002.

Wang, J., Hossain, M. S., Lyu, Z., Schmutz, J., Stacey, G., Xu, D., & Joshi, T. (2019). SoyCSN: Soybean context-specific network analysis and prediction based on tissue-specific transcriptome data. *Plant Direct, 3*(9), e00167.

Wang, X., & Komatsu, S. (2017). Improvement of soybean products through the response mechanism analysis using proteomic technique. *Advances in Food and Nutrition Research, 82*, 117–148. https://doi.org/10.1016/bs.afnr.2016.12.006.

Wang, Z., Libault, M., Joshi, T., Valliyodan, B., Nguyen, H. T., Xu, D., et al. (2010). SoyDB: A knowledge database of soybean transcription factors. *BMC Plant Biology, 10*(1), 14. https://doi.org/10.1186/1471-2229-10-14.

Wu, D. D., Hu, X., Park, E., Wang, X., Feng, J., & Wu, X. (2010). Exploratory analysis of protein translation regulatory networks using hierarchical random graphs. *BMC Bioinformatics, 11*(S3), S2. https://doi.org/10.1186/1471-2105-11-s3-s2.

Xanthopoulou, A., Montero-Pau, J., Picó, B., Boumpas, P., Tsaliki, E., Paris, H. S., et al. (2021). A comprehensive RNA-Seq-based gene expression atlas of the summer squash (*Cucurbita pepo*) provides insights into fruit morphology and ripening mechanisms. *BMC Genomics, 22*(1), 341. https://doi.org/10.1186/s12864-021-07683-2.

Xiong, J. (2006). Introduction to biological databases. In *Essential Bioinformatics* (pp. 10–28). Cambridge University Press. https://doi.org/10.1017/cbo9780511806087.003.

Xu, Y., Guo, M., Liu, X., Wang, C., & Liu, Y. (2014). SoyFN: A knowledge database of soybean functional networks. *Database, 2014,* bau019. https://doi.org/10.1093/database/bau019.

Yu, J., Zhang, Z., Wei, J., Ling, Y., Xu, W., & Su, Z. (2014). SFGD: A comprehensive platform for mining functional information from soybean transcriptome data and its use in identifying acyl-lipid metabolism pathways. *BMC Genomics, 15*(1), 271. https://doi.org/10.1186/1471-2164-15-271.

Zdobnov, E. M., & Apweiler, R. (2001). InterProScan–An integration platform for the signature-recognition methods in InterPro. *Bioinformatics, 17*(9), 847–848. https://doi.org/10.1093/bioinformatics/17.9.847.

Zhang, P., Dreher, K., Karthikeyan, A., Chi, A., Pujar, A., Caspi, R., et al. (2010). Creation of a genome-wide metabolic pathway database for *Populus trichocarpa* using a new approach for reconstruction and curation of metabolic pathways for plants. *Plant Physiology, 153*(4), 1479–1491. https://doi.org/10.1104/pp.110.157396.

Zou, D., Ma, L., Yu, J., & Zhang, Z. (2015). Biological databases for human research. *Genomics, Proteomics & Bioinformatics, 13*(1), 55–63. https://doi.org/10.1016/j.gpb.2015.01.006.

PART II

Improvements in Soybean

CHAPTER 6

Oil Quality Improvement in Soybean

RAMAN PUSHPA,[1,*] PALANIVELU SATHYA,[2]
ARAVINDAN SHANMUGAM,[2] KALAIARASAN MANIVELAN,[1] and
BALAMURUGAN MOHANAPRIYA[2]

[1]*Tamil Nadu Rice Research Institute, Tamil Nadu Agricultural University, Aduthurai, Tamil Nadu, India*

[2]*Department of Genetics and Plant Breeding, Tamil Nadu Agricultural University, Coimbatore, Tamil Nadu, India*

*Corresponding author

ABSTRACT

Over the past 25 years, oil crops have considerably gained importance in world agriculture alongside food crops. Soybean is a multipurpose crop known for its feed (yield), food (protein and oil), and fuel (biodiesel) uses, cultivated over 90.1 million hectares. Soybean oil consists mainly of linoleic (52%), oleic (25%), palmitic (11%), linolenic (8%), and stearic (4%) fatty acids (FAs). A reduction in saturated FAs is preferred for lowering cardiovascular risk and improving heart health, while increased unsaturated FA content benefits human health. Among the unsaturated FAs, increasing oleic acid and reducing linoleic and linolenic acids improves both human health and oil stability. The poor keeping quality of soybean oil is primarily due to linolenic FA. The hydrogenation process was once used to reduce linolenic FA content; however, it produces trans FAs, which are undesirable for food purposes due to their negative impact on health. The major goal of breeders is to develop altered FA compositions in soybean oil that better meet consumer needs. This chapter discusses breeding strategies, from mutation breeding (conventional) to genome editing (GE) tools (modern breeding), targeting soybean oil quality in

Soybeans, Volume 2: Genome, Biotechnology, Stress, and Disease.
R. Ambika Rajendran, Ayyagari Ramlal, Sanjay Kumar Lal, and Godfree Chigeza (Eds.)

DOI: 10.1201/9781998511730-8

terms of FA composition, inheritance of modified traits, and their impact on other agronomic traits to support sustainable oil production.

6.1 INTRODUCTION

Soybean (*Glycine max* (L.) Merr.) is a major oilseed as well as a major protein source grown worldwide. The crop soybean was domesticated from its ancient progenitor (*G. soja* Sieb. and Zucc.) in East Asia (Zhang et al., 2020). Due to its high oil content, soybean is widely used for cooking purposes and in various food industries. Soybean-based nutritious food products such as margarine, miso, mayonnaise, soybean milk, sauce, and tofu have been formulated for food purposes. The oil-extracted soybean is used as a nourishing feed for cattle. Soybean oil is mostly used in industries such as producing cosmetics, pesticides, plastics, pharmaceuticals, paper, inks, paints, and varnishes. And now, soybean oil used as biodiesel is another chance for renewable sources of energy for industrial purposes. Soybean comes under the family of legume crops; it fixes atmospheric nitrogen to the soil through biological nitrogen fixation. Hence, it is less dependent on synthetic nitrogen fertilizers. Soybean has approximately 40% protein and 6% ash (Pratap et al., 2012). The human body is not able to produce the essential amino acids (AAs). Soybean, as a good protein source, contains 18 AAs, including all essential AAs (Jeong et al., 2018). Soybean oil contains both saturated and unsaturated fatty acids (FAs). The FA compositions of soybean are shown in Figure 6.1. Soybean oil is composed of two FA groups, namely saturated and unsaturated FAs. Palmitic acid (16:0) and stearic acid (18:0) fall under the group of saturated FAs. Oleic acid (18:1), linoleic acid (18:2), and linolenic acid (18:3) fall under the group of unsaturated FAs. Soybean oil has 10% palmitic acid, 4% stearic acid, 18% oleic acid, 55% linoleic acid, and 13% linolenic acid (Clemente et al., 2009). The polyunsaturated FAs, linoleic and linolenic acids, are essential for mammals, including humans, because they cannot be synthesized in the human body. Hence, the essentiality of polyunsaturated FAs in humans is significant. Raw soybean oil contains about 8–13% linolenic FA and 53% linoleic FA. Partially hydrogenated oil contains about 3% linolenic acid and 23% linoleic acid. Among the polyunsaturated FAs, linolenic acid is responsible for the poor keeping quality of oil. The availability of 8–13% of linolenic FA is accountable for low oxidative stability. The hydrogenation process may lower the linolenic acid content to a favorable amount, but trans FAs are produced during the hydrogenation process, making it inadvisable for human consumption (Fehr, 2007). Soybeans have excellent protein quality (40%). Soybeans serve as a meat substitute for vegetarians, maintaining dietary balance for them. The protein content in soybean seeds is controlled by quantitative genes. The protein content and oil content are negatively correlated (Brim and Burton, 1979), making it difficult to modify soybean protein and oil composition at the same time.

The major breeding objective in soybeans was developing soybean lines with high oleic and low linolenic FAs. The development of lines with less linolenic FA through natural conditions is a major breeding objective. The development of high oil-yielding lines without affecting protein availability is another breeding objective to improve the soybean crop. This chapter discusses the oil quality improvement of soybean crops.

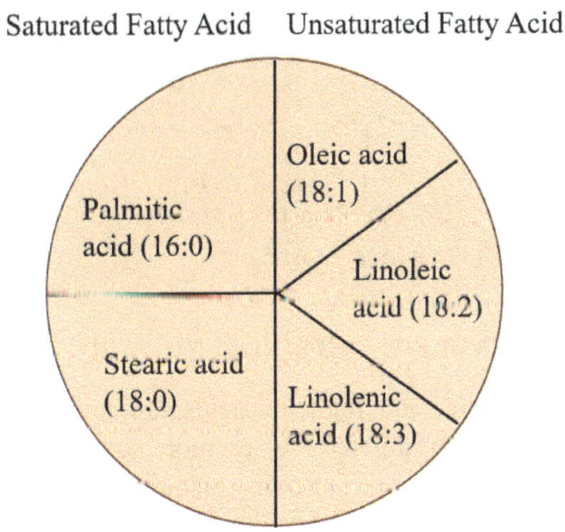

FIGURE 6.1 Fatty acid composition of soybean.

6.2 OIL QUALITY IMPROVEMENT

FAs were classified according to their presence or absence of double bonds as saturated and unsaturated FAs. Again, the unsaturated FAs are classified into two groups, namely monounsaturated FAs and polyunsaturated FAs. The monounsaturated FA has one double bond, and the polyunsaturated FA has two to six double bonds (Figure 6.2). Soybean oil, containing considerable quantities of linolenic acid, is capable of off-flavor and rancidity. The linolenic FA content in cooking oil usually ranges between 7% and 8%. Hence, the improvement of the oil quality of soybean, without affecting the availability of unsaturated FA content, is only achieved by increasing the oleic FA content and lowering the linolenic and linoleic acid content. The monounsaturated FA is good for human health, and it also increases the oil's oxidative stability.

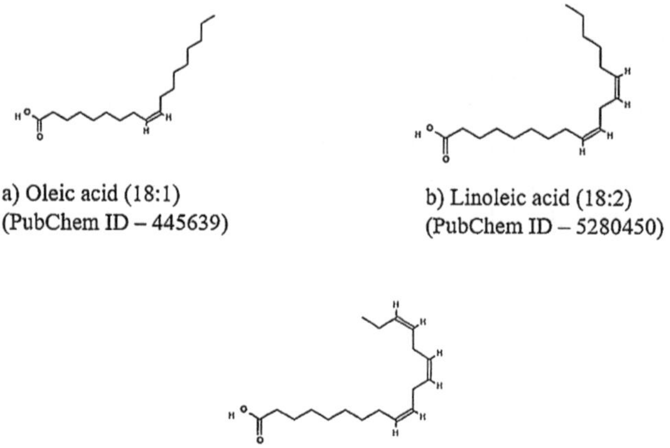

a) Oleic acid (18:1)
(PubChem ID – 445639)

b) Linoleic acid (18:2)
(PubChem ID – 5280450)

c) Linolenic acid (18:3)
(PubChem CID – 5280934)

FIGURE 6.2 MUFA and PUFA in soybean oil.

6.3 OLEIC V/S LINOLEIC AND LINOLENIC ACID

All three oleic, linoleic, and linolenic acids fall under the category of unsaturated FAs. Monounsaturated FAs are more suitable for human health as well as the stability of the oil. The mutation or elimination of the enzyme/ factor responsible for linolenic acid production is effective in reducing the linolenic FA content without the hydrogenation process. Linolenic acid is an essential component of chloroplasts, and all photosynthetic organisms have linolenic FA in the photosynthetic pathway. This clearly explains the importance of linolenic FA in the plant life cycle. Several plants manage to maintain a very low linolenic acid content in their oil. The compositions of FAs in several crop species have been evaluated during maturation. During the maturation process, the linolenic FA content is reduced. This also occurs in soybeans, but the level at the maturity stage still has 7 to 8% linolenic acid. Reduction of this 7–8% to less than 1% and increasing the oleic acid is important. The oleic acid content is recorded to have a significant and negative association with linoleic acid. The correlation coefficient between these associations was –0.75. It concludes that increasing the linoleic acid content decreases the oleic FA content (Liu et al., 2020). The monounsaturated FA (oleic) shows a highly significant and negative correlation with saturated FA (stearic acid) and polyunsaturated FA (linolenic). In another study, linoleic FA and linolenic FA had a significant and positive correlation. Simply put, the monounsaturated FA (oleic) had a significant and negative correlation with polyunsaturated FAs (linoleic and linolenic). The polyunsaturated FAs had a significant and positive correlation. Oleic acid increases the oil's oxidative stability and also decreases the production of trans FAs during hydrogenation, making it most desired for human consumption.

6.4 GENETICS OF LINOLENIC ACID AND BREEDING METHODOLOGY TO IMPROVE OIL QUALITY

The development of any trait depends on the genetics of that particular trait. The selection of breeding methodology, alteration in chemical pathways, and parental material selection all depend on the genetics of the specific trait. The "FAD2" and "FAD3" are major enzymes responsible for the production of linoleic and linolenic FAs (Povkhova et al., 2020). The FAD2 enzyme has two identical enzyme copies, that is, FAD2-1 and FAD2-2. These two enzyme copies have different gene family members. The enzyme "FAD2-1" had two enzyme members, that is, FAD2-1A and FAD2-1B Another enzyme, "FAD2-2," had three members, that is, FAD2-2A, FAD2-2B, and FAD2-2C. The FAD2-2 family gene members were found to share 78 to 95% similarity, a similar structure, and contain eight exons. However, significant dissimilarity is observed between the introns of *"FAD3"* genes. These are the major genes accountable for the production of linoleic and linolenic FAs. Changes in this gene complex may help to alter the oil quality. Based on this oil quality relationship and genes, conventional and molecular breeding methodologies are used to develop superior quality soybean oil.

6.5 CONVENTIONAL METHOD-MUTATION BREEDING

Mutation breeding helps to change the trait of interest by altering gene expression. The mutation breeding method is more acceptable than the transgenic method, and the improvement of the trait of interest takes less time compared to the conventional breeding method.

Hammond and Fehr (1975) mutated the M1 of parental and M1F2 of crosses using X-ray and EMS mutagens. They reported that the physical mutagen (X-ray treatment) may induce physiological changes. It does not affect the genetic constituents of the M_1 generation. The combination of the chemical mutagen (EMS) with the physical mutagen (X-radiation) was efficient in increasing the variability of the FA composition in soybean oil (Patil et al., 2007).

The enzymes GmFAD2-1A (omega-6-desaturases) and GmFAD2-1B (omega-6-desaturases) convert oleic FA into linoleic FA. *Glyma.10g278000* and *Glyma.20g111000* are the candidate genes linked with GmFAD2-1A and GmFAD2-1B. The omega-3-desaturases GmFAD3A, GmFAD3B, and GmFAD3C are responsible for the conversion of linoleic FA into linolenic FA. The candidate genes *Glyma.14g194300*, *Glyma.02g227200*, and *Glyma.18g062000* are linked with the enzymes GmFAD3A, GmFAD3B, and GmFAD3C. Pham et al. (2012) produced low polyunsaturated FA and high monounsaturated lines using different sources of mutations in the omega-3 desaturases gene *FAD2-1A*. The mutant lines had high oleic acid content, and low linoleic acid was constant across the two testing locations, but the composition of linolenic FA content changed based on the combination of mutant "*FAD3*" genes with mutant "*FAD2-1*" genes. Increased oleic and decreased linolenic mutant soybean lines are utilized in the crossing program to develop a diverse mapping population. RNAi silencing techniques, mutagens, and gene editing tools are utilized to suppress the action of *FAD2* and *FAD3* genes. Using the mutating agent, Patil et al. (2007) concluded that the combination of chemical and physical mutagens would be effective in achieving higher variation in oleic acid content. Higher concentrations of the chemical mutagen EMS combined with the physical mutagen (gamma radiation) effectively increase the variability for the unsaturated FA. The frequency distribution of the M_2 population was skewed towards low levels of linoleic acid and high levels of oleic acid. Hence, enhancing the oil quality (oxidative stability) of soybean oil without changing the nutritional availability is achieved through the natural process. During this process, it does not produce any trans FAs.

Mutant lines having downregulated omega-6 desaturases (GmFAD2 gene family) and omega-3 desaturases (GmFAD3A) were adequate to produce soybean lines with linolenic FA (1.8–2.6%), linoleic FA (1.5–4.0%), and oleic acid (80%) (Pham et al., 2012).

6.6 GENES RESPONSIBLE FOR FATTY ACID COMPOSITION

The first promising candidate gene linked to soybean oleic FA content is "*Glyma.04G116500.1*" (*GmWRI14*), which was detected by GWAS (Di et al., 2021). This candidate gene was confirmed using three years of repeated GWAS studies. The expression of *GmWRI14* in different plant tissues, Viz., seed, leaf, stem, and root, was investigated by Northern blot analysis. The gene expression was high in seed (26.12–52.32) compared to other plant tissues. Additionally, the candidate gene "*GmWRI14*" was negatively associated with the linoleic FA content (–0.901). The development of transgenic lines with the *GmWRI14* gene is an easy way to reduce polyunsaturated FA and increase monounsaturated FA. Di et al. (2021) transferred the *GmWRI14* gene into JN 38 lines. In the T_2 plant generation, an increased oleic acid content and decreased linoleic acid content were recorded. The positive strains of JN38 lines have a linoleic FA content of 11.42% and an oleic FA content of 41.02% (Lakhssassi et al., 2017). *FAD2* is a key gene that catalyzes the production of linoleic FA. RNA interference (RNAi) technology targeting *FAD2* and b-ketoacyl-acyl carrier protein synthase (KASII) increases the composition of oleic acids and decreases the amount of linoleic acid (Yang et al., 2018). The investigation showed that the loss of the role of *FAD3* increased the composition of linoleic FA to 67% (Thapa et al., 2018). The candidate gene *GmWRI14* down-regulates the function of the *FAD2* gene (Held et al., 2019). Due to the downregulation action, *FAD2* does not convert oleic acid into linoleic acid. This report clearly explains that the candidate gene *WRI1* plays a major positive role in controlling oleic acid content and a negative role in linoleic acid content. The decreasing level of linoleic FA content is stimulated by *WRI1*. It is due to the dysfunction of a soybean *FAD2* gene family. The overall gene expressions are presented in Figure 6.3.

6.7 CRISPR/CAS9-MEDIATED GENOME EDITING

In plants, the CRISPR-Cas9 gene-editing tool has been efficaciously used in some species such as *Arabidopsis thaliana, Zea mays, Nicotiana tabacum, Oryza sativa, Sorghum bicolor,* and *Triticum aestivum* (Sun et al., 2015). In soybean, the first successful GE was done for the traits of hairy roots and flowering (Cai et al., 2015; Han et al., 2019), followed by plant structure and yield (Bao et al., 2019), plant height (PH) (Jacobs et al., 2015), and seed protein genes (Li et al., 2019). Based on these successful GE techniques, researchers have started using this GE technology to improve the oil quality in soybeans. Do et al. (2019) developed two gRNAs to guide Cas9 to cleave two sites at the same time. The two sites are spaced 1 kb apart from each other. High oleic and low linoleic lines were observed in the transgenic first-generation seeds derived from CRISPR/Cas9 genome-edited plants. GE in the *GmFAD2* gene family increases the production of oleic acid by up to 73.5% and decreases linoleic acid by up to 12.20%. Based on these results, the GE technique CRISPR/Cas9 is a speedy and highly effective method to develop soybean lines with increased oleic FA content and decreased levels of linoleic and linolenic FA composition.

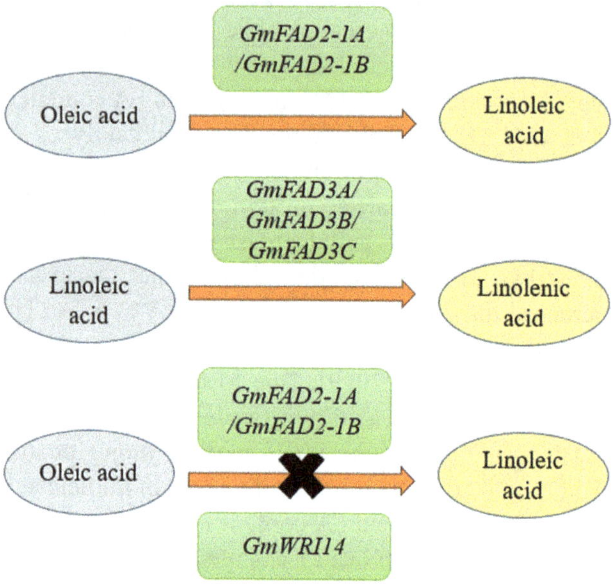

FIGURE 6.3 Major gene responsible for fatty acid conversion.

6.8 HIGH OLEIC VARIETIES/LINES DEVELOPED THROUGH VARIOUS BREEDING METHODOLOGIES

The multinational seed company Monsanto has newly released the soybean variety MON 87705 (Vistive Gold) using gene silencing technology (RNAi). The genes involved in this research are *FAD2* and *FatB*. Vistive Gold is a genetically altered variety that has a high level of oleic

FA and a low level of linolenic FA. Pham et al. (2012) developed soybean lines with 2% linolenic acid and 85% oleic acid content using mutations in genes responsible for the conversion of oleic into linoleic (*GmFAD2-1A*, *GmFAD2-1B*) and linoleic into linolenic FA (*GmFAD3A* and *GmFAD3C*). Silva et al. (2018) developed CS303TNKCA lines through the deletion of *GMFAD3A* (1 bp deletion). The line has linolenic FA in the range of 3.5–4%. DuPont Pioneer Seed Company developed another soybean variety having low linolenic FA of less than 3% (Waltz, 2010). Patil et al. (2007) observed that the combination of irradiation with chemical mutagens produces effective mutant lines with less than 3% linolenic acid. Yang et al. (2018) developed a soybean line having 80% oleic acid and less than 3% linoleic acid using RNAi-mediated knockdown methodology. In this method, they developed the mutants of Williams 82 and Shennong 9 lines.

6.9 QUANTITATIVE TRAIT LOCUS STUDY FOR IMPROVING OIL QUALITY

The QTL region *GqOil20* was an environmentally stable and significant quantitative trait locus (QTL) associated with oil content in soybean. It accounts for 23.70% of the total phenotypic variance for seed oil across multiple environments. This is the region responsible for increasing oil content. The candidate gene *GmOLEO1* had a significant and positive correlation with seed oil content. This oleosin protein-encoding gene was identified using haplotype material through genome-wide association studies (GWAS). This gene was localized in the oil bodies in maturing seeds. Predominant expression of *GmOLEO1* is recorded during seed maturation using expression analysis. Up-regulation of *GmOLEO1* significantly enriched smaller oil bodies of soybean seeds and increased the oil content by 10.60% compared with that of control seeds. This positive enrichment of smaller oil bodies is achieved by the downregulation of TAG metabolism. Transcriptomic studies were utilized to study the function of the candidate gene *GmOLEO1*. Hence, the selection of the promoter region of *GmOLEO1* leads to an increase in high seed oil content lines. These lines have higher oil content than that of wild ones. Therefore, utilizing this candidate gene along with another candidate gene responsible for increasing oleic acid content will help to improve oil quality and oil content.

Based on synteny between candidate genes and QTLs, Qiu et al. (2013) identified 12 candidate genes accountable for increasing the oil content. The sulfur-containing AA synthase gene (113 genes) and related QTLs (33 QTLs) were utilized for the identification of candidate genes. Liu et al. (2020) identified two candidate genes on chromosomes 11 and 4 that are related to oleic acid content in soybean. The candidate genes "*Glyma.11G229600.1*" (linkage group 11) and "*Glyma.04G102900.1*" (linkage group 4) were detected in a 2-year-long genome-wide association study. The candidate gene "*Glyma.11G229600.1*" in linkage group 11 showed a significant and positive association (0.98) with oleic FA content. The second candidate gene "*Glyma.04G102900.1*" on linkage group 4 showed a significant and negative association (–0.964) with oleic acid content.

Zhang et al. (2018) studied the 13,433 germplasm entries, and they also explored the allelic variation. They identified a total of 138 QTL associated with soybean seed composition. They also explored the nine candidate genes responsible for FA composition. Major candidate genes and QTL linked to soybean FAs are presented in Table 6.1. The enzyme delta-9-stearoyl-acyl carrier protein desaturase is another major enzyme associated with the conversion of saturated FAs into monounsaturated FAs. The candidate genes, *viz.*, *GmSACPD-A*, *GmSACPD-B*, and *GmSACPD-C*, are responsible for the conversion of stearic acid into oleic acid (Gillman et al. 2014; Vogel et al. 2019; Weselake et al. 2009).

TABLE 6.1 QTL Region and Candidate Gene Linked to Soybean Oil

SL. No.	Character	QTL /Candidate Gene	Chromosome	R^2 Value/ Expression Percent	References
1.	Linolenic acid	*qlin-14*	14	70.37	Silva et al. (2021)
2.	Oleic acid	*qole-19*	19	12.13	Silva et al. (2021)
3.	Linoleic acid	*qlol-19*	19	12.72	Silva et al. (2021)
4.	Oil	*Glyma05g245000*	5	11	Zhang et al. (2018)
5.	Oleic acid	*AK-HSDH*	8	9	Zhang et al. (2018)
6.	Linoleic acid	*FAD 3A*	14	34	Zhang et al. (2018)
7.	Linoleic acid	*SAT 1*	15	33	Zhang et al. (2018)
8.	Oil content	*qoil-10-2-2*	10	30.57	Karikari et al. (2019)
9.	Oil content	*qoil-10-2-4*	10	21	Karikari et al. (2019)
10.	Oil content	*qoil-10-4-4*	10	27.49	Karikari et al. (2019)

6.10 CONCLUSION

Soybean oil quality enhancement has been achieved by using conventional breeding methods and molecular breeding techniques. This compiled information could help to reveal the genetic architecture of soybean genetics and breeding. Altering the omega-6 FA and omega-3 FA composition of soybean oil has been a major goal for soybean breeders. Increasing the oleic FA and reducing the linolenic FA is a major breeding objective in soybean breeding to improve oil quality. Therefore, breeding methods according to specific requirements are needed. Being an important oil crop, attempts are still needed to conduct more novel research studies to improve oil quality. The studies should focus on the genes that play a major role in the conversion of MUFA into PUFA. Hence, mutation in a particular gene improves oil quality. RNAi interference technology, gene transfer technology, and CRISPR/Cas9 technologies are major methods used to improve soybean oil quality. Identified improved soybean lines may be utilized in a breeding program to develop a better oil-improved variety.

KEYWORDS

- **CRISPR/Cas9**
- **fatty acids**
- **genome editing**
- **molecular breeding**
- **mutation breeding**
- **oleic acid**
- **soybean oil quality**

REFERENCES

Bao, A., Chen, H., Chen, L., Chen, S., Hao, Q., & Guo, W. (2019). CRISPR/Cas9-mediated targeted mutagenesis of *GmSPL9* genes alters plant architecture in soybean. *BMC Plant Biology, 19*(1), 1–12.

Brim, C. A., & Burton, J. W. (1979). Recurrent selection in soybeans. II. Selection for increased percent protein in seeds. *Crop Science, 19*(4), 494–498.

Cai, Y., Chen, L., Liu, X., Sun, S., Wu, C., & Jiang, B. (2015). CRISPR/Cas9-mediated genome editing in soybean hairy roots. *PLoS One, 10*(8), 1–10.

Clemente, T. E., & Cahoon, E. B. (2009). Soybean oil: Genetic approaches for modification of functionality and total content. *Plant Physiology, 151*(3), 1030–1040.

Di, Q., Piersanti, A., Zhang, Q., Miceli, C., & Liu, X. (2021). Soybean AP2-domain transcription factor *GmWRI4* optimizes the ratio of oleic acid to linoleic acid in seed. *Research Square*, 1–15.

Do, P. T., Nguyen, C. X., Bui, H. T., Tran, L. T., Stacey, G., & Gillman, J. D. (2019). Demonstration of highly efficient dual gRNA CRISPR/Cas9 editing of the homeologous *GmFAD2-1A* and *GmFAD2-1B* genes to yield a high oleic, low linoleic and α-linolenic acid phenotype in soybean. *BMC Plant Biology, 19*(1), 311.

Fehr, W. R. (2007). Breeding for modified fatty acid composition in soybean. *Crop Science, 47*, 72.

Gillman, J. D., Tetlow, A., Hagely, K., Boersma, J. G., Cardinal, A., Rajcan, I., & Bilyeu, K. (2014). Identification of the molecular genetic basis of the low palmitic acid seed oil trait in soybean mutant line RG3 and association analysis of molecular markers with elevated seed stearic acid and reduced seed palmitic acid. *Molecular Breeding, 34*(2), 447–455.

Hammond, E. G., & Fehr, W. R. (1975). Oil quality improvement in soybeans (*Glycine max* L. Merr.). *Fette, Seifen, Anstrichmittel, 77*(3), 97–101.

Han, J., Guo, B., Guo, Y., Zhang, B., Wang, X., & Qiu, L. (2019). Creation of early flowering germplasm of soybean by CRISPR/Cas9 technology. *Frontiers in Plant Science, 10*, 1446.

Held, J. P., Carrero-Colón, M., & Hudson, K. A. (2019). Combination of novel mutations in *FAD3C* and *FAD3A* for low linolenic acid soybean. *Agrosystems, Geosciences & Environment, 2*(1), 1–4.

Jacobs, T. B., LaFayette, P. R., Schmitz, R. J., & Parrott, W. A. (2015). Targeted genome modifications in soybean with CRISPR/Cas9. *BMC Biotechnology, 15*(1), 1–10.

Jeong, N., Kim, K. S., Jeong, S., Kim, J. Y., Park, S. K., Lee, J. S., Jeong, S. C., Kang, S. T., Ha, B. K., & Kim, D. Y. (2019). Korean soybean core collection: Genotypic and phenotypic diversity, population structure, and genome-wide association study. *PLoS One, 14*, 1–10.

Karikari, B., Li, S., Bhat, J. A., Cao, Y., Kong, J., Yang, J., & Zhao, T. (2019). Genome-wide detection of major and epistatic effect QTLs for seed protein and oil content in soybean under multiple environments using a high-density bin map. *International Journal of Molecular Sciences, 20*(4), 979.

Lakhssassi, N., Zhou, Z., Liu, S., Colantonio, V., AbuGhazaleh, A., & Meksem, K. (2017). Characterization of the *FAD2* gene family in soybean reveals the limitations of gel-based TILLING in genes with high copy number. *Frontiers in Plant Science, 8*, 324.

Li, C., Nguyen, V., Liu, J., Fu, W., Chen, C., & Yu, K. (2019). Mutagenesis of seed storage protein genes in soybean using CRISPR/Cas9. *BMC Research Notes, 12*(1), 176.

Liu, X., Qin, D., Piersanti, A., Zhang, Q., Miceli, C., & Wang, P. (2020). Genome-wide association study identifies candidate genes related to oleic acid content in soybean seeds. *BMC Plant Biology, 20*(1), 1–14.

Patil, A., Taware, S. P., Oak, M. D., Tamhankar, S. A., & Rao, V. S. (2007). Improvement of oil quality in soybean (*Glycine max* L. Merrill) by mutation breeding. *Journal of the American Oil Chemists' Society, 84*(12), 1117–1124.

Pham, A. T., Shannon, J. G., & Bilyeu, K. D. (2012). Combinations of mutant *FAD2* and *FAD3* genes to produce high oleic acid and low linolenic acid soybean oil. *Theoretical and Applied Genetics, 125*, 503–515.

Povkhova, L., Pushkova, E., Dmitriev, A., Kezimana, P., Novakovskiy, R., Melnikova, N., & Rozhmina, T. (2020). Molecular markers based on SNPs in *FAD3* genes for determination of linolenic acid content in flax seed. *Bioinformatics of Genome Regulation and Structure/Systems Biology*.

Pratap, A., Gupta, S. K., Kumar, J., & Solanki, R. K. (2012). Soybean. In S. Gupta (Ed.), *Technological Innovations in Major World Oil Crops, Volume 1* (pp. 293–321). Springer, New York, NY.

Qiu, L. J., Xing, L. L., Guo, Y., Wang, J., & Chang, R. Z. (2013). A platform for soybean molecular breeding: The utilization of core collections for food security. *Plant Molecular Biology, 83*, 41–50.

Silva, L. C. C., Bueno, R. D., Matta, L. B., Pereira, P. H. S., Mayrink, D. B., Piovesan, N. D., & Dal-Bianco, M. (2018). Characterization of a new *GmFAD3A* allele in Brazilian CS303TNKCA soybean cultivar. *Theoretical and Applied Genetics, 131*(5), 1099–1110.

Silva, L., Matta, L., Ramos, P. G., Bueno, R., Piovesan, N., Cardinal, A., & Ribeiro, C. (2021). Association studies and QTL mapping for soybean oil content and composition. *Euphytica, 217*, 24.

Sun, X., Hu, Z., Chen, R., Jiang, Q., Song, G., & Zhang, H. (2015). Targeted mutagenesis in soybean using the CRISPR-Cas9 system. *Scientific Reports, 5*(1), 1–10.

Thapa, R., Carrero-Colón, M., Addo-Quaye, C., Held, J., Dilkes, B., & Hudson, K. A. (2018). New alleles of *FAD3A* lower the linolenic acid content of soybean seeds. *Crop Science, 58*(2), 713–718.

Vogel, P. A., Bayon de Noyer, S., Park, H., Nguyen, H., Hou, L., Changa, T., & Clemente, T. E. (2019). Expression of the *Arabidopsis WRINKLED 1* transcription factor leads to higher accumulation of palmitate in soybean seed. *Plant Biotechnology Journal, 17*(7), 1369–1379.

Weselake, R. J., Taylor, D. C., Rahman, M. H., Shah, S., Laroche, A., McVetty, P. B., & Harwood, J. L. (2009). Increasing the flow of carbon into seed oil. *Biotechnology Advances, 27*(6), 866–878.

Yang, J., Xing, G., Niu, L., He, H., Guo, D., Du, Q., & Qian, X. (2018). Improved oil quality in transgenic soybean seeds by RNAi-mediated knockdown of *GmFAD2-1B*. *Transgenic Research, 27*(2), 155–166.

Zhang, J. P., Wang, X. Z., Lu, Y. M., Bhusal, S. J., Song, Q. J., Cregan, P. B., Yen, Y., Brown, M., & Jiang, G. L. (2018). Genome-wide scan for seed composition provides insights into soybean quality improvement and the impacts of domestication and breeding. *Molecular Plant, 11*(3), 460–472.

Zhang, P., Du, H., Wang, J., Pu, Y., Yang, C., Yan, R., Yang, H., Cheng, H., & Yu, D. (2020). Multiplex CRISPR/Cas9-mediated metabolic engineering increases soybean isoflavone content and resistance to soybean mosaic virus. *Plant Biotechnology Journal, 18*, 1384–1395.

CHAPTER 7

Herbicide-Tolerant Soybean: Past, Present, and Future

PREETI GUPTA,[1] ANJANA RUSTAGI,[2] and PREM KUMAR GUPTA[3,*]

[1]*Independent Researcher, New Delhi, India*

[2]*Department of Botany, Gargi College, Delhi University, Delhi, India*

[3]*Modern Diagnostics and Research Center, Gurgaon, Haryana, India*

*Corresponding author

ABSTRACT

Herbicide-tolerant (HT) soybean has played an important role in increasing soybean yields, reducing weed management costs, and mitigating harmful environmental impacts. The introduction of genetically modified (GM) glyphosate-tolerant (GT) soybean, corn, and canola in 1996–97 marked a turning point in weed control practices. This technology was quickly adopted by farmers due to its effectiveness in weed control, lower production costs, increased yields, and reduced environmental impact. The use of GM HT soybean is now widespread, with over 95% of soybeans cultivated in the United States genetically modified for herbicide tolerance. The first HT soybean variety, Roundup Ready, was released in 1996 by Monsanto. Since then, several soybean varieties tolerant to different herbicides have been developed. Major HT varieties cultivated globally include Roundup Ready (RR), Roundup Ready 2 Yield® (RR2), Roundup Ready 2 Xtend® (RRX), Liberty Link® (LL), XtendFlex®, Enlist E3™, and Liberty Link GT27. On the flip side, reliance on glyphosate has led to the emergence of at least 47 glyphosate-resistant (GR) weed species. Moreover, studies indicate that

Soybeans, Volume 2: Genome, Biotechnology, Stress, and Disease.
R. Ambika Rajendran, Ayyagari Ramlal, Sanjay Kumar Lal, and Godfree Chigeza (Eds.)

DOI: 10.1201/9781998511730-9

the use of GM HT soybean has resulted in an overall increase in herbicide use, raising environmental concerns. This has sparked debate over its long-term sustainability and the need for alternative weed control technologies.

7.1 INTRODUCTION

Sustainable agriculture is the need of the hour due to the increasing human population and the ever-deteriorating soil and environmental resources. Genetically modified (GM) organisms can potentially reduce the use of pesticides and thus contribute to achieving agricultural sustainability goals (Brookes and Barfoot, 2020b). GM crops have been well received by farmers mainly due to various farm-level economic benefits obtained from them (Ali et al., 2021; ISAAA brief, 2017). Currently, GM soybean, GM cotton, and GM corn account for about 95% of these crops planted in the USA (USDA, 2022). In Argentina, 100% of soybean and 97% of corn grown in 2018 were GM (Klein and Vidal Luna, 2021).

Insect resistance (corn, cotton, soybeans, and potatoes) and herbicide tolerance (HT) (canola and soybean) are the two most common traits in GM crops. GM HT crops engineered for tolerance to glyphosate-based herbicides were introduced in 1996 and have since become widespread in major agricultural producing countries (Kumar et al., 2020). A total of 353 million metric tons of soybean were produced in the world in 2020, with Brazil being the largest producer (34%), followed by the USA (32%) (Bonea, 2021; Meier et al., 2020). Available data indicate that in 1997, one year after the commercialization of GM HT soybeans, approximately 17% of soybean acres were planted with these varieties. The adoption rate grew rapidly, and by 2001, 68% of planted soybeans were GM. After 2001, the adoption rate of GM HT soybeans continued to rise, eventually reaching a plateau of 94% in 2014 (ISAAC brief, 2014). In 2021–22, the adoption rate of GM HT soybeans in the USA was 95% (USDA, 2022). Monsanto, the pioneer and current leader in soybean GM HT technology, released the first commercial GM HT soybean variety in 1996 (Brookes and Barfoot, 2005). This variety, named Roundup Ready (RR), was tolerant to glyphosate and heralded a revolution in the chemical weed management of crops. GT varieties completely dominated the HT soybean plantation until a few years back, when the development of glyphosate tolerance by many weed species in soybean-producing regions started to pose challenges for weed control using this technology (Baek et al., 2021; Duke, 2018; EFSA GMO Panel, 2019; Gage et al., 2019; Heap and Duke, 2018). This prompted seed companies to introduce tolerance against additional herbicides in soybean and other crops, mainly in the form of stacked traits, conferring tolerance to more than one herbicide (Brookes and Barfoot, 2020a). In this chapter, we chart the GM HT technology from its initial commercial launch to its present situation and projected future direction. We also discuss the socioeconomic and environmental effects of this technology. GM HT technology currently plays an important role in world soybean production. In the future, it is probably going to be crucial for feeding the constantly expanding human population.

7.2 SOYBEAN TRANSGENIC HERBICIDE TOLERANCE TECHNOLOGY

Early efforts to develop HT crops, including soybean, were centered around conventional breeding methods. Site-directed mutagenesis and mutagenesis in tissue culture were employed to develop soybean varieties tolerant to sulfonylurea herbicides (Duke et al., 2002) and triazine herbicides (Barrentine et al., 1982), respectively. These non-transgenic herbicide-tolerant soybean varieties were modestly successful in fulfilling the expectations of the farmers, and the quest for the development of soybean varieties tolerant to broad-spectrum herbicides for more comprehensive weed control continued.

Choosing a crop variety appropriate for a particular field is one of the most important and difficult choices made by a farmer. Maturity group, yield potential, disease resistance, insect resistance, and HT are some of the aspects that need to be considered while choosing a variety for planting. Farmers have several options for HT soybean varieties, with up to seven major traits available. These include Roundup Ready (RR), Roundup Ready 2 Yield® (RR2), Roundup Ready 2 Xtend® (RRX) Liberty Link® (LL), XtendFlex®, Enlist E3™, and Liberty Link GT27 (Table 7.1) (Jhala, 2021).

TABLE 7.1 Different Herbicide-Tolerant Varieties of Soybean

Soybean Trait	Tolerance to Herbicide(s)
Roundup Ready	Glyphosate
Roundup Ready 2 Yield®	Glyphosate
Roundup Ready 2 Xtend®	Glyphosate/Dicamba
LibertyLink®	Glufosinate
Enlist E3™	Glyphosate/glufosinate/2,4-D choline
XtendFlex®	Glyphosate/glufosinate/Dicamba
LibertyLink GT27	glufosinate, glyphosate, and 4-hydroxyphenylpyruvate dioxygenase (HPPD) /Group 27 herbicides

7.2.1 ROUNDUP READY TECHNOLOGY

7.2.1.1 ROUNDUP READY

In 1996, Monsanto introduced the first widely used GM crop, the RR soybean (Brookes, 2003). RR soybean varieties are tolerant to the active ingredient glyphosate, which inhibits the 5-enolpyruvylshikimate-3-phosphate (S3P) synthase enzyme (EPSPS) (EC 2.5.1.19), catalyzing the penultimate step of the shikimate pathway: the transfer of the enolpyruvyl moiety from phosphoenol pyruvate (PEP) to S3P (Figure 7.1), producing EPSP and inorganic phosphate. EPSP is required for the synthesis of aromatic amino acids (AAs) and other aromatic products such as alkaloids, flavonoids, and hormones, and its deficiency results in plant death (Arango, 2009).

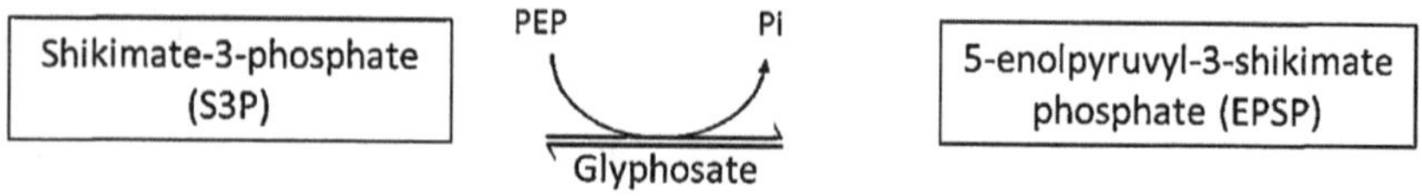

FIGURE 7.1 Reaction catalyzed by EPSP synthase, inhibited by glyphosate.

However, the ESPS enzyme found in some microorganisms is resistant to glyphosate inhibition. The glyphosate-resistant *ESPS* gene used for creating RR soybean was isolated from *Agrobacterium* sp. strain CP4 (CP4 EPSPS) (Padgette et al., 1995). The *CP4 EPSPS* was cloned, and the 5' end of the gene was fused to the sequence of a chloroplast transit peptide derived from the EPSPS from *Petunia hybrida*. This transit peptide had been previously shown to deliver bacterial EPSPS to the chloroplasts of other plants. The gene was cloned in the plant expression vector PV-GMGT04 with beta-glucuronidase (β-GUS) from *Escherichia coli* as a selection marker. The *CP4 EPSPS* was overexpressed in all plant tissues using the strong cauliflower mosaic virus (CaMV) 35S promoter. The DNA was delivered into the A5403 cultivar of soybean using the gene-gun method, and the transformed plants were tested for glyphosate tolerance over many generations. This particular event of insertion of *CP4 EPSPS* in the soybean genome was named event MON88978 (Padgette et al., 1995; Watson, 2019).

The use of RR soybean technology allowed the replacement of multiple applications of various selective post-emergence herbicides. Glyphosate has low toxicity for humans and other animals, and by using glyphosate, farmers could control weeds with just one or two over-the-top applications, reducing the need for multiple herbicides. Moreover, since glyphosate is an effective broad-spectrum weedicide, it extended the herbicide application window. Over many production cycles, adopting RR soybeans resulted in higher yields due to more effective weed control and reduced production risk (Brookes and Barfoot, 2018).

RR was well received by the farmers in the U.S. In just four seasons, the RR soybean planting area went from representing less than 5% of the total soybean plantation area to more than 80%. A total of 87% of soybeans planted in the U.S. in 2005 was RR, which rose to 90% by the year 2009, before the Roundup Ready 2 soybean release in 2010. The patent on Roundup technology expired in 2015, and even to this date, some academic breeding programs continue to develop generic soybean varieties using the Roundup technology (ISAAA Briefs, 2014). The wide-spread use of GM HT soybeans, particularly RR soybeans, has been a major trend in American agriculture.

7.2.1.2 ROUNDUP READY 2 YIELD®

Roundup ready soybean made available an effective and convenient chemical weed control strategy. However, various studies performed since the launch of RR soybean revealed that it suffered from a yield drag of up to 7% (Zilberman et al., 2010). To address this issue, Monsanto released the RR2 GM HT variety in 2008. This variety contained the same exogenous *CP4 EPSPS* gene; however, it differed from RR soybean in the following important aspects (Padgette et al., 1995):

1. A CTP2 transit peptide from *Arabidopsis thaliana* L. was used to target the *CP4 EPSPS* to the chloroplasts;
2. CaMV was replaced with a 35S promoter obtained from the figwort mosaic virus;
3. An E9 3' untranslated sequence from ribulose-1,5-biphosphate carboxylase small subunit from *Pisum sativum* L. was used to attach the Poly-A tail in place of the untranslated sequence from nopaline synthase gene;
4. The DNA construct was inserted into the PV-GMGOX20 plasmid instead of the PV-GMGT04 plasmid;
5. Gene transfer was performed using *Agrobacterium*-mediated transformation rather than the gene-gun method; and
6. The soybean cultivar A3244 was used for the transformation process in place of the A5403 cultivar used in the development of the original RR soybeans.

The transformation event was named MON89788 (Nandula, 2010). Southern blotting analysis showed that a single copy of the EPSPS gene was inserted into the genome of the A3244 cultivar. No antibiotic resistance markers from the transformation cassette were inserted into the genome. Monsanto claimed a yield improvement of 7–11% for RR2 soybeans, with mixed results in various field trials (Padgette et al., 2018).

7.2.1.3 ROUNDUP READY 2 XTEND®

The widespread use of glyphosate as the primary herbicide for GM glyphosate-tolerant (GT) crops has led to the emergence of GT weeds. This necessitated the use of another herbicide with the GM herbicide-tolerant (HT) technology to tackle the GR weeds. RRX was the first soybean variety with stacked HT traits that offered resistance to dicamba in addition to glyphosate (Hartzler and Owen, 2016). The glyphosate resistance event in the RRX soybean was the same as in the RR2 soybean. However, the RRX soybean additionally contained the dicamba monooxygenase (*DMO*) gene from the soil bacterium *Pseudomonas maltophilia* (strain DI-6). DMO is a Rieske non-heme monooxygenase that confers tolerance to Dicamba by converting it to 3,6-dichloro salicylic acid (DSCA) (Behrens et al., 2007) (Figure 7.2). *DMO* gene was expressed under the FLt36 promoter from the Peanut chlorotic streak virus, and the expression was enhanced using a translational enhancer from the tobacco etch virus. The gene was targeted to chloroplasts using a transit peptide coding region from *P. sativum,* and the transcription was terminated using a terminator sequence from the Rubisco small subunit gene (rbcS3′) of *P. sativum* (Behrens et al., 2007).

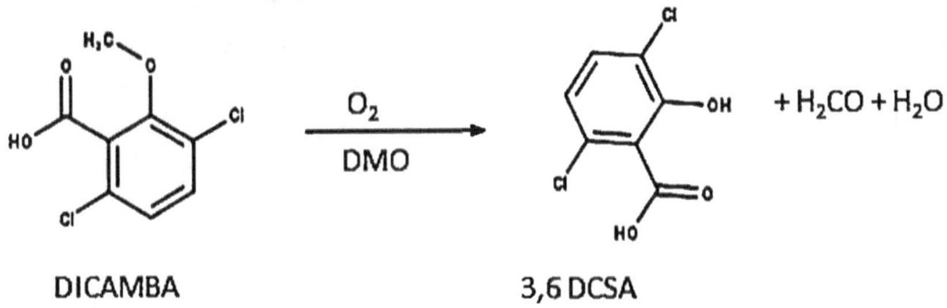

FIGURE 7.2 Reaction catalyzed by dicamba monooxygenase (DMO) that converts dicamba to 3,6-dichloro salicylic acid (3,6DSCA).

The inclusion of dicamba tolerance in addition to glyphosate tolerance in RRX soybeans allows farmers to apply dicamba-based herbicides without harming the soybean crop, protecting against both GR weeds and off-target dicamba injury (Jhala, 2021). RRX soybean was planted on 20 million acres in the first year of its commercial availability in 2017. This doubled to 40 million acres in 2018 and rose to 50 million acres in 2020, which was 66% of total soybean acreage in the U.S. this year (Rose, 2020).

7.2.2 *LIBERTY LINK TECHNOLOGY*

7.2.2.1 LIBERTY LINK®

Liberty Link® (LL) soybean trait provides resistance to glufosinate (phosphinothricin). Glufosinate inhibits the activity of the enzyme glutamine synthetase (GS) required for the assimilation of ammonia in plants (Nandula, 2019). The inhibition of GS results in the depletion of glutamine (Gln) and glutamate (Glu) and the accumulation of ammonia produced during photorespiration. Moreover, Glu depletion prevents the formation of glycine from glyoxalate by Glu glyoxalate aminotransferase, resulting in the accumulation of glyoxalate and glycolate. The accumulating glycolate is acted upon by glycolate oxidase (GO), producing glyoxalate and hydrogen peroxide (H_2O_2) (a source of free radicals) (Duke, 2014; Takano and Dayan, 2020). The accumulation of glycolate, a byproduct of incomplete carbon fixation during photorespiration, can lead to several effects, including feedback inhibition of the photorespiration pathway and inhibition of the enzyme Rubisco activase. Since the photorespiration pathway acts as a sink for excess electrons generated by the photosystems, after its inhibition, the extra electrons reduce the molecular oxygen produced by the PSII water-splitting complex, causing a buildup of lethal amounts of reactive oxygen species (ROS). The plants are killed due to the accumulation of ROS and the buildup of ammonia, which is phytotoxic (Figure 7.3) (Carbonari et al., 2016; Takano et al., 2020).

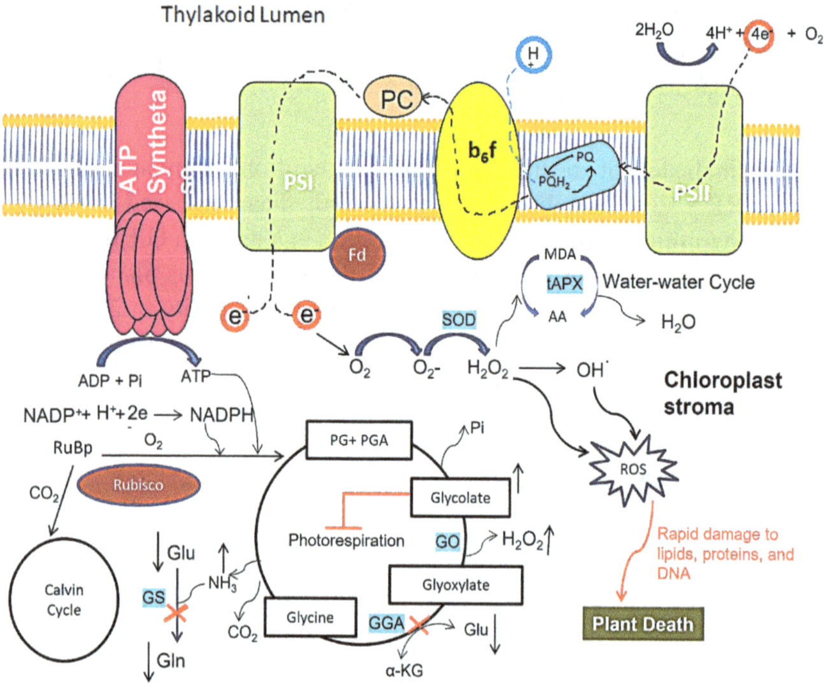

FIGURE 7.3 Mechanism of herbicide action of glufosinate.

Abbreviations: AA: ascorbic acid; GGA: glutamate glyoxalase aminotransferase; Gln: glutamine; Glu: glutamate; GO: glycolate oxidase; GS: glutamine synthase; MDA: monodehydroascorbate; PGA: 3-phosphoglycerate; PG: phosphoglycolate; SOD: superoxide dismutase; and tAPX: ascorbate peroxidase.

Source: Adapted from Takano et al. (2020).

The gene encoding the bacterial enzyme phosphinothricin-N-acetyltransferase (PAT) is often found in only one copy in transgenic plants engineered to exhibit the glufosinate tolerance trait, such as LL plants. The PAT enzyme confers glufosinate tolerance by inactivating glufosinate. It achieves this by catalyzing the acetylation of glufosinate, converting it into non-toxic acetylphosphinothricin (Figure 7.4). The gene was inserted into the soybean genome using the gene-gun method (Que et al., 2010). The gene was cloned in the plasmid pB2/35SAcK, which has the backbone of a plasmid cloning vector. The vector contains a β-lactamase (bla) gene from *E. coli* that can be used as a selection marker for gene cloning in bacteria. The plasmid was linearized by a single cut within the *bla* gene (thereby disrupting it) by the restriction enzyme PvuI before the transformation. The *pat* gene from *Streptomyces viridosporus* was optimized for expression in plants by site-directed mutagenesis without changing the AA sequence of the PAT protein. The *pat* gene was overexpressed in all plant tissues using the strong CaMV 35S promoter (Huang et al., 2015).

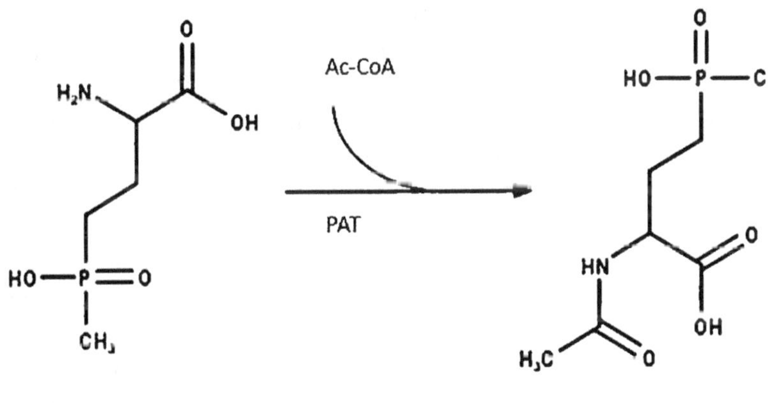

Phosphinothricin

Acetylphosphinothricin

FIGURE 7.4 Catalytic reaction of phosphinothricin-N-acetyltransferase.

Glufosinate should be applied to the LL soybean in the presence of sunlight for maximum herbicide action. Also, the temperature should not drop below 10°C following the application of glufosinate. The activity of PAT protein is significantly reduced below this temperature, and the transgenic plants may sustain injury at temperatures lower than 10°C (Meyer and Norsworthy, 2020).

LL soybean gained good acceptance in the market, approaching 20% of the soybean market share in 2018, up from 6% in 2006. It is particularly popular in the southern and midwestern U.S., where GT weeds pose a formidable challenge. The widespread infestation of GR *Amaranthus palmeri* L. in soybean fields led to increased adoption of LL soybean in regions affected by this weed (Riar et al. 2013a, b).

In 2019, BASF Corp. released Liberty Link GT27 (LL GT27) soybeans with triple-stacked traits for resistance against glyphosate, glufosinate, and isoxaflutol (an HPPD inhibitor) (Nandula, 2019). The LL GT27 soybean was created by crossing the LL soybean with the FG72 soybean, which was GM to express a modified maize *EPSPS* gene and the *hppd* gene (*hppdpfW336*) from *Pseudomonas fluorescens* that confers resistance to HPPD inhibitors such as isoxaflutol. The gene responsible for glyphosate resistance in LL GT27 soybean is the maize *2mEPSPS*, which is different from the *EPSPS* gene used to develop the RR trait. The plasmid pSF10 was used to clone the expression cassettes for the *2mepsps* and *hppdPfW336* genes. The *2mepsps* expression cassette consisted of the Ph4a748 promoter and intron 1 of the histone H3.3 of *A. thaliana*, which directs high-level constitutive expression of the *2mEPSPS* gene. The TPotp C transit peptide, a fusion peptide containing sequences from the Rubisco (ribulose-1,5-bisphosphate carboxylase/oxygenase) small subunit genes of corn and sunflower, was used to target the *2mEPSPS* enzyme to the chloroplast. The 3' untranslated region of the histone H3 gene from *A. thaliana* was used for gene transcription termination. The *hppdPfW336* is a mutated version of the wild-type *hppd* from *P. fluorescens* that encodes a 358-AA-long protein with a molecular weight of 40 kDa. The *hppdPfW336* was expressed under the same promoter as used for expressing the *2mEPSPS*, but an internal promoter sequence was replicated to enhance the promoter activity, and the leader sequence of the tobacco etch virus was utilized to further increase the expression of *hppdPfW336*. This design helped to achieve high expression levels of the hppdPfW336 protein. The TPotp C transit peptide directed the synthesized protein to the chloroplast. The 3' untranslated region of the nopaline synthase from *A. tumefaciens* was used for transcription termination and polyadenylation of the *hppdPfW336* gene. The fragment of the plasmid containing both expression vectors was obtained by cleaving the vector pSF10 with *SalI* enzyme and purified by HPLC. The "Jack" variety of soybean was transformed with this purified fragment by the gene-gun method (OECD Unique Identifier MST-FGØ72-2).

HPPD inhibitors prevent plant growth by inhibiting the enzyme HPPD, which catalyzes the breakdown of the AA tyrosine into molecules that act as building blocks for other molecules essential for plant survival. One of the breakdown products of tyrosine is homogentisic acid, which is a precursor for the synthesis of both tocopherols and plastoquinone. Plastoquinone is a critical cofactor for the synthesis of carotenoids. Lack of tocopherol leads to oxidative damage, and lack of carotenoids results in severe damage to chlorophyll by UV light.

The plant turns white due to the loss of chlorophyll and dies (Reade and Cobb, 2002) (Figure 7.5). The isoxaflutole herbicide formulation for use with LL GT27 soybeans was approved by the EPA only in 2020, and only in a small number of counties in 25 states across the USA, representing only 15% of the total soybean acreage (Nandula, 2019).

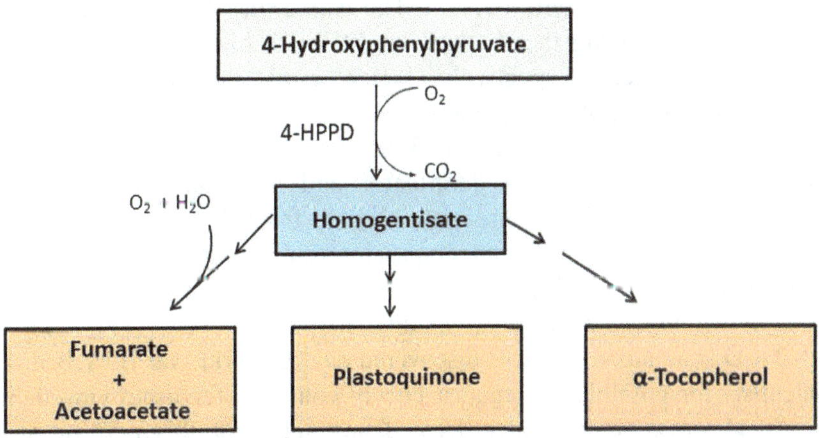

FIGURE 7.5 Biochemical pathways dependent on 4-hydroxyphenylpyruvate dioxygenase (4-HPPD). (i) catabolism of tyrosine; (ii) biosynthesis of plastoquinone; and (iii) biosynthesis of tocopherol.

7.2.3 ENLIST TECHNOLOGY

7.2.3.1 ENLIST E3™

Enlist E3™ soybean is a GM soybean variety that is tolerant to three different herbicides: 2,4-D, glufosinate, and glyphosate. When these herbicides are applied either pre-emergence or post-emergence up to the R2 growth stage, the plants can tolerate the herbicide treatments without significant damage or yield loss (Robinson et al., 2015). Enlist E3™ soybean varieties that express the *aad-2* gene from *Delphia acidovorans*, the *2mESPS* gene from maize, and the *pat* gene from *Streptomyces viridochromogenes*. These genetic modifications allow Enlist E3™ soybean varieties to withstand applications of 2,4-D, glyphosate, and glufosinate herbicides, providing farmers with multiple options for effective weed control in their soybean fields (Frene et al., 2018). The *aad2* gene encodes the enzyme aryloxyalkanoate dioxygenase-12 (AAD-12), which can degrade 2,4-D to 2,4-dichlorophenol (DCP), a compound lacking herbicide action. The publicly available soybean cultivar "Maverick" was used to generate Enlist E3™ soybean. The cotyledonary node was used for transformation using *A. tumefaciens* bacteria carrying a binary vector containing all three genes within the T-DNA region. The transformed shoots were selected by including glufosinate in the growth medium. The *aad*, *2mEPSPS*, and *pat* genes were expressed using the *A. thaliana* AtUbi10, histone H4A748, and CsVMV promoters, respectively (Wright et al., 2010). Transcription termination and polyadenylation of *aad*, *2mEPSPS*, and *pat* were performed using Histone H4A748 3′ UTR, AtuORF23 3′ UTR (3′ UTR encompassing transcription terminator and polyadenylation site of open reading frame 23 of plasmid pTi15955 from *A. tumefaciens*), and AtuORF1 3′ UTR (3′ UTR encompassing transcription terminator and polyadenylation site of open reading frame 1 of plasmid pTi15955 from *A. tumefaciens*), respectively (Nandula, 2019).

2,4-D is an auxin that was discovered during World War II by British scientists for possible use against enemy countries for destroying their crops. However, it was later deemed unsuitable for this purpose and released as a commercial herbicide. It soon became one of the most widely used herbicides worldwide because of its selectivity for dicotyledonous plants (making it suitable for weed control in crops such as wheat, maize, rice, etc.). At low concentrations, 2,4-D acts as an auxin, regulating plant growth (Tu et al., 2001), but at higher concentrations, the compound can cause growth irregularities that kill sensitive plants. The application of 2,4-D leads to the induction of auxin-responsive genes, which, in turn, upregulates the biosynthesis of abscisic acid (ABA) and ethylene hormones. Both ethylene and ABA stimulate leaf senescence and increase ROS production, leading to tissue damage and plant death (Figure 7.6) (Song, 2014).

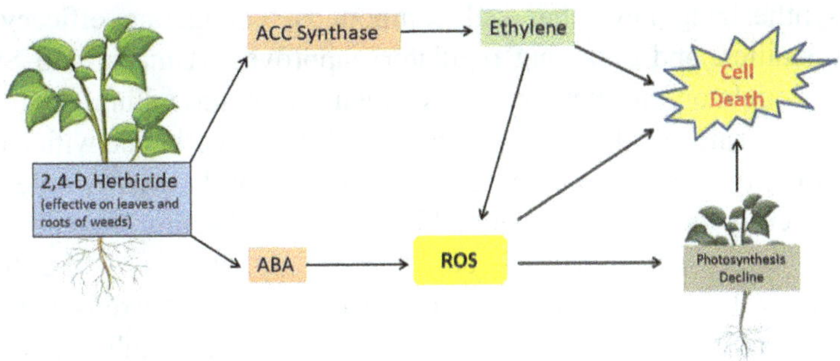

FIGURE 7.6 Mode of action of herbicide 2,4-D.

Abbreviations: ACC: 1-aminoacyclopropane 1-carboxylate synthase; ABA: abscisic acid; and ROS: reactive oxygen species.

Enlist HT technology has been favorably received by farmers due to its ability to control GT weeds and the non-volatility of 2,4-D, glyphosate, and glufosinate, which minimizes damage to adjacent non-target crops. About 20% of soybeans planted in the USA in 2020 were Enlist E3®, which increased to 35% in 2021 (Bagemann, 2021). The Enlist E3 is currently the fastest-growing weed control system in the U.S., and the trend is likely to continue (Johnston, 2022).

7.2.4 XTEND FLEX TECHNOLOGY

Xtendflex® soybean is a GM soybean variety that was developed through traditional breeding methods. It combines three independent genetic modifications from soybean lines MON 87708, MON 89788, and A5547-12 developed by Monsanto (now Bayer CropScience Ltd.), whereas the event A5547-127 is owned by BASF Agriculture Solutions Seed U.S. LLC. Protein expression analysis indicated that the expression of these proteins in the triple-stacked variety was comparable to that in the varieties with individual events (EFSA GMO Panel, 2021). Xtendflex® was planted on approximately 15 million acres in 2021.

7.3 FUTURE OF GENETICALLY MODIFIED (GM) HERBICIDE TOLERANCE (HT) SOYBEAN TECHNOLOGIES

The development of a new herbicide with a novel mode of action requires significant investment in research and development, including identifying and synthesizing new chemical compounds, conducting efficacy and safety studies, and obtaining regulatory approvals (Umetsu and Shirai, 2020). Regulatory hurdles can pose challenges for agricultural companies seeking to bring new herbicides to market. No new herbicide with a novel mode of action has been commercially released in the past two decades. Only recently have a few novel herbicide modes of action been described that are being developed for commercial applications in the future. Tetraflupyrolimet is an aryl pyrrolidinone anilide that exerts herbicide action by inhibiting *de novo* pyrimidine biosynthesis via inhibition of the dihydroorotate dehydrogenase enzyme. It is particularly effective against grasses. BASF Corp is developing Cinmethylin as a herbicide that works by inhibiting the enzyme FA thioesterase, which is required for fatty acid (FA) biosynthesis. It is also particularly effective against grasses (Campe et al., 2018). Another herbicide with a novel site of action, Bixlozone, is under development by FMC Corp., which works by inhibiting carotenoid synthesis in plants, leading to photo-oxidative chlorophyll damage (Jeanmart et al., 2021). These herbicide technologies are expected to be important elements of weed management strategies for the later part of this decade.

As weeds develop resistance to existing herbicides, and many to multiple herbicides, the fight against the weeds is increasingly using the strategy of stacking existing herbicide traits. In 2022, Bayer announced the development of soybeans with stacked resistance against five different herbicides (glyphosate, glufosinate, dicamba, 2,4-D, and HPPD inhibitors) with the expected availability of seeds in the middle of this decade (Bayer Crop Science Annual R&D Pipeline Update, 16th February 2022). Another approach that agricultural companies are taking to maximize the yield potential of various herbicide-tolerant traits is to combine herbicide-tolerance traits with other agronomically important traits, such as nematode resistance. Soybean designated as event GMB151 has been developed by BASF to provide resistance against the soybean cyst nematode (SCN) and tolerance to HPPD-4 inhibitors. This soybean trait has been deregulated and is expected to be available shortly (EFSA Panel on GMO, 2021).

On the technology front, gene editing holds great promise for developing novel herbicide-tolerant crops, enabling more sustainable and effective weed management strategies in the future. Crops with useful traits developed using gene editing technology do not have foreign genetic elements and thus can potentially find greater public acceptance due to their non-GMO nature. The recently developed CRISPR/Cas9-mediated gene editing techniques are particularly effective in generating crops with herbicide-tolerant traits and have been used to generate various herbicide-tolerant crops, including rice, maize, soybean, wheat, rapeseed, tomato, and potato, by introducing target site mutations in the enzymes affected by the herbicides (Dong et al., 2021). Since no foreign genetic element is introduced in the plants using the *CRISPR/Cas* gene editing, such products are also likely to face fewer regulatory hurdles. The browning-resistant mushrooms, false flax with enhanced omega-3 content, and high amylopectin waxy corn, developed using gene editing technology, were approved by the U.S. Department of Agriculture (USDA) in record time (USDA, 2018; Zaidi et al., 2020).

7.4 SOCIOECONOMIC IMPACT OF GENETICALLY MODIFIED (GM) HERBICIDE-TOLERANT SOYBEAN

It is estimated that soybean farmers generated an additional income of 4.78 billion USD in 2018 by using GM HT soybean varieties. Between 1996 and 2018, GM HT technology delivered 64.3 billion USD in additional income to the farmers (Brookes and Barfoot, 2014, 2020b). An analysis by Gardner et al. (2009) revealed that households cultivating GM HR soybeans saved an average of 14.5% on the cost of weed management. However, an important motivation for the adoption of GM HT soybeans by farmers has been the increased off-farm time due to reduced labor requirements for weed management. This additional off-farm time can also be used to generate additional family income. The farm-level income increase from GM HT soybeans has been due to both reductions in the cost of weed control and yield gains due to effective weed control. Furthermore, no-tillage farming with GM HT soybeans can contribute to shorter production cycles. Without the need for pre-plant tillage, farmers can save time and labor, allowing for more efficient and timely planting. This allows farmers in many countries, such as Argentina and Paraguay, to plant an additional crop per year, generating substantial additional income (Brookes and Barfoot, 2014; 2020a). Research also indicates that various indirect benefits, such as cost reduction on equipment, better operator and worker safety, and increased overall usability, may be important factors in explaining the widespread adoption of HR GM soybean technology in the major soybean-producing countries (Marra and Piggott, 2006).

7.5 ENVIRONMENTAL IMPACT OF HERBICIDE-TOLERANT SOYBEAN

HT technology has had a significant impact on the goal of achieving agricultural sustainability. Brookes and Barfoot (2020a) estimated that an additional 12 million hectares of land would have been required to match the current levels of soybean production using non-GMO varieties, and about one-third of this land would have come from deforestation (Mahaffey et al., 2016). Moreover, no-till farming of GM HT soybean allows farmers in many countries, such as Paraguay and Argentina, to plant an additional soybean crop immediately after the wheat crop, which also reduces pressure to expand land use for agriculture (Brookes and Barfoot, 2020b).

The GM HT technology is expected to benefit the environment by reducing the amounts of herbicides applied to the fields. Nevertheless, various studies indicate that GM HT crops have not contributed significantly to reductions in the amounts of herbicides applied to the fields worldwide, which is in contrast to the initial expectations from this technology. Some studies indicate that herbicide use on GM HT crops has increased over time (Perry et al., 2016), whereas others indicate a 30% reduction (Smyth et al., 2011). A small net increase of 0.1% in herbicide use on GM HT soybean compared to what would have been used on conventional crops has been reported. However, significant environmental benefits have accrued due to a shift towards less toxic herbicides such as glyphosate (Brookes and Barfoot, 2020b).

The adoption of soybean GM HT technology has, however, contributed significantly to the reduction in greenhouse gas (GHG) emissions. It is estimated that in 2018, GM HT technology made possible a reduction in fuel usage of 920 million liters, which would have contributed 2456 million kg of CO_2 emissions to the atmosphere. Overall, between 1996 and 2018, the practice of no-tillage farming made possible by GM HT soybean technology contributed to 78% of the total GHG emission reduction by all GM HT crops combined (Brookes, 2021). Widespread adoption of no-tillage farming, especially in soybeans, has also contributed to the sequestration of 5606 million kg of soil carbon, a significant contribution towards mitigating climate change and promoting sustainable agricultural practices (Brookes, 2022, 2021).

7.6 CONCLUSION

The GM HT crop technology has revolutionized how weeds are controlled in crops such as soybean, maize, and cotton. The socioeconomic benefits of this technology are immense, as a majority of the cultivated area of these crops consists of HT varieties. The development of HT weeds has necessitated the stacking of HT traits in the crops, and crops tolerant to multiple herbicides are available. Despite some environmental concerns, this technology is expected to be more widely adopted in the future as the pressure to feed the growing human population mounts.

KEYWORDS

- **environmental impact**
- **genetically modified crops**
- **glyphosate-resistant weeds**
- **herbicide stacking**
- **herbicide-tolerant soybean**
- **roundup ready soybean**
- **weed management**

REFERENCES

Ali, S., Nawaz, M. A., Ghufran, M., Hussain, S. N., & Hussein Mohammed, A. S. (2021). GM trust shaped by trust determinants with the impact of risk/benefit framework: The contingent role of food technology neophobia. *GM Crops & Food, 12*(1), 170–191.

Arango Isaza, L. M. (2009). *Impact of Glyphosate Application to Transgenic Roundup Ready® Soybean on Horizontal Gene Transfer of the EPSPS Gene to Bradyrhizobium japonicum and on the Root-Associated Bacterial Community* (Doctoral dissertation, Ludwig Maximilians Universität, München).

Baek, Y., Bobadilla, L. K., Giacomini, D. A., Montgomery, J. S., Murphy, B. P., & Tranel, P. J. (2021). Evolution of glyphosate-resistant weeds. In J. B. Knaak (Ed.), *Reviews of Environmental Contamination and Toxicology* (Vol. *5*, pp. 93–128).

Bagemann, S. (2021). Corn and soy seed production on track, 2021 Liberty soybean acreage jump estimated at 15 percentage points. *Seed World.*

Barrentine, W. L., Hartwig, E. E., Edwards, C. J., & Kilen, T. C. (1982). Tolerance of three soybean (*Glycine max*) cultivars to metribuzin. *Weed Science, 30*(4), 344–348.

Bayer Crop Science. (2022). *Annual R&D Pipeline Update: Advancing Tomorrow's Innovations Today.*

Behrens, M. R., Mutlu, N., Chakraborty, S., Dumitru, R., Jiang, W. Z., LaVallee, B. J., Herman, P. L., Clemente, T. E., & Weeks, D. P. (2007). Dicamba resistance: Enlarging and preserving biotechnology-based weed management strategies. *Science, 316*(5828), 1185–1188.

Bonea, D. (2021). Evolution and global distribution of genetically modified soybean area in the period 2014–2018. *Evolution, 21*(4).

Bonny, S. (2011). Herbicide-tolerant transgenic soybean over 15 years of cultivation: Pesticide use, weed resistance, and some economic issues. The case of the USA. *Sustainability, 3*(9), 1302–1322.

Brookes, G. (2003). The farm level impact of using Roundup Ready soybeans in Romania. *PG Economics.*

Brookes, G. (2021). Environmental impacts of genetically modified (GM) crop use: Impacts on pesticide use and carbon emissions. In A. Ricroch, S. Chopra, & M. Kuntz (Eds.), *Plant Biotechnology* (pp. 87–101). Springer, Cham.

Brookes, G. (2022). Genetically modified (GM) crop use 1996–2020: Impacts on carbon emissions. *GM Crops & Food, 13*(1), 242–261.

Brookes, G., & Barfoot, P. (2005). GM crops: The global economic and environmental impact—The first nine years 1996–2004. *AgBioForum, 8*(2, 3), 187–196. Retrieved from: https://mospace.umsystem.edu/xmlui/handle/10355/121 (accessed on 25 March 2025).

Brookes, G., & Barfoot, P. (2014). Economic impact of GM crops: The global income and production effects 1996–2012. *GM Crops & Food, 5*(1), 65–75.

Brookes, G., & Barfoot, P. (2018). Farm income and production impacts of using GM crop technology 1996–2016. *GM Crops & Food, 9*(2), 59–89.

Brookes, G., & Barfoot, P. (2020a). GM crop technology use 1996-2018: Farm income and production impacts. *GM Crops & Food, 11*(4), 242–261.

Brookes, G., & Barfoot, P. (2020b). Environmental impacts of genetically modified (GM) crop use 1996–2018: Impacts on pesticide use and carbon emissions. *GM Crops & Food, 11*(4), 215–241.

Campe, R., Hollenbach, E., Kämmerer, L., Hendriks, J., Höffken, H. W., Kraus, H., Lerchl, J., Mietzner, T., Tresch, S., Witschel, M., & Hutzler, J. (2018). A new herbicidal site of action: Cinmethylin binds to acyl-ACP thioesterase and inhibits plant fatty acid biosynthesis. *Pesticide Biochemistry and Physiology, 148*, 116–125.

Carbonari, C. A., Velini, E. D., Silva Jr, J. F., Tropaldi, L., & Gomes, G. L. G. C. (2016). Absorption velocity of glufosinate and its effects on weeds and cotton. *Agrociencia, 50*(2), 239–249.

Dong, H., Huang, Y., & Wang, K. (2021). The development of herbicide-resistant crop plants using CRISPR/Cas9-mediated gene editing. *Genes, 12*(6), 912.

Duke, S. O. (2014). Biotechnology: Herbicide-resistant crops. In *Encyclopedia of Agriculture and Food Systems* (Vol. *2*, pp. 94–116).

Duke, S. O. (2018). The history and current status of glyphosate. *Pest Management Science, 74*(5), 1027–1034.

Duke, S. O., Scheffler, B. E., Dayan, F. E., & Dyer, W. E. (2002). Genetic engineering crops for improved weed management traits. In K. Rajasekaran, T. J. Jacks, & J. W. Finley (Eds.), *Crop Biotechnology* (Vol. *829*, pp. 52–66). ACS Publications, US.

EFSA Panel on Genetically Modified Organisms (EFSA GMO Panel), Naegeli, H., Bresson, J. L., Dalmay, T., Dewhurst, I. C., Epstein, M., Firbank, L. G., Guerche, P., Hejatko, J., Moreno, F. J., & De Sanctis, G. (2019). Assessment of genetically modified soybean MON 87708× MON 89788× A5547-127, for food and feed uses, under Regulation (EC) No 1829/2003 (application EFSA-GMO-NL-2016-135). *EFSA Journal, 17*(7), e05733.

EFSA Panel on Genetically Modified Organisms (GMO), Naegeli, H., Bresson, J. L., Dalmay, T., Dewhurst, I. C., Epstein, M. M., Firbank, L. G., Guerche, P., Hejatko, J., Moreno, F. J., & De Sanctis, G. (2021). Assessment of genetically modified soybean GMB 151 for food and feed uses, under Regulation (EC) No 1829/2003 (application EFSA-GMO-NL-2018-153). *EFSA Journal, 19*(4), e06424.

Frene, R. L., Simpson, D. M., Buchanan, M. B., Vega, E. T., Ravotti, M. E., & Valverde, P. P. (2018). Enlist E3™ soybean sensitivity and Enlist™ herbicide-based program control of Sumatran fleabane (*Conyza sumatrensis*). *Weed Technology, 32*(4), 416–423.

Gage, K. L., Krausz, R. F., & Walters, S. A. (2019). Emerging challenges for weed management in herbicide-resistant crops. *Agriculture, 9*(8), 180.

Gardner, J. G., Nehring, R. F., & Nelson, C. H. (2009). Genetically modified crops and household labor savings in U.S. crop production. *AgBioForum, 12*, 303–312.

Hartzler, R., & Owen, M. (2016). Dicamba and dicamba-resistant soybean varieties. *Integrated Crop Management News,* 2339.

Heap, I., & Duke, S. O. (2018). Overview of glyphosate-resistant weeds worldwide. *Pest Management Science, 74*(5), 1040–1049.

Huang, J., Ellis, C., Hauge, B., & Qi, Y. (2015). Herbicide tolerance. In K. Azhakanandam, A. Silverstone, H. Daniell, & M. Davey (Eds.), *Recent Advancements in Gene Expression and Enabling Technologies in Crop Plants*. Springer

ISAAA. (2014). *Global Status of Commercialized Biotech/GM Crops: 2014*.

ISAAA. (2017). *Global Status of Commercialized Biotech/GM Crops in 2017: Biotech Crop Adoption Surges as Economic Benefits Accumulate in 22 Years* (ISAAA Brief No. 53, pp. 25–26).

Jeanmart, S., Edmunds, A. J., Lamberth, C., Pouliot, M., & Morris, J. A. (2021). Synthetic approaches to the 2015–2018 new agrochemicals. *Bioorganic & Medicinal Chemistry, 39*, 116162.

Jhala, A. (2021). Managing soybeans with multiple herbicide-resistant traits.

Johnston, P. (2022). Exclusive, new Pioneer® brand A-Series Enlist E3® soybeans deliver the total performance package. *Corteva Agriscience (Iowa)*.

Klein, H., & Vidal Luna, F. (2021). The growth of the soybean frontier in South America: The case of Brazil and Argentina. *Revista de Historia Económica–Journal of Iberian and Latin American Economic History, 39*(3), 427–468. https://doi.org/10.1017/S0212610920000269.

Kumar, K., Gambhir, G., Dass, A., Tripathi, A. K., Singh, A., Jha, A. K., Yadava, P., Choudhary, M., & Rakshit, S. (2020). Genetically modified crops: Current status and future prospects. *Planta, 251*, 1–27.

Mahaffey, H., Taheripour, F., & Tyner, W. E. (2016). Evaluating the economic and environmental impacts of a global GMO ban (No. 333-2016-14338, pp. 1–34).

Marra, M., & Piggott, N. (2006). The value of non-pecuniary characteristics of crop biotechnologies: A new look at the evidence. In R. E. Just & J. M. Alston (Eds.), *Regulating Agricultural Biotechnology: Economics and Policy* (Vol. *30*, pp. 145–178). Springer.

Meier, C., Sampson, G., Larrea, C., Schlatter, B., Voora, V., Dang, D., Bermudez, S., Wozniak, J., & Willer, H. (2020). *The State of Sustainable Markets 2020: Statistics and Emerging Trends*.

Meyer, C. J., & Norsworthy, J. K. (2020). Timing and application rate for sequential applications of glufosinate are critical for maximizing control of annual weeds in LibertyLink® soybean. *International Journal of Agronomy, 2020*.

Nandula, V. K. (2010). *Glyphosate Resistance in Crops and Weeds: History, Development, and Management*. John Wiley & Sons.

Nandula, V. K. (2019). Herbicide resistance traits in maize and soybean: Current status and future outlook. *Plants, 8*(9), 337.

Padgette, S. R., Kolacz, K. H., Delannay, X., Re, D. B., LaVallee, B. J., Tinius, C. N., Rhodes, W. K., Otero, Y. I., Barry, G. F., Eichholtz, D. A., & Peschke, V. M. (1995). Development, identification, and characterization of a glyphosate-tolerant soybean line. *Crop Science, 35*(5), 1451–1461.

Padgette, S. R., Re, D. B., Barry, G. F., Eichholtz, D. E., Xavier, D., Fuchs, R. L., Kishore, G. M., & Fraley, R. T. (2018). New weed control opportunities: Development of soybeans with a Roundup Ready™ gene. In *Herbicide-Resistant Crops* (pp. 53–84). CRC Press.

Perry, E. D., Moschini, G., & Hennessy, D. A. (2016). Testing for complementarity: Glyphosate-tolerant soybeans and conservation tillage. *American Journal of Agricultural Economics, 98*(3), 765–784.

Que, Q., Chilton, M. D. M., de Fontes, C. M., He, C., Nuccio, M., Zhu, T., Wu, Y., Chen, J. S., & Shi, L. (2010). Trait stacking in transgenic crops: Challenges and opportunities. *GM Crops, 1*(4), 220–229.

Reade, J. P., & Cobb, A. H. (2002). Herbicides: Modes of action and metabolism. *Weed Management Handbook, 9*, 134–170.

Riar, D. S., Norsworthy, J. K., Steckel, L. E., Stephenson, D. O., & Bond, J. A. (2013). Consultant perspectives on weed management needs in midsouthern United States cotton: A follow-up survey. *Weed Technology, 27*(4), 778–787.

Riar, D. S., Norsworthy, J. K., Steckel, L. E., Stephenson, D. O., Eubank, T. W., & Scott, R. C. (2013b). Assessment of weed management practices and problem weeds in the midsouth United States—soybean: A consultant's perspective. *Weed Technology, 27*(3), 612–622.

Robinson, A. P., Simpson, D. M., & Johnson, W. G. (2015). Response of aryloxyalkanoate dioxygenase-12 transformed soybean yield components to postemergence 2,4-D. *Weed Science, 63*, 242–247.

Rose, J. (2020). *Sensitivity of Enlist™ and Roundup Ready Xtend® Technologies to Auxin Herbicides and Comparison of Resistance to Susceptible Cotton and Soybean Cultivars*. University of Arkansas.

Smyth, S. J., Gusta, M., Belcher, K., Phillips, P. W., & Castle, D. (2011). Environmental impacts from herbicide-tolerant canola production in Western Canada. *Agricultural Systems, 104*(5), 403–410.

Song, Y. (2014). Insight into the mode of action of 2,4-dichlorophenoxyacetic acid (2,4-D) as an herbicide. *Journal of Integrative Plant Biology, 56*(2), 106–113.

Takano, H. K., & Dayan, F. E. (2020). Glufosinate-ammonium: A review of the current state of knowledge. *Pest Management Science, 76*(12), 3911–3925.

Takano, H. K., Beffa, R., Preston, C., Westra, P., & Dayan, F. E. (2020). A novel insight into the mode of action of glufosinate: How reactive oxygen species are formed. *Photosynthesis Research, 144*(3), 361–372.

Tu, M., Hurd, C., & Randall, J. M. (2001). Weed control methods handbook, The Nature Conservancy. *Bulletin of Environmental Contamination and Toxicology, 7*(1).

Umetsu, N., & Shirai, Y. (2020). Development of novel pesticides in the 21st century. *Journal of Pesticide Science, 45*(2), 54–74.

United States Department of Agriculture (USDA). (2018). *Recent Trends in GE Adoption.* United States Department of Agriculture, Economic Research Service. Retrieved from: https://www.ers.usda.gov/data-products/adoption-of-genetically-engineered-crops-in-the-us/recent-trends-in-ge-adoption (accessed on 25 March 2025).

United States Department of Agriculture (USDA). (2022). *Adoption of Genetically Engineered Crops in the US.* United States Department of Agriculture, Economic Research Service. Retrieved from: https://www.ers.usda.gov/data-products/adoption-of-genetically-engineered-crops-in-the-u-s/recent-trends-in-ge-adoption/ (accessed on 25 March 2025).

Watson, D. (2019). *Genetically Modified Crops in Agriculture*. Scientific e-Resources.

Wright, T. R., Shan, G., Walsh, T. A., Lira, J. M., Cui, C., Song, P., Zhang, Z. M., Arnold, N. L., Lin, G., Yau, K., & Russell, S. M. (2010). Robust crop resistance to broadleaf and grass herbicides provided by aryloxyalkanoate dioxygenase transgenes. *Proceedings of the National Academy of Sciences, 107*(47), 20240–20245.

Zaidi, S. S. E. A., Mahas, A., Vanderschuren, H., & Mahfouz, M. M. (2020). Engineering crops of the future: CRISPR approaches to develop climate-resilient and disease-resistant plants. *Genome Biology, 21*(1), 1–19.

Zilberman, D., Sexton, S. E., Marra, M. C., & Fernandez-Cornejo, J. (2010). The economic impact of genetically engineered crops. *Choices, 25*(316-2016-6937), 1–13.

CHAPTER 8

Molecular Tools for the Improvement of Oil Quality in Soybeans

VISWABHARATHY SAKTHIVEL,[1] CHANDRASEKARAN DEEPIKA,[1] PONNUCHAMY MAGUDEESWARI,[2] SUBRAMANIAN DIVYA,[1] PRANAY GUNDALA,[3] RAMAN PUSHPA,[4] MURUGESAN DHANDAPANI,[4] and ARAVINDAN SHANMUGAM[1,*]

[1]*Department of Genetics and Plant Breeding, Tamil Nadu Agricultural University, Coimbatore, Tamil Nadu, India*

[2]*School of Crop Improvement, College of Post Graduate Studies in Agricultural Sciences, Central Agricultural University, Imphal, Manipur, India*

[3]*Department of Genetics and Plant Breeding, Indira Gandhi Krishi Vishwavidyalaya, Raipur, Chhattisgarh, India*

[4]*Tamil Nadu Rice Research Institute, Tamil Nadu Agricultural University, Aduthurai, Tamil Nadu, India*

*Corresponding author

ABSTRACT

Soybean (*Glycine max* (L.) Merr.) is an important commercial crop with low input responsiveness, grown for protein-rich vegetable meat and oil purposes across the world. Protein content, fatty acids (FAs), and moisture levels are key quality parameters of soybean. Structural genomic studies on QTLs governing these traits have been conducted. Precise mapping of genomic regions involved in oil synthesis, using SNPs that regulate unsaturated FA proportions, has revealed the complexity of these traits

Soybeans, Volume 2: Genome, Biotechnology, Stress, and Disease.
R. Ambika Rajendran, Ayyagari Ramlal, Sanjay Kumar Lal, and Godfree Chigeza (Eds.)

DOI: 10.1201/9781998511730-10

through genome-wide association studies (GWAS). These loci are found across diverse cultivar groups in the germplasm lines for oleic acid (OA, 18:1), linoleic acid (LLA, 18:2), and linolenic acid (LNA, 18:3). Germplasm lines vary widely in the concentration and proportion of OA, LLA, and LNA, and their synthesis is often suppressed in high-accumulating lines of each FA type. GWAS has explained the genetic architecture of unsaturated fatty acid synthesis (FAS). Integrated omics analysis has confirmed candidate genes, their transcription factors, and microRNAs, along with phylogenetic and co-expression studies of the *PEPC* gene family. QTLs for oil synthesis are not stable due to environmental and genotypic variation. Therefore, integrating metabolomics, transcriptomics, and both structural and functional genomics is vital for breeding high-yielding soybean cultivars with enhanced oil content, unsaturated FAs, and improved storability.

8.1 INTRODUCTION

A major problem in the 21st century will be the development of a sustainable food supply following the increasing global population. By 2050, the predicted global population will be above 9 billion, necessitating a doubling of food production to meet the growing demand (Varshney et al., 2013; Zhou et al., 2015). Soybean is referred to as the miracle bean, super legume, golden bean, or crop of the globe due to its wide range of applications as a crucial source of protein (40–42%), vitamins, minerals, and edible oil (18–22%). All of the necessary amino acids (AAs) that the human body cannot produce can be found in soybeans (Kanchana et al., 2015; Nandakishor et al., 2017). The protection against kidney disease is greatly aided by soy (Messina, 1995). Bioactive compounds from soybean extracts serve as a potential ingredient in many drug formulations against diabetes, obesity, and spinal cord-related disorders (Ramlal et al., 2022, 2023). It is a significant oilseed crop in the international market, contributing more than 50% to the production of vegetable oil. Comparatively speaking, to competitive oils like olive (75%) and canola (61%), soybean oil has much less monounsaturated fatty acids (FAs) (around 24%) (Teres et al., 2008; White, 2007). Monounsaturated fats are thought to be healthier, and oils high in them often have a longer shelf life and better oxidative stability. For developing better soybean varieties in the future, it is essential to comprehend the process of genetic improvement (Zhou et al., 2015). Increased soybean production worldwide is a result of technological advancements. With a natural outcrossing of 0.5–1%, soybeans are classified as self-pollinated plants. Given their autogamous nature, plant breeding techniques like single pod descent, pedigree, population breeding, and single pod descent have been used to genetically improve quality traits and create high-yielding varieties of this significant oilseed crop (Tripathi et al., 2016). The development of soybean seed qualities, which are difficult to manage because of their multi-gene complexity and environmental influence, would benefit especially from the use of molecular tools. In hybrid development programs, molecular markers are extremely useful for estimating the genetic diversity and organization of the available genotype and for predicting heterosis based on the polymorphism and genetic distance between the parents. In addition to traditional breeding techniques, scientific advancements in the field of structural and functional plant genomics will result in the emergence of novel soybean varieties with the necessary distinctiveness to improve agronomic qualities. Bringing together biotechnology and plant breeding can undoubtedly bring about a "yellow revolution" in oilseed crops like soybeans, improving seed quality in the process. This chapter discusses the importance of soybean as a global commercial crop, genomic studies on quality parameters, the complexity of oil synthesis through SNP mapping and GWAS, the diverse variations in unsaturated FAs, and the future goal of sustainable oil production through omics approaches and breeding high-yield, unsaturated FA-rich soybean cultivars.

8.2 BREEDING METHODS FOR OIL QUALITY IMPROVEMENT

Improving the oil content and quality traits in soybean through the conventional breeding approach is one of the oldest and best-known methods. There is a huge demand for non-GMO soybean cultivars that are high-yielding and early-maturing, which consist of high oleic content (Chen et al., 2020). In general, the soybean oil quality is determined by the FA composition, which affects low oxidative stability, off-flavor, and rancidity. Modification of soybean oil composition to reduce polyunsaturated FAs can be a viable alternative to improving oil stability, flavor, and eliminating the need for hydrogenation. Soybean oil content is governed by 2–3 major genes that show the multiple allelism mode of inheritance of altered FA traits, though in this case, they have no additive effects but act independently (Velasco et al., 2013).

The most common plant breeding methods used are pedigree selection, single seed descent method, hybridization, recurrent selection, and mutational breeding (Kulkarni et al., 2021). Still, the traditional pedigree method and the single-seed descent method (SSD) have been used successfully and most often in soybean breeding (Goulden, 1939). The highest content of oleate that has been achieved by conventional breeding is greater than 70% (Alt et al., 2005).

A primary method is the screening of soybean lines from the germplasm collection of *Glycine* species that contain elevated levels of oleic acid (Pham et al., 2010). Further, the best elite lines are crossed to enhance genetic variation. Progenies of these crosses segregate genetically in the course of subsequent generations of selfing, and new genotypes are thus formed. The progenies with the most advantageous combinations of the desired features are identified using a variety of selection techniques. The most common of these methods are pedigree selection, which involves visually selecting the best-looking families in each generation, followed by within-family selection of one or more plants to advance to the next generation; and single-seed descent, which entails advancing one seed or pod from each plant to the next generation to develop nearly homozygous lines that still preserve the majority of the original genetic variation in a population (Miladinovic et al., 2011).

Hybridization between species of the same genus or even between species from closely related genera is a useful approach for incorporating traits that are not present in the crop species. Wild relatives are the sources of new genes and harbor many useful genes, which are a potential reservoir of diversity. The perennial *Glycine tomentella* and the annual *Glycine soja* have been hybridized successfully with the domesticated soybean to produce breeding lines suitable for yield testing (Ma and Nelson, 2012).

In mutational breeding, physical and chemical mutagens are treated, which induce the alteration of FA composition. The stable soybean mutants that have high oleic content, oxidative stability, and nutritional traits are identified by recurrent selection and evaluation. The most common mutagens are gamma radiation and ethyl methane sulfonate (EMS) (Patil et al., 2007). Physical mutagens tend to cause larger genomic abnormalities than some chemical mutagens, and more dominant or more easily observable traits could be created at a higher frequency of gamma rays (Jankowicz Cieslak and Till, 2015).

8.3 QUANTITATIVE TRAIT LOCI (QTL) MAPPING APPROACHES

The identification, isolation, and even transfer of advantageous genes into elite germplasm without passing on unfavorable loci have been made feasible by the use of molecular markers in plant genetics and breeding (Tanksley and McCouch, 1997). In addition, quantitative trait loci (QTL) or chromosomal regions associated with seed oil concentration and other important agronomic and seed composition traits in soybean have been discovered using molecular markers over the past two decades. These QTLs may be utilized in marker-assisted selection (MAS) programs (Qi et al., 2011). Seed yield, protein concentration, seed size, and maturity are just a few of the agronomic and seed composition characteristics that have been linked to the co-localization of many seed oil QTLs in soybean (Palomeque et al., 2009). Mapping populations derived from parental lines with large differences in target traits or from plant introductions and exotic germplasm were used in the majority of studies aimed at detecting QTLs associated with important agronomic and seed traits (Hyten et al., 2004). Major QTLs for the trait under investigation may be detectable in populations with large parental differences, but minor QTLs that may be obscured by major QTLs may not be as easily found in such populations (Asins, 2002).

Identifying the number of stable QTL for these quality traits in soybean is critical to understanding the genetic factors; however, prior studies had two shortcomings: a low-density genetic map and a lack of QTL integration. In previous studies, the majority of QTL related to soybean isoflavone, protein, or oil content were based on low-density genetic maps with low-throughput molecular markers like restriction fragment length polymorphism (RFLP), variable number of tandem repeats (VNTR), and simple sequence repeat (SSR). This made QTL mapping inefficient and inaccurate. With the advancement of science and DNA sequencing technologies, QTL mapping can now be carried out using high-density genetic maps and high-throughput molecular markers. A powerful method for building a high-density genetic map based on the large-scale identification of single nucleotide polymorphism (SNP) markers is high-throughput sequencing, a next-generation sequencing (NGS) technology (Depristo et al., 2011), and it is a useful tool for mapping QTL. The integration of QTL is an effective method for obtaining stable QTL through comparing and combining the QTL from various studies because the QTL are primarily derived from diverse populations and environments in various studies. It is also crucial to comprehend the complex quantitative characters.

Researchers in the medical, social, and behavioral sciences have primarily utilized meta-analysis, a technique for combining data from multiple sources into a single study (Borenstein et al., 2009). A modified Akaike criterion that could be used to determine which QTL was represented by the QTL acquired from various studies was presented by Goffinet and Gerber (2000), who described a method for combining QTL mapped from multiple independent experiments. Meta-QTL, the QTL that was combined from multiple experiments through the use of meta-analysis and has shorter confidence intervals than the original QTL, may be more accurate and representative.

8.4 QUANTITATIVE TRAIT LOCI FOR OIL CONTENT AND QUALITY TRAITS

As of right now, the Soybase database (Grant et al., 2010) indicates that 315 QTLs are connected to soybean oil content. Additionally, 5 QTLs are linked to soybean oil and protein (Chen et al., 2007).

Akond et al. (2014), using 5376 SNP markers from the Illumina Infinium SoySNP6K Bead Chip array, identified the QTL controlling the content of protein, oil, and FAs in a set of F5:8 RILs derived from a cross between the lines "MD 96-5722" and "Spencer." On LG-B2 (Chr_14), one notable QTL for protein (*qPro001*) was found with LOD scores of 4.13. Around 11 QTLs for oil content (*qOil001* to *qOil011*) were found. Two of these QTLs were on LG-N (Chr_3), one on LG-A1 (Chr_5), one on LG-K (Chr_9), and one on LG-F (Chr_13). Among the remaining six, three QTLs were found on LG-B2 (Chr_14) and LG-J (Chr_16). The LOD scores for oil QTL varied from 2.51 to 4.67. There were 16 identified QTLs for FAs: one for stearic acid (*qSte001* on LG-J/Chr_16), three for oleic acid (*qOle001* on LG-F/Chr_13), and two for palmitic acid (*qPal001* on LG-N/Chr_3 and *qPal002* on LG-G/Chr_18); three for linoleic acid (*qLin001* on LG-N/Chr_3; *qOle002* and *qOle003* on LG-J/Chr_16); and seven for linolenic acid (*qLinn001* and *qLinn002* on LG-F/Chr_13; *qLinl002* and *qLinl003* on LG-J/Chr_16) on LG-B2/Chr_14 by *qLinn003*; on LG-E/Chr_15, *qLinn004* and *qLinn005*; and on LG-J/Chr_16, *qLinn006* and *qLinn007*.

Using a 181 RIL population derived from a wild soybean and a cultivated soybean, Yao et al. (2020) constructed a high-density genetic map with 11,398 SNP markers and identified 24 stable QTLs for soybean seed oil content and FA composition using SLAF-seq technology. One of these QTLs, *qPA10_1*, is a novel locus because it did not contain any QTNs associated with palmitic acid and did not overlap with or come close to any QTLs that had been previously reported.

8.5 META QUANTITATIVE TRAIT LOCI FOR OIL QUALITY TRAITS

Important quality characteristics in soybean breeding include isoflavone, soy protein, and soybean oil because these compounds are essential to human health. The whole-genome re-sequencing method was used by Chen et al. (2021) to create a high-density genetic map and plot the QTLs with short intervals. The public transcriptome data were used for the expression analysis, and several important genes were identified. They looked into the molecular origins of isoflavones for the first time using the meta-analysis technique. Based on 1034 QTL gathered from earlier studies, 212 meta-QTL comprising 122 stable and consentaneous QTL were mapped. In their research, 70 meta-QTL linked to isoflavones were mapped for the first time. Additionally, it was discovered that 69 and 73 meta-QTLs were linked to protein and oil, respectively.

8.6 MODERN MOLECULAR TOOLS TO IMPROVE OIL QUALITY IMPROVEMENT

8.6.1 GENE EDITING AND GENETIC ENGINEERING APPROACHES

Oil crops, such as sunflower, soybean, rapeseed, and oil palm, have attained the highest status in global agriculture over the past decades by providing valuable resources for mankind (Vollmann and Rajcan, 2010). With the increase in the demand for oil crop products, one of the promising ways to alleviate the supply pressure of oil crop products is to bring elite varieties with modified agronomic characteristics under cultivation. Although traditional breeding strategies are available, advanced biotechnological approaches could yield fruitful results in creating genetic variation as well as in developing crop varieties with significant improvements in quality characteristics. One such approach is gene editing technology, which has been widely employed in oil crops (He et al., 2021).

It has been used in elevating monounsaturated FAs and saturated FAs while decreasing the long-chain FA content. The *FAD2* gene desaturates 18:1 to 18:2 (Subedi et al., 2020). Hence, mutating the gene would result in the accumulation of 18:1, which is a desirable oil quality trait. Recently, using genome editing (GE), *FAD2* genes were mutated precisely to achieve increased levels of 18:1 in seed oils. Soybean with enriched 18:1 content has been derived from TALEN-based GE, which is already available in the United States market (Subedi et al., 2020). To lower the long-chain FAs, the *FAE1* gene has been mutated, which catalyzes the elongation of 18:1 to 22:1. RNAi-mediated silencing of two genes, *viz., FAE1* and *FAD2*, effectively lowered the 22:1 and elevated the 18:1 in oilseed crops (Li et al., 2016; Peng et al., 2010).

In soybean, a gene-knocking system was employed to create mutants with increased oleic acid content and decreased linoleic acid by editing the *FAD2-2* gene (Al Amin et al., 2019). *GmFATB1* has been modified to reduce the amount of saturated FAs in seeds (Ma et al., 2021). In rapeseed, oleic acid content was increased by editing *BnFAD2* (Huang et al., 2020; Okuzaki et al., 2018). The gene *BnLPAT2/5* caused wizened and enlarged oil bodies (Zhang et al., 2019), while Yan et al. (2021) observed a reduction in C18:1 and an increase in C18:2 content in the oil. This was made possible by employing a gene editing strategy for *BnaLEC1.*

Engineering agronomically significant oilseeds to produce non-native FAs with nutritional or industrial value, such as n-3 VLC-PUFA, HFA, and epoxy FAs, is gaining popularity in addition to increasing oil content and modifying the levels of existing FAs. Even though some plants have the biosynthetic pathways to create these uncommon FAs in high concentrations on a natural basis, most are not appropriate for large-scale agronomic production, which has restricted the supply of their seed oils. As a result, a lot of research is being done now to try and create these novel FAs in species where they are not generally found. For instance, the health-promoting VLC-PUFA docosapentaenoic acid has been successfully bred into Arabidopsis and camelina to yield levels (>12%) comparable to fish oil. The number of genes that would need to be altered to produce sufficiently high levels of the desired FA would make this technique technically difficult, but the rate at which the CRISPR/Cas toolkit is now increasing may open the door for such initiatives. Despite the challenges in efficiently obtaining the CRISPR-based knockout lines in some oil crops, especially sesame, sunflower, and groundnut, with ever-increasing newly developed CRISPR/Cas systems and their modified derivative editors, and delivery methods such as carbon nanomaterials-based delivery of genetic elements (Demirer et al., 2019), GE to improve the oil content and quality will flourish shortly.

8.6.2 *NEXT GENERATION SEQUENCING*

Three expressed *CaFAD2* genes were subjected to EMS mutagenesis in Crambe, followed by Illumina sequencing, to test for mutations (Cheng et al., 2015). To find mutation locations, two original analysis techniques were applied. In the first one, mutations targeted specific sequence motifs. The second method involved searching for alterations at each nucleotide location in a *CaFAD2* fragment. In 1100 one-dimensional pools (11,000 individuals), 17 new mutations were found in the three expressed *CaFAD2* genes, including nonsense and missense variants in *CaFAD2*-C1, -C2, and -C3. In comparison to the wild type, the homozygous *CaFAD2*-C3 nonsense mutants had a 25% greater level of C18:1 and a 25% lower concentration of PUFA. The research concluded that the use of targeted mutation detection using NGS in a polyploid revealed that even a single nonsense mutation in the *CaFAD2* gene can alter the composition of Crambe oil. Further alterations in C18:1 and PUFA concentrations may result from stacking the mutations in various *CaFAD2* genes (Cheng et al., 2015). If none exhibit detrimental agronomic features, crossings between *CaFAD2*-C1 and -C3 nonsense mutants can further improve oil quality for Crambe. This was clear in the soybean double *FAD2* mutant, which had more C18:1 than the single-gene mutants (Pham et al., 2010).

8.7 OMICS APPROACH

8.7.1 *TRANSCRIPTOMICS*

It is essential to understand the current developments in the omics-level understanding of de novo fatty acid synthesis (FAS), FA buildup, and protein storage during seed development in oilseed crops like rapeseed/canola (*Brassica napus*) and soybean. Transcriptome data from multiple soybean seed developmental stages were analyzed to identify lipid biosynthesis-related gene expression patterns (O'Rourke et al., 2014). These analyses revealed that genes like *FAD2-2B* and *FAD2-2C* were highly expressed in the early stages of seed formation and that *FAD2-1B* and *FAD2-1A* were highly expressed in the later stages of seed development. In a different study, RNA-Seq was used to examine potential genes in pod tissues from different developmental stages of *B. napus* cultivars implicated in biological processes associated with oil (5–7, 15–17, and 25–27 DAF). The genes were discovered to be involved in lipid, AA, and carbohydrate metabolism (Xu et al., 2015).

8.7.2 PROTEOMICS

According to Niu et al. (2009), *B. napus* (Huyou 15) transcriptomes had more copies of the genes involved in the glycolysis pathway than Arabidopsis. In contrast to 3-ketoacyl-ACP synthase, which had only two copies, one of the genes for the FAS pathway's acyl carrier protein 1 was present in 16 copies. The results suggested that FAS genes are transcriptionally regulated. The observations revealed that *B. napus* had a higher flux of carbon from sucrose to glycolysis and ultimately to de novo FAS. This discrepancy may be the cause of the higher oil content of rapeseed than soybean (Agrawal et al., 2008; Niu et al., 2009). Also, it was shown that the rapeseed seeds in development had significantly higher levels of the Rubisco large subunit (Agrawal et al., 2008). To investigate differences in protein composition and oil body features, the proteomes of ripe seeds from two winter-type *B. napus* lines, low (36.49%; LO) and high (55.19%; HO) in oil content, were studied (Gan et al., 2013). The findings demonstrated that late embryogenesis proteins (LEA; Dong et al., 2004) were expressed differently in HO and LO lines and may be associated with the quantity and size of oil bodies.

8.7.3 METABOLOMICS

During leaf detachment, phloem peeling, and silique darkening stages, a total of 443 potential metabolites were found in the oil accumulation stage in the seeds and silique walls of canola. The metabolite profiles demonstrated that a subset of genes involved in FA synthesis (FAS) and sugar metabolism were activated by high concentrations of metabolites in seeds when they were transferred from silique walls, increasing the metabolic flux and, ultimately, the amount of seed oil (Tan et al., 2015). It is also possible to use metabolomics methods to compare the metabolomes of wild-type and transgenic (or mutagenesis-derived) canola seeds with seeds that have different FA compositions. Databases have been created for integrating, processing, and displaying diverse omics and phenotypic data from soybeans (http://soymetDB.org; http://soykb.org; Joshi et al., 2010, 2012). The possible results from numerous metabolic investigations conducted on diverse soybean tissues will serve as a foundation for increasing the production and quality of soybean seeds. There are well-established metabolic pathways for the de novo production of oil. There has been progress in identifying important enzymes or families that are numerous and abundant in higher oil crops. Yet, to genetically engineer the phenotype, the regulatory mechanisms of these pathways must be understood. Furthermore, in gaining insight into gene function and regulation in the context of the genome, omics research must be integrated with other genetic methodologies, such as gene mapping and GWAS.

8.7.4 GWAS APPROACHES

Genome-wide scans in soybean identified six key SNPs and the genes that were involved in various traits. Annotation of these genes has been made. For example, the lead SNP ss715619179 (*FAD3A* gene) was identified on chromosome 14 at position 45971865, which was associated with linolenic acid (Zhang et al., 2018). The major-effect loci contained the putative *MtN21* nodulin, *FATB*, and steroid-5-alpha-reductase involved in N_2 fixation, AA biosynthesis, and FA metabolism, as well as the candidate genes *GmSAT1*, *AK-HSDH*, *SACPD-C*, and *FAD3A* with known functions. Hence, GWAS studies offer useful genes and markers for improving soybean nutrient content both quantitatively and qualitatively.

Xiao et al. (2019) discovered 385,692 high-quality SNPs with a minor allele frequency (MAF) > 0.05 through high-throughput genome resequencing. They identified 17 loci that were significantly correlated with the amount of seed oil, of which 12 SNPs were distributed on the A3 (11 loci) and A1 (one locus) chromosomes, and four novel significant SNPs on the C5 (one locus) and C7 (four loci), respectively. A total of 64

DEGs associated with lipid metabolism were discovered, 14 of which are essential for the production and assembly of TAGs. Between high- and low-oil content *B. napus* accessions, insights have been made to reveal differences in the transcription levels of key genes involved in *de novo* FA production in the plastid, TAG assembly, and lipid droplet packaging in the endoplasmic reticulum (ER). Seven potential genes that were within the confidence interval of significant SNPs were discovered by combining the results of GWAS and transcriptome investigations (Xiao et al., 2019).

8.8 CONCLUSION AND FUTURE PROSPECTS

Soybean is a significant source of plant protein and oil for the consumption of humans and animals. Native soybeans' (20%) low seed oil content needs to be increased, and its FA profile falls short of various dietary or industrial requirements. It has taken a lot of work to bioengineer soybean oil, and substantial advancements have been made in these genetic alterations, such as the improvement of soybean oil that has been enhanced with oleic acid, linolenic acid, saturated FAs, or uncommon FAs. Combined multiple "omics" approaches, efficient gene editing techniques, and pangenomic sequencing will further shed light on the intricate mechanisms of TAG production and its control. This will greatly speed up genetic modification for oil improvement and increase the predictability of their positive and negative effects on soybean agricultural traits.

KEYWORDS

- **gene editing**
- **genome-wide association studies**
- **integrated omics**
- **oil synthesis**
- **oleic acid**
- **quantitative trait loci**
- **unsaturated fatty acids**

REFERENCES

Agrawal, G. K., Hajduch, M., Graham, K., & Thelen, J. J. (2008). In-depth investigation of the soybean seed-filling proteome and comparison with a parallel study of rapeseed. *Plant Physiology, 148*, 504–518.

Akond, M., Liu, S., Boney, M., Kantartzi, S. K., Meksem, K., Bellaloui, N., & Kassem, M. A. (2014). Identification of quantitative trait loci (QTL) underlying protein, oil, and five major fatty acids' contents in soybean. *American Journal of Plant Sciences, 5*(1), 158–167.

Al Amin, N., Ahmad, N., Wu, N., Pu, X., Ma, T., Du, Y., & Wang, P. (2019). CRISPR-Cas9 mediated targeted disruption of FAD2-2 microsomal omega-6 desaturase in soybean (*Glycine max* L.). *BMC Biotechnology, 19*, 9. https://doi.org/10.1186/s12896-019-0501-2.

Asins, M. J. (2002). Present and future of quantitative trait locus analysis in plant breeding. *Plant Breeding, 121*(4), 281–291.

Borenstein, M., Hedges, L. V., Higgins, J. P., & Rothstein, H. R. (2021). *Introduction to Meta-Analysis* (pp. 1–7). John Wiley & Sons.

Chen, H., Pan, X., Wang, F., Liu, C., Wang, X., Li, Y., & Zhang, Q. (2021). Novel QTL and Meta-QTL mapping for major quality traits in soybean. *Frontiers in Plant Science, 12*, 774270.

Chen, Q. S., Zhang, Z. C., Liu, C. Y., Xin, D. W., Qiu, H. M., & Shan, D. P. (2007). QTL analysis of major agronomic traits in soybean. *Agricultural Sciences in China, 6*, 399–405.

Cheng, J., Salentijn, E. M., Huang, B., Denneboom, C., Qi, W., Dechesne, A. C., Krens, F. A., Visser, R. G., & Loo, E. N. (2015). Detection of induced mutations in *Ca* FAD2 genes by next-generation sequencing leading to the production of improved oil composition in *Crambe abyssinica*. *Plant Biotechnology Journal, 13*(4), 471–481.

Demirer, G. S., Zhang, H., Matos, J. L., Goh, N. S., Cunningham, F. J., Sung, Y., Chang, R., Aditham, A. J., Chio, L., & Cho, M. J. (2019). High aspect ratio nanomaterials enable delivery of functional genetic material without DNA integration in mature plants. *Nature Nanotechnology, 14*, 456–464.

DePristo, M. A., Banks, E., Poplin, R., Garimella, K. V., Maguire, J. R., Hartl, C., & Daly, M. J. (2011). A framework for variation discovery and genotyping using next-generation DNA sequencing data. *Nature Genetics, 43*(5), 491–498.

Dong, J., Keller, W. A., Yan, W., & Georges, F. (2004). Gene expression at early stages of *Brassica napus* seed development as revealed by transcript profiling of seed-abundant cDNAs. *Planta, 218*, 483–491.

Gan, L., Zhang, C., & Wang, X. (2013). Proteomic and comparative genomic analysis of two *Brassica napus* lines differing in oil content. *Journal of Proteome Research, 12*, 4965–4978.

Goffinet, B., & Gerber, S. (2000). Quantitative trait loci: A meta-analysis. *Genetics, 155*, 463–473. https://doi.org/10.1093/genetics/155.1.463.

Goulden, C. H. (1939). Problems in plant selection. *Proceedings of the 7th International Genetics Congress*, Edinburgh, Scotland, 132–133.

Grant, D., Nelson, R. T., Cannon, S. B., & Shoemaker, R. C. (2010). SoyBase, the USDA-ARS soybean genetics and genomics database. *Nucleic Acids Research, 38*, D843–D846.

He, J., Zhang, K., Tang, M., Zhou, W., Chen, L., Chen, Z., & Li, M. (2021). CRISPR-based genome editing technology and its applications in oil crops. *Oil Crop Science, 6*(3), 105–113.

Huang, H., Cui, T., Zhang, L., Yang, Q., Yang, Y., Xie, K., Fan, C., & Zhou, Y. (2020). Modifications of fatty acid profile through targeted mutation at *BnaFAD2* gene with CRISPR/Cas9-mediated gene editing in *Brassica napus*. *Theoretical and Applied Genetics, 133*, 2401–2411.

Hyten, D. L., Pantalone, V. R., Sams, C. E., Saxton, A. M., Landau-Ellis, D., Stefaniak, T. R., & Schmidt, M. E. (2004). Seed quality QTL in a prominent soybean population. *Theoretical and Applied Genetics, 109*, 552–561.

Jankowicz-Cieslak, J., & Till, B. J. (2015). Forward and reverse genetics in crop breeding. In *Advances in Plant Breeding Strategies: Breeding, Biotechnology and Molecular Tools* (pp. 215–240).

Joshi, T., Patil, K., & Fitzpatrick, M. R. (2012). Soybean Knowledge Base (SoyKB): A web resource for soybean translational genomics. *BMC Genomics, 42*, D1245–D1252. https://doi.org/10.1186/1471-2164-13-S1-S15.

Joshi, T., Yao, Q., & Franklin, L. D. (2010). SoyMetDB: The soybean metabolome database. In *2010 IEEE International Conference on Bioinformatics and Biomedicine* (pp. 203–208). IEEE.

Kanchana, P., Santha, M. L., & Raja, K. D. (2015). A review on *Glycine max* (L.) Merr. (soybean). *World Journal of Pharmacy and Pharmaceutical Sciences, 5*(1), 356–371.

Kulkarni, K. P., Tayade, R., Jo, H., Song, J. T., & Lee, J. D. (2021). Breeding strategy for improvement of omega-3 fatty acid through conventional breeding, genetic mapping, and genomics in soybean. *Plant Breeding–Current and Future Views*.

Li, X., Mei, D., Liu, Q., Fan, J., Singh, S., Green, A., & Zhu, L. H. (2016). Down-regulation of *Crambe* fatty acid desaturase and elongase in *Arabidopsis* and *Crambe* resulted in significantly increased oleic acid content in seed oil. *Plant Biotechnology Journal, 14*, 323–331. https://doi.org/10.1111/pbi.12386.

Ma, J. (2012). *Evaluation of Fertile Lines Derived from the Hybridization of Glycine max and G. tomentella* (Doctoral dissertation, University of Illinois at Urbana-Champaign). University of Illinois at Urbana-Champaign.

Ma, J., Sun, S., Whelan, J., & Shou, H. (2021). CRISPR/Cas9-mediated knockout of *GmFATB1* significantly reduced the amount of saturated fatty acids in soybean seeds. *International Journal of Molecular Sciences, 22*, 3877.

Messina, M. (1995). Modern applications for an ancient bean: Soybeans and the prevention and treatment of chronic disease. *Journal of Nutrition, 125*(3), 567S–569S.

Miladinovic, J., Burton, J. W., Tubic, S. B., Miladinovic, D., Djordjevic, V., & Djukic, V. (2011). Soybean breeding: Comparison of the efficiency of different selection methods. *Turkish Journal of Agriculture and Forestry, 35*(5), 469–480.

Nandakishor, H., Kumar, P., & Mane, S. (2017). Transmission studies of soybean mosaic virus. *International Journal of Current Microbiology and Applied Sciences, 6*(4), 867–869.

Niu, Y., Wu, G. Z., & Ye, R. (2009). Global analysis of gene expression profiles in *Brassica napus* developing seeds reveals a conserved lipid metabolism regulation with *Arabidopsis thaliana*. *Molecular Plant, 2*, 1107–1122.

O'Rourke, J. A., Bolon, Y. T., Bucciarelli, B., & Vance, C. P. (2014). Legume genomics: Understanding biology through DNA and RNA sequencing. *Annals of Botany, 113*, 1107–1120.

Okuzaki, A., Ogawa, T., Koizuka, C., Kaneko, K., Inaba, M., Imamura, J., & Koizuka, N. (2018). CRISPR/Cas9-mediated genome editing of the fatty acid desaturase 2 gene in *Brassica napus*. *Plant Physiology and Biochemistry, 131*, 63–69.

Palomeque, L., Li-Jun, L., Li, W., Hedges, B., Cober, E. R., & Rajcan, I. (2009). QTL in mega-environments: II. Agronomic trait QTL co-localized with seed yield QTL detected in a population derived from a cross of high-yielding adapted × high-yielding exotic soybean lines. *Theoretical and Applied Genetics, 119*, 429–436.

Patil, A., Taware, S. P., Oak, M. D., Tamhankar, S. A., & Rao, V. S. (2007). Improvement of oil quality in soybean (*Glycine max* (L.) Merrill) by mutation breeding. *Journal of the American Oil Chemists' Society, 84*(12), 1117–1124.

Peng, Q., Hu, Y., Wei, R., Zhang, Y., Guan, C., Ruan, Y., & Liu, C. (2010). Simultaneous silencing of *FAD2* and *FAE1* genes affects both oleic acid and erucic acid contents in *Brassica napus* seeds. *Plant Cell Reports, 29*, 317–325. https://doi.org/10.1007/s00299- 010-0823-y.

Pham, A. T., Lee, J. D., Shannon, J. G., & Bilyeu, K. D. (2010). Mutant alleles of *FAD2-1A* and *FAD2-1B* combine to produce soybeans with the high oleic acid seed oil trait. *BMC Plant Biology, 10*, 195.

Qi, Z., Wu, Q., Han, X., Sun, Y., Du, X., Liu, C., Jiang, H., Hu, G., & Chen, Q. (2011). Soybean oil content QTL mapping and integrating with meta-analysis method for mining genes. *Euphytica, 179*(3), 499–514.

Ramlal, A., Bhat, I., Nautiyal, A., Baweja, P., Mehta, S., Kumar, V., Tripathi, S., Mahto, R. K., Saini, M., Mallikarjuna, B. P., & Saluja, S. (2023). *In silico* analysis of angiotensin-converting enzyme inhibitory compounds obtained from soybean (*Glycine max* (L.) Merr.). *Frontiers in Physiology, 14*, 1172684.

Ramlal, A., Nautiyal, A., Baweja, P., Kumar, V., Mehta, S., Mahto, R. K., Tripathi, S., Shanmugam, A., Pujari Mallikarjuna, B., Raman, P., & Lal, S. K. (2022). Angiotensin-converting enzyme inhibitory peptides and isoflavonoids from soybean (*Glycine max* (L.) Merr.). *Frontiers in Nutrition, 9*, 1068388.

Subedi, U., Jayawardhane, K. N., Pan, X., Ozga, J., Chen, G., Foroud, N. A., & Singer, S. D. (2020). The potential of genome editing for improving seed oil content and fatty acid composition in oilseed crops. *Lipids, 55*(5), 495–512.

Tan, H., Xie, Q., & Xiang, X. (2015). Dynamic metabolic profiles and tissue-specific source effects on the metabolome of developing seeds of *Brassica napus*. *PLOS ONE, 10*(4), e0124794. https://doi.org/10.1371/journal.pone.0124794.

Tanksley, S. D., & McCouch, S. R. (1997). Seed banks and molecular maps: Unlocking genetic potential from the wild. *Science, 277*(5329), 1063–1066.

Teres, S., Barcelo-Coblijn, G., Benet, M., Alvarez, R., Bressani, R., Halver, J. E., & Escriba, P. V. (2008). Oleic acid content is responsible for the reduction in blood pressure induced by olive oil. *Proceedings of the National Academy of Sciences, 105*, 13811–13816.

Tripathi, N., & Khare, D. (2016). Molecular approaches for genetic improvement of seed quality and characterization of genetic diversity in soybean: A critical review. *Biotechnology Letters, 38*(10), 1645–1654.

Varshney, R. K., Roorkiwal, M., & Nguyen, T. (2013). Legume genomics: From genomic resources to molecular breeding. *The Plant Genome, 6*, 1–7. https://doi.org/10.3835/plantgenome2013.12.0002in.

Velasco, L., & Fernández-Martínez, J. M. (2002). Breeding oilseed crops for improved oil quality. *Journal of Crop Production, 5*(1, 2), 309–344.

Vollmann, J., & Rajcan, I. (2010). *Oil Crop Breeding and Genetics* (pp. 1–30). Springer New York.

White, P. J. (2007). Fatty acid in oilseeds (vegetable oils). In *Fatty Acids in Foods and Their Health Implications* (pp. 210–263). CRC Press.

Xiao, Z., Zhang, C., Tang, F., Yang, B., Zhang, L., Liu, J., Huo, Q., Wang, S., Li, S., Wei, L., & Du, H. (2019). Identification of candidate genes controlling oil content by combination of genome-wide association and transcriptome analysis in the oilseed crop *Brassica napus*. *Biotechnology for Biofuels, 12*(1), 1–16.

Xu, H. M., Kong, X. D., Chen, F., Huang, J. X., Lou, X. Y., & Zhao, J. Y. (2015). Transcriptome analysis of *Brassica napus* pod using RNA-Seq and identification of lipid-related candidate genes. *BMC Genomics, 16*, 858. https://doi.org/10.1186/s12864-015-2062-7.

Yan, G., Yu, P., Tian, X., Guo, L., Tu, J., Shen, J., Yi, B., Fu, T., Wen, J., & Liu, K. (2021). DELLA proteins *BnaA6.RGA* and *BnaC7.RGA* negatively regulate fatty acid biosynthesis by interacting with *BnaLEC1s* in *Brassica napus*. *Plant Biotechnology Journal, 19*(10), 2011–2026. https://doi.org/10.1111/pbi.13628.

Yao, Y., You, Q., Duan, G., Ren, J., Chu, S., Zhao, J., & Jiao, Y. (2020). Quantitative trait loci analysis of seed oil content and composition of wild and cultivated soybean. *BMC Plant Biology, 20*, 1–13.

Zhang, J., Wang, X., Lu, Y., Bhusal, S. J., Song, Q., Cregan, P. B., Yen, Y., Brown, M., & Jiang, G. L. (2018). Genome-wide scan for seed composition provides insights into soybean quality improvement and the impacts of domestication and breeding. *Molecular Plant, 11*(3), 460–472.

Zhang, K., Nie, L., Cheng, Q., Yin, Y., Chen, K., Qi, F., Zou, D., Liu, H., Zhao, W., & Wang, B. (2019). Effective editing for lysophosphatidic acid acyltransferase 2/5 in allotetraploid rapeseed (*Brassica napus* L.) using CRISPR-Cas9 system. *Biotechnology for Biofuels, 12*, 1–18.

Zhou, Z., Jiang, Y., Wang, Z., Gou, Z., Lyu, J., & Li, W. (2015). Resequencing 302 wild and cultivated accessions identifies genes related to domestication and improvement in soybean. *Nature Biotechnology, 33*, 408–414. https://doi.org/10.1038/nbt.3096.

CHAPTER 9

In vitro Intervention in Soybean (*Glycine max* (L.) Merr.)

SHIVAM RAJ,[1] SUMIT KUMAR KASHYAP,[2] RAYEES AHMAD BHAT,[3] AYYAGARI RAMLAL,[4,5,*] SANJAY KUMAR LAL,[5] DHANDAPANI RAJU,[6] SREERAMANAN SUBRAMANIAM,[4,7,8] and AMBIKA RAJENDRAN[5,*]

[1]*Department of Life Sciences, Hansraj College, University of Delhi, New Delhi, Delhi, India*

[2]*Department of Molecular and Cellular Engineering, Jacob Institute of Biotechnology and Bioengineering, Sam Higginbottom University of Agriculture, Technology and Sciences, Naini, Prayagraj, Uttar Pradesh, India*

[3]*Department of Botany, University of Delhi, New Delhi, Delhi, India*

[4]*School of Biological Sciences, Universiti Sains Malaysia (USM), Georgetown, Penang, Malaysia*

[5]*Division of Genetics, Indian Council of Agricultural Research (ICAR)–Indian Agricultural Research Institute (IARI), Pusa Campus, New Delhi, India*

[6]*Division of Plant Physiology, ICAR–Indian Agricultural Research Institute, Pusa Campus, New Delhi, India*

[7]*Chemical Center Biology (CCB), Universiti Sains Malaysia (USM), Bayan Lepas, Penang, Malaysia*

[8]*Department of Biology, Faculty of Science and Technology, Universitas Airlangga, Surabaya, Indonesia*

*Corresponding author

Soybeans, Volume 2: Genome, Biotechnology, Stress, and Disease.
R. Ambika Rajendran, Ayyagari Ramlal, Sanjay Kumar Lal, and Godfree Chigeza (Eds.)

DOI: 10.1201/9781998511730-11

ABSTRACT

In vitro techniques and approaches in soybean play an essential role in crop research. Soybean is an agriculturally, industrially, and economically important wonder legume. Therefore, the methodologies of *in vitro* and genetic approaches will be critical in determining the trajectory of soybean agriculture in the future. The chapter aims to provide an overview of different *in vitro* approaches that have been employed in soybean so far, such as micro-propagation, production of doubled haploids (DHs), and genetic engineering. The chapter highlights the significance of these techniques in crops, which will eventually be helpful in crop advancement and improvement programs.

9.1 INTRODUCTION

Soybeans (*Glycine max* (L.) Merr.), a key member of the family Fabaceae, are a multipurpose food crop with 40% protein, 20% oil, and several macro- and micro-molecules (Muthu et al., 2023; Rajendran and Lal, 2020; Ramlal et al., 2022a, b; Rizzo and Baroni, 2018; Tiwari et al., 2021). Its global importance is apparent since it substantially contributes to the production of 25% of edible oil globally and two-thirds of the world's protein concentrate for livestock feed (Maranna et al., 2021; Vasil and Thorpe, 2013). Soybeans are used extensively in cooking and industries for making tofu, and so forth, containing bioactive substances with potential health benefits (Ramlal et al., 2022a, c, 2023a; Swallah et al., 2023). In order to choose desirable features and improve crops, haploids and doubled haploids (DHs) are essential in plant breeding (Dirks et al., 2009; Ferrie and Mollers, 2011; Germanà, 2011; Hale et al., 2022; Khan et al., 2010; Pratap and Gupta, 2007; Pratap et al., 2016; Ramlal et al., 2023b; Rudolf-Pilih et al., 2019; Seeja and Sreekumar, 2020; Singh et al., 2020; Wijnker et al., 2012; Zulkarnain, 2014).

Micropropagation, also known as tissue culture propagation, is a method employed for the asexual reproduction of plants. This technique involves taking minute portions of a plant, such as the shoot tips or leaves, and introducing them into a nutrient medium (Skrzypek et al., 2011). The nutrient-rich environment serves as a conducive setting for the stimulated growth of these plant segments, allowing them to develop into complete, independent plants. Through careful control of environmental factors like temperature, light, and nutrient concentrations, researchers can optimize conditions for the successful cultivation of these plant tissues (Amarasinghe and Yuesheng, 2005; Komatsuda, 1989; Muthu et al., 2023). This precise control allows for the rapid multiplication of plants, generating clones of the genetically identical parents. Micropropagation is particularly valuable in the mass production of elite plant varieties, ensuring consistency in desirable traits. Previous works by Tiwari et al. (2021) and Vasil and Thorpe (2013) likely provide further insights into the specific methodologies and advancements in the field of micropropagation. Another essential component of plant biotechnology is androgenesis, which is the process by which immature pollen or microspores develop into full-fledged plants. It contributes significantly to the advancement of plant science and technology by rerouting pollen development along an embryogenic pathway (Ramlal et al., 2024). On the other hand, gynogenesis, a portion of the female gametophyte that has not yet undergone fertilization, gives rise to sporophytes, which eventually produce haploid plants. This technique includes the culture of female gametophytes, ovules, and non-fertilized ovaries (He et al., 2014). Various factors influence the *in vitro* activities in soybean, including plant growth regulators (PGRs) such as auxins and cytokinins, physical parameters (light intensity, quality, and duration), carbon and nitrogen sources, and media composition (Franklin et al., 2004; Kim et al., 2001; Ramlal et al., 2024a).

This chapter will delve deeper into the intricacies of androgenesis, micropropagation, and genetic approaches, specifically exploring their applications in the context of soybean crop enhancement. By unraveling the unique advantages and challenges posed by these methodologies in relation to soybean cultivation, we seek to understand and harness the synergies between them.

9.2 MICROPROPAGATION IN SOYBEAN

Micropropagation refers to the asexual propagation of the plant using tissue culture techniques. This method involves taking small portions of the plant, such as the shoot tip or a leaf, and placing them in a nutrient medium to stimulate their growth into new plants (Tiwari et al., 2021; Vasil and Thorpe, 2013). The process typically consists of several stages, including the initiation of the culture, multiplication of the plant material, rooting of the plantlets, and acclimatization of the new plants to *in vitro* conditions. Micropropagation is an important tool for the rapid and large-scale production of disease-free and genetically uniform plants. It has been successfully used for the propagation of soybean through the regeneration of plantlets (Muthu et al., 2023; Tiwari et al., 2020). Additionally, it contributes to the augmentation of various advantageous qualitative traits in soybeans (Tiwari et al., 2021). In the context of soybean, the direct application of tissue culture methodologies, encompassing techniques like embryo rescue, creation of homozygous lines by the use of somatic hybridization, haploid technology, and the purposeful use of *in vitro* selection in conjunction with the deliberate exploration of somaclonal variations, is revealing substantial advantages. These sophisticated techniques offer targeted approaches for the screening of mutant lines characterized by heightened tolerance or resistance (Amarasinghe and Yuesheng, 2005; Komatsuda, 1989; Muthu et al., 2023). The integration of tissue culture and molecular biology is pivotal in advancing the enhancement of cultivated soybean. This cooperative strategy is crucial for the production of transgenic soybean varieties that exhibit tolerance or resistance to diverse biotic and abiotic factors (Muthu et al., 2023; Vasil and Thorpe, 2013). The primary challenge hindering these applications is the absence of standardized protocols for plantlet regeneration. The substantial body of research conducted worldwide underscores the interest in employing *in vitro* techniques to enhance soybean improvement (Table 9.1).

9.2.1 EMBRYO CULTURE

Embryo culture has evolved into a well-established field within plant tissue culture (PTC). Notably, in soybean, successful regeneration has been achieved through the utilization of intact isolated zygotic embryos, as well as immature and mature embryos, along with their respective sections (Lazzeri et al., 1987a, b, 1988; Shan et al., 2005). This has proven to be an effective strategy for plant regeneration, involving both shoot morphogenesis and somatic embryogenesis (SE). Genetic variations were mostly observed for embryo culture in soybeans (Amarasinghe and Yuesheng, 2005; Vasil and Thorpe, 2013).

9.2.2 SOMATIC EMBRYOGENESIS (SE)

Somatic embryogenesis (SE) is a remarkable phenomenon wherein embryos are regenerated from somatic cells, tissues, or organs, either *de novo* or directly from the tissues (adventive origin) (Komatsuda, 1990;

TABLE 9.1 Summary of Experiments Conducted by Various Investigators on Soybean Micropropagation

Cultivars	Nature of Explant	Media	Remarks	References
Cotyledon/Mature/ Immature	Immature cotyledons	NAA (6.25, 12.5, 25.0, or 50.0 mg L^{-1}) and Suc (0.5, 1, 2, or 4%)	Better somatic embryo formation with low to moderate levels of Suc	Lazzari et al. (1988)
–	Cotyledons	MS + 90.4 µM 2,4-D + B5 + 0.6 µM IBA + 0.36 µM GA	Embryoid formation, maturation, and plantlet regeneration	Ferreira et al. (1990)
–	Cotyledons	Basal medium + BA + NAA + 2,4-D	Frequency of embryoids increased with 2,4-D instead of 10.0 mg L^{-1} NAA	Komatsuda (1989)
–	Cotyledons	Basal medium + 2,4-D	Genotype and auxin proved important factors	Griga (1993)
Northern-adapted soybean cultivars	Cotyledons	MS + 2,4-D + B5 vitamins + Suc (43.5 mM)	Somatic embryogenesis affected	Tian et al. (1994)
–	Immature cotyledons	MS + 100 mg L^{-1} myo-inositol + 30.0 g L^{-1} Suc with: L1 – 5.0 mg L^{-1} 2,4-D; L2 – 9.3 mg L^{-1} NAA; L3 – 10.0 mg L^{-1} NAA + 0.1 mg L^{-1} BA	L1 was found most effective for somatic embryo development	Nawaracala and Konieczny (1996)
Bragg, IAS-5, RS-7	Cotyledons	Basal medium + ethylene inhibitors	Ethylene inhibitors may aid plant recovery in soybean genotypes	Santos et al. (1997)
–	Mature cotyledons	MS + NAA, organic nitrogen supplements, and IBA	Higher IBA enhances frequency of somatic embryogenesis	Fu et al. (1997)
–	Immature cotyledons	MS salts + B5 vitamins – 6% Suc + 40 mg L^{-1} 2,4-D; +15 µM $AgNO_3$	Highest embryo induction from abaxial side; $AgNO_3$ accelerated secondary embryo formation	Santarem et al. (1997)
Kaohsiung 10 (cv. K10)	Immature cotyledons	MS + 164.8 µM 2,4-D, 5 mM asparagine, 6% Suc, 684 µM glutamine	Plants were morphologically normal and fertile	Loganathan et al. (2010)

TABLE 9.1 *(Continued)*

Cultivars	Nature of Explant	Media	Remarks	References
Cheongja 3, Jinpumkong 2, Taekwangkong, Uram	Cotyledon and hypocotyl segments	MS + 0–4 mg L^{-1} BA and TDZ	*Taekwangkong* showed highest regeneration via organogenesis and callus induction	Kim et al. (2016)
CO1, CO2, CO3	Cotyledon	MS + auxin and cytokinin: 2,4-D (5 μM), kinetin (10 μM)	Callus induction from cotyledonary segments, esp. CO3	Muthu et al. (2023)
Embryo Culture	Immature embryos	MS + NAA, 2,4-D, BA, IBA (alone or combined)	Somatic embryogenesis and plant regeneration	Lazzeri et al. (1987a)
–	Immature embryos	MS + B5 + NAA + Suc	SE increased with decreased sugar concentration	Lazzeri et al. (1987b)
–	Immature zygotic embryo	Basal medium + 40 mg L^{-1} 2,4-D + 6% Suc	Somatic embryo initiated after 4 weeks	Finer (1988)
–	Immature embryos (3–30 DAP)	MS + B5 + 0.1 mg L^{-1} NAA + 0.1 mg L^{-1} IBA	Formation of callus, shoots, roots, and embryos	Yeh (1989)
–	Immature zygotic embryo	L2 medium + various growth regulators	Somatic embryogenesis observed	Sellars et al. (1990)
–	Immature embryonic axes	Basal medium + various Suc levels	Culture response affected by genotype × Suc interaction	Komatsuda et al. (1991)
–	Immature embryonic axes	½ MS + BA + IBA	Multiple shoot formation	Yeh and Chyuan (1991)
–	Embryos	Basal medium + various Suc concentrations	Genotype × Suc interaction significant; 60 mg L^{-1} Suc required for embryo maturation	Komatsuda (1992)
–	Mature cotyledons	MS basal medium supplemented with 5.0 mg. L^{-1} PCPA. 3.0 mg. L^{-1} PCPA + 0.5 mg. L^{-1} BAP 3.0 mg L^{-1} BAP + 0.5 mg L^{-1} NAA	High *in vitro* morphogenesis and plantlet regeneration depending on genotype	Tripathi and Tiwari (2004)

TABLE 9.1 *(Continued)*

Cultivars	Nature of Explant	Media	Remarks	References
–	Mature/Immature cotyledons, Cotyledonary node	MS + 0.29 μM CA + 2.69 μM NAA	Synergistic effect of BAP and TDZ on regeneration	Franklin et al. (2004)
Various Explants	Shoot apices	B5 + 0.5–1.0 mg L^{-1} IBA; B5 + 6% Suc (no growth regulators)	High-frequency shoot formation, low callusing in 2nd passage	Kothari et al. (1991)
–	Anthers	B5 + 0.5 mg L^{-1} BAP + 0.5 mg L^{-1} 2,4-D	Embryoid formation observed	Zhuang et al. (1991)
–	Immature embryonic axes + cotyledons	MS + 2.5 mg L^{-1} 2,4-D + 30 g L^{-1} Suc + 8.0 g L^{-1} agar	Casein hydrolysate promoted callus/SE; axes more responsive	Yeh and Chyuan (1992)
–	Young inflorescence	B5 + 2.0 mg L^{-1} 2,4-D + 0.5 mg L^{-1} BA	Somatic embryos resembled zygotic ones	Kaltchuk-Santos et al. (1997)
–	Mature cotyledons + hypocotyls (5–6 day-old seedlings)	MS + 0.25 mg L^{-1} TDZ (induction); MS + 1.5 mg L^{-1} BA + 0.5 mg L^{-1} kinetin (germination)	Mature cotyledons formed more embryoids; 63–83% germination	Gai and Guo (1997)
–	Half seed	B5 + BAP, 2,4-D, KIN, IBA, NAA (various)	Callus induction and regeneration highest with BAP (13.3 μM) + 2,4-D (13.5 μM)	Radhakrishnan and Ranjithakumari (2007)
–	Hypocotyls, cotyledons, cotyledonary nodes	MS/B5 + TDZ (0.9–5.4 μM)	TDZ functioned as cytokinin for shoot induction	Radhakrishnan et al. (2009)

Komatsuda and Kao, 1990). SE, apart from organogenesis, results in the formation of bipolar embryos having a closed radial end, in contrast to the monopolar structure observed in organogenesis. Originating from a single cell, the somatic embryo lacks any vasculature of the cultivated explants or the mother tissue (Komatsuda, 1992; Muthu et al., 2023). Somatic embryos, also referred to as embryoids, are intricate, small, bipolar structures that develop at the callus surface. Notably, they share a multitude of structural similarities with sexual or zygotic embryos found in the natural development of seeds. This process not only showcases the versatility of plant regeneration but also provides a fascinating alternative to traditional reproduction, offering insights into the intricate world of plant development (Amarasinghe and Yuesheng, 2005; Komatsuda et al., 1991; Muthu et al., 2023; Nawracala and Konieczny, 1996; Tiwari et al., 2021). There are several aspects related to SE in soybean, *viz.*, the influence of media components along with other physical and chemical parameters, nutritional status, carbon source (sucrose, etc.), PGRs, and pH (Cheng et al., 1980; Hofmann et al., 2004a, b; Lazzeri et al., 1987a, b, 1988).

9.2.3 PROTOPLAST CULTURE

The application of protoplast culture in soybean stands as a valuable tool. The enzymatic removal of cell walls from soybean plant cells yields isolated protoplasts. These protoplasts can then be cultured in a nutrient medium, facilitating their regeneration into whole plants or the development of Calli (Lazzeri et al., 1987a, b, 1988; Wright et al., 1987). Protoplast culture in soybean finds utility in diverse applications, encompassing genetic transformation and investigations into cell biology. This methodology enhances the efficiency of manipulating soybean cells, contributing to advancements in soybean breeding and the creation of genetically modified (GM) soybean plants with desired traits (Komatsuda and Ohyama, 1986, 1988). It finds extensive use in physiological research combined with genetic engineering and the direct introduction of foreign DNA. Plant protoplasts, in particular, open up exciting possibilities in somatic cell genetics and crop enhancement. Somatic hybridization is a technique wherein somatic protoplasts are fused in an *in vitro* environment, leading to the subsequent development of the plantlets.

9.2.4 HYPOCOTYL AND EPICOTYL CULTURE

Hypocotyl and epicotyl cultures (H/EC) have been regarded as beneficial for *in vitro* selection, transformation studies, and conserving the genetic pool. Kimball and Bingham (1973) provided the first report on soybean HC; however, they did not include shoot initiation. Subsequently, effective regeneration from soybean hypocotyl and epicotyl portions has been reported in various studies by Amarasinghe and Yang (2006), Cristea and Cachita-Cosma (1992), Dan and Reichert (1998), Kameya and Widholm (1981), Li et al. (1989), Tripathi and Tiwari (2003), Wright et al. (1987), and Yoshida (2002).

9.2.5 MERISTEM CULTURE

Meristem culture stands as a pivotal technique in soybean plant biotechnology. This method, involving the meticulous isolation and cultivation of meristematic tissue from shoot or root tips, plays a critical role in the production of virus-free soybean plants. Its application in clonal propagation ensures the replication of specific and desirable genetic traits essential for soybean crop improvement (Freytag et al., 1989; Hepher et al., 1988). Moreover, meristem culture is instrumental in the conservation of soybean germplasm, serving as a mechanism for preserving the genetic diversity of this economically significant crop. Its rejuvenating impact on aging soybean plants, promoting the development of vigorous shoots, underscores its utility (Freytag et al., 1989; Kartha et al., 1981; Lippmann and Lippmann, 1993). With its demonstrated efficiency in rapid multiplication, meristem culture assumes a strategic position in large-scale propagation initiatives, contributing to the advancement of soybean agriculture (Droste et al., 1993; Freytag et al., 1989; Hepher et al., 1988; Kartha et al., 1981; Lippmann and Lippmann, 1993).

9.2.6 ORGANOGENESIS

In PTC, a process referred to as organogenesis leads undifferentiated cells to grow into organs, including roots, shoots, and leaves. Using the aforementioned method, plant tissue is manipulated in a nutritional medium in order to promote the differentiation of cells into distinct organ structures. Contrary to SE, which produces embryos, organogenesis causes adult organs to develop straight from the cultured tissue (Barwale et al., 1986; Sairam et al., 2003; Shan et al., 2005). In the context of soybean, organogenesis assumes a crucial role in plant breeding initiatives by facilitating the clonal propagation of elite varieties and ensuring the stable maintenance of specific genetic traits (Kim et al., 2004; Sairam et al., 2003; Texeira et al., 2011). Within the domain of research and development, this technique has a pivotal role in the regeneration of transformed plants incorporating genetically engineered modifications. This contribution significantly drives advancements in crop improvement, playing a key role in the development of soybean varieties meticulously tailored to meet specific and evolving agricultural demands.

9.3 ANDROGENESIS AND GYNOGENESIS IN SOYBEAN

Androgenesis, a key component of plant biotechnology, is the process by which immature pollen or microspores develop into plants. Rerouting pollen along an embryogenic pathway contributes significantly to the advancement of plant science and technology for the development of haploids and DHs (Ramlal et al., 2024). Gynogenesis involves the development of sporophytes from an unfertilized element of the female gametophyte, resulting in haploid plants. This approach encompasses the culture of non-fertilized ovaries, ovules, or female gametophytes (He et al., 2014). Androgenesis in soybean and the formation of callus were first reported by Tang et al. (1973), which marked the beginning of the historical evolution of haploids in soybeans to date (Hale et al., 2022). Numerous investigations documented androgenesis in soybeans, aiming to create procedures for isolating (Rodrigues et al., 2006) and cultivating microspores (de Moraes et al., 2004; Rodrigues et al., 2004; Skinner and Liang, 1996; Tiwari et al., 2004; Zhao et al., 1998) (Tables 9.2 and 9.3). Androgenesis is influenced by several factors, such as temperature stress (Liu and Zhao, 1986; Rodrigues et al., 2005; Zhuang et al., 1991) and ultrasonication as a pretreatment of buds during sterilization for 20 minutes (Deswal, 2018; Hu et al., 1996), among others (for more details, refer to the review by Ramlal et al. (2024)). Figure 9.1 represents the factors affecting androgenesis and gynogenesis.

Soybean androgenesis has limitations that prevent its widespread use. These limitations include erratic activity in microspores, a preference for microspore embryogenesis, and morphological problems. Haploid technology is still in its infancy in soybeans, and its lack of standardization and uniform protocols makes it difficult to use effectively in breeding projects (Maranna et al., 2021; Ramlal et al., 2024; Swallah et al., 2023).

TABLE 9.2 Summary of Androgenesis in Soybean Using Anthers

Genotypes (Cultivars)	Media*	Remarks	References
Hark	Miller's + 20 mg L^{-1} NAA + 1 mg L^{-1} Kin + coconut milk; also prepared several other media and hormone combinations	Profuse callus formation with root initiation and downward movement of callus; no embryoids or plantlets observed	Deswal (2018); Ivers et al. (1974)
Chinese cultivars	–	Callus production from anthers; sporadic induction; uninucleate pollen	Yin (1981)
Chinese cultivars	B5 + 2 mg L^{-1} 2,4-D + 12% Suc	Callus developed from microspores with few shoots	Deswal (2018); Yin et al. (1982)
85 Chinese cultivars	Modified B5 + 2 mg L^{-1} 2,4-D + 2 mg L^{-1} BAP + 0.5 mg L^{-1} Kin + 12% Suc	Induction of haploid callus (36%) with shoot formation in 1 genotype (1% regeneration)	Deswal (2018); Jian et al. (1986)
Unknown	–	Callus and embryo-like structures (no data shown)	Kadlec et al. (1991)
Williams 82, Asgrow, A1929	B5 long + 2 mg L^{-1} 2,4-D + 0.5 mg L^{-1} BA + 9% Suc + 0.3% agarose	Embryo-like structures; abnormal	Deswal (2018); Zhuang et al. (1991)
Chinese cultivars and IUS line PI486355	Modified MS and B5 + 2 mg L^{-1} 2,4-D + 12% Suc; S: B5 + 0.5 mg L^{-1} NAA + 1 mg L^{-1} Kin + 1% Suc; S: Modified MS + 0.5 mg L^{-1} IBA + 0.5 mg L^{-1} BAP + 0.5 mg L^{-1} Kin + 0.5 mg L^{-1} zeatin + 5% Suc + 1% maltose	Callus induced (36%) in the best responsive genotypes; 84% were haploids when 15% Suc was supplied	Deswal (2018); Ye et al. (1994)
–	B5 long + 2 mg L^{-1} 2,4-D + 0.5 mg L^{-1} BAP + 9–12% Suc + 0.35% agarose	–	Deswal (2018); Hu et al. (1996)
IAS5, RS7	B5 long + 2 mg L^{-1} 2,4-D + 0.5 mg L^{-1} BAP + 9% Suc + 0.8% agarose	Callus frequency with embryos was 2%	Deswal (2018); Kaltchuk-Santos et al. (1997)

TABLE 9.2 *(Continued)*

Genotypes (Cultivars)	Media*	Remarks	References
110 tested genotypes (64 cultivated soybeans from Heilongjiang and Liaoning, 18 wild soybeans, 28 cultivated soybeans from America)	M1: MS + 2 mg L^{-1} 2,4-D + 2 mg L^{-1} NAA + 1 mg L^{-1} 6-BA + 0.01 mg L^{-1} TDZ + 12% Suc; M2: MS + 0.5 mg L^{-1} Kin + 0.5 mg L^{-1} 6-BA + 0.15 mg L^{-1} TDZ + 6% Suc (cultured in light); ½M3: ½ MS + 0.025 mg L^{-1} IBA + 0.01 mg L^{-1} 6-BA + 0.005 mg L^{-1} TDZ + 0.3 mg L^{-1} GA_3 + 0.5% AC + 2% Suc	26–68.1% callus induction; 4–18% embryos; 0.5–2% plant differentiation. Haploid plants from genotypes: LA88–408814, PI-9501, Buckshot 723, Changnong No. 4, Jilin Nos. 20, 32, 36, and 70203–I	Zhao et al. (1998)
Decada, IAS5, RS7	–	Variation in developmental stages of microspores within the same buds	Lauxen et al. (2003)
IAS5	B5 or B5 long + YSaa + 2 mg L^{-1} 2,4-D + 0.5 mg L^{-1} BAP + 9% Suc + 0.3% phytagel	B5 suggested as basal medium for culturing soybean anthers	Cardoso et al. (2004); Deswal (2018); Yeung and Sussex (1979)
IAS5, Bragg, Decada, RS7, BR4, BRS 133	Embryo Induction I: B5 + YSaa + 2 mg L^{-1} 2,4-D + 0.5 mg L^{-1} BAP + 9% Suc + 0.25% phytagel; II: As above + 1 mg L^{-1} 2,4-D + 1 mg L^{-1} BAP; Maturation I: MSM6; Maturation II: MS + B5 vitamins + 3% Suc + 0.25% phytagel; III: MSO + 1% Suc	Embryogenic calluses observed after 4 days	de Moraes et al. (2004); Deswal (2018); Yeung and Sussex (1979)
IAS5, F2 seeds (BRQ96–3065 × BRSMG–Liderança)	B5 long + YSaa + 2 mg L^{-1} 2,4-D + 0.5 mg L^{-1} BAP + 9% Suc + 0.8% agarose; SM: B5 + 1 mg L^{-1} 2,4-D + 3 mg L^{-1} BAP + 3% Suc	Higher auxin concentration + dark conditions led to embryogenic callus emergence	Deswal (2018); Rodrigues et al. (2004a, b)

TABLE 9.2 *(Continued)*

Genotypes (Cultivars)	Media*	Remarks	References
10 genotypes	B5DBIG, B5DB, B5DK, B5BKN (all with 9% Suc + 0.7% agar); RM: MS – 0.4 mg L^{-1} NAA + 0.4 mg L^{-1} BAP + 2% Suc + 0.8% agar	JS 90–41 had best response with B5DBIG; B5DBIG superior to other media	Deswal (2018); Tiwari et al. (2004)
Bragg, IAS5, MG/ BR-46–Conquista, BRS 133	B5 long + 10 ml YSaa + 2 mg L^{-1} 2,4-D + 0.5 mg L^{-1} BAP + 9% Suc + 0.25% phytagel	No positive response observed	Deswal (2018); Rodrigues et al. (2005a)
IAS5, BRS 133	B5 and B5 long + YSaa + 2 mg L^{-1} 2,4-D + 0.5 mg L^{-1} BAP + 9% Suc + 0.3% phytagel or modified PTA-15	Calluses and embryogenic calluses formed after 45 days; gelling agents showed no significant difference	Cardoso et al. (2007); Deswal (2018)
Merubetiri, Wilis	MS + IAA + 2,4-D/NAA + Kin/BAP under 50 µmol m^{-2} s^{-1}, 16 h/day at 25 ± 1°C (fluorescent light)	Callus proliferation observed in 5–18 days; Merubetiri more responsive	Zulkarnain (2014)

Abbreviation: 2,4-D: 2,4-Dichlorophenoxyacetic acid; AC: Activated charcoal; BAP/BA: 6-Benzylaminopurine; B5: Gamborg medium; B5DBIG: B5 + 2,4-D + IBA + myo-inositol + glutamine; Kin: Kinetin; IBA: Indole-3-butyric acid; IAA: Indole-3-acetic acid; GA_3: Gibberellic acid; MS/MSO: Murashige and Skoog medium; MSM6: Finer and McMullen medium; NAA: Naphthalene acetic acid; Suc: Sucrose; TDZ: Thidiazuron; YSaa: Yeung and Sussex amino acid supplement.

Source: Adapted from Ramlal et al. (2024a);

TABLE 9.3 Summary of Androgenesis in Soybean Using Microspores

Genotypes (Cultivars)	Media*	Remarks	References**
Chinese cultivars	Enriched B5 + 0.5–1.0 mg L^{-1} NAA, 0.1–0.5 mg L^{-1} zeatin	No data shown for callus	Deswal (2018); Liu and Zhao (1986)
Decada, IAS-5, Ipagro 21, BR4	Studied for pollen dimorphism (not cultured)	Pollen dimorphism observed; most were normal, some were P-pollen (smaller, lightly stained, with one or two similar nuclei)	Kaltchuk-Santos et al. (1993)
Bragg, MG/BR-46 Conquista, IAS-5, BRSMT Uirapuru	Modified PTA-15 + 8 mg L^{-1} agarose gel	Calluses were formed	Deswal (2018); Rodrigues et al. (2006)
IAS-5, Embrapa-1	BNN (initial cultural medium); macronutrients and vitamins of NN medium; micronutrient salts of B5; 2% Suc + 2% sorbitol; pH 5.8; various hormone concentrations (2,4-D + NAA + Pic and/or BA, ABA)	Callus formed; heart-shaped embryo formation also observed	Hale et al. (2019)
IAS-5, Maverick, Williams 82	BNN (initial cultural medium); 2% Suc + 2% sorbitol; also used coconut water and ABA	Globular, rough-surfaced pro-embryos formed; some developed into heart-shaped embryos with bilateral symmetry	Hale et al. (2020)
Pioneer 93Y21	BNN induction medium + 2 mg L^{-1} 2,4-D + 0.1 mg L^{-1} BA + 2% Suc + 2% sorbitol	Microspores and gynoecium-derived trichomes used; trichomes led to pseudo-embryo formation (false positive in androgenic response)	Hale et al. (2021)

*Different groups used different media for culturing.

** For complete details, refer to Ramlal et al. (2024).

Abbreviations: 2,4-D: 2,4-Dichlorophenoxyacetic acid; ABA: Abscisic acid; BA/BAP: 6-Benzylaminopurine; B5: Gamborg medium; BNN: Nitsch and Nitsch medium with reduced nitrogen (modification of Nitsch and Nitsch (1969) medium); °C: Degree Celsius; d: days; hr: hours; IBA: Indole-3-butyric acid; GA_3: Gibberellic acid (A_3); Kin: Kinetin; mg L^{-1}: milligrams per liter; NAA: Naphthalene acetic acid; NN: Nitsch and Nitsch medium; Pic: Picloram; PTA-15: Skinner and Liang medium; Suc: Sucrose.

Source: Adapted from Ramlal et al. (2024a); ©2024 Society for In Vitro Biology. Reproduced with permission.

Furthermore, the small number of research teams studying soybean androgenesis limits the field's potential progress. Novel approaches for *in vitro* organogenesis can be employed through pollen irradiation and CRISPR. However, in this context, gynogenesis in soybean has been explored recently and needs much attention (Ramlal et al., 2023b). To highlight the necessity of standardized protocols in soybean for various applications, the ultimate goal is to benefit soybean breeding programs and related investigations, such as backcross breeding, bulked segregant analysis, quantitative trait loci (QTL) analysis, and cultivar development (Ferrie and Mollers, 2011; Germanà, 2011; Kaneda et al., 1997; Zulkarnain, 2014).

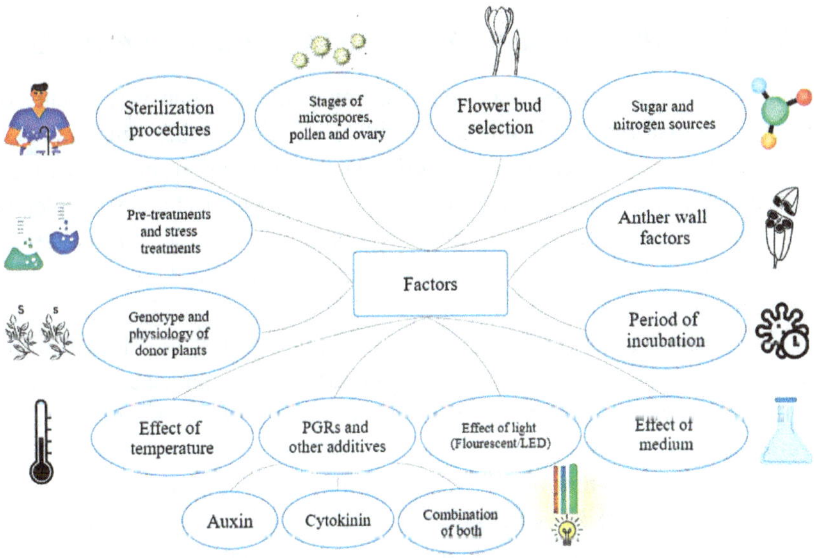

FIGURE 9.1 Factors affecting androgenesis and gynogenesis in soybean.

9.4 CROP IMPROVEMENT IN SOYBEAN: A TRANSGENIC APPROACH

On a global scale, soybean oil and food have a huge demand, and the demand is increasing day by day. On the other hand, factors such as climate change, abiotic and biotic stresses lead to a decline in yield. Therefore, to meet the demands, the development of improved cultivars with enhanced traits is needed (Amarasinghe and Yuesheng, 2005; Komatsuda, 1989; Muthu et al., 2023; Tiwari et al., 2020). In the last few decades, there has been significant development in the field of plant biotechnology. This has led to innovations in agricultural practices at a global level to improve crop yields. This technology, apart from making an impact on the agricultural market, also opens up new avenues for agriculture-related industrial applications of soybean (Barwale et al., 1986; Shan et al., 2005). As a result, soybean has received considerable attention in the last 20 years for genetic improvements, and genes have been manipulated to obtain desired traits. However, compared to crops like rice, it is still not as successful. This is because the production of transgenic soybean has been affected by limitations such as genotype specificity, low transformation efficiency, and complex protocols (Table 9.4).

9.4.1 HISTORICAL PERSPECTIVE AND LATEST DEVELOPMENTS

It has been three decades since the first transgenic soybean was produced, but stable transformation is still not considered routine. Stable transformation depends on the capability to combine efficient transformation and regeneration techniques. Two methods have shown relative success in producing completely and stably transformed plants: particle bombardment and the *Agrobacterium tumefaciens* method. Apart from these, transformation by *A. rhizogenes* has been used as a tool for functional studies (Homrich et al., 2012).

The genetic transformation of the soybean was first developed in the late 1980s. Using the meristems of immature soybean seeds, the first fertile transgenics were believed to be either produced by the particle bombardment method (McCabe et al., 1988) or by the regeneration of cotyledonary nodes that were infected with *A. tumefaciens* (Hinchee et al., 1988). Over the last 30 years, these methods have been continuously used to improve the soybean. The first herbicide-resistant transgenic soybean was commercialized in the mid-1990s. Since then, this transgenic soybean has been cultivated and grown commercially and has also been continuously improved using modern plant biotechnology. This has become one of the major commercially grown transgenics around the globe. Transgenic soybean, particularly the GM Roundup Ready® (which is resistant to glyphosate herbicides), has been grown in different countries, including the United States, Argentina, and Brazil (Pagano and Miransari, 2016), making this a leading transgenic crop. GM Roundup Ready soybean is resistant to herbicides, allowing farmers to spray the herbicide, killing the weeds without affecting the soybean. In total, about 105 million hectares of GM soybean were grown in 2017, and about 272 million metric

TABLE 9.4 Transgenic Approaches Used for the Enhancement of Potential Traits in Soybean

Factor/Trait	Targeted Gene	Result	Method Used	Reference
Salt stress / tolerance	Overexpression of AtMYB44 transcription factor	Dwarf phenotype with Improved salt and drought tolerance	*Agrobacterium tumefaciens* strain NTL4 mediated	Seo et al. (2012)
	Overexpression of *MSWRKY11* of alfalfa	Increased salt tolerance of soybean seedlings	*Agrobacterium*-mediated (strain EHA101)	Wang et al. (2018)
Drought tolerance	Overexpression of the ABA dependent *AREB1* transcription factor from *Arabidopsis thaliana*	Improved drought tolerance	Biolistic transformation methods	Barbosa et al. (2012)
	Overexpression of *LOS5/ABA3*	At least a 21% increase in seed yield of transgenic lines under drought stress	Agrobacterium mediated transformation	Li et al. (2013)
	Introduction of a transcription factor Dehydration responsive element binding *(DREB)* and ABA responsive element binding (*AREB*)	Showed improvement in the total number of pods, number of pods with seeds, and number of seeds at field level under drought stress	*Agrobacterium tumefaciens* EHA105 strain mediated	Pagliarini et al. (2020)
Modification in Oil and fatty acid content	*Δ6desaturase*	Converts LA and α-linolenic; GLA to ~27% and SDA to ~3% in seed oil (ALA) to GLA and SDA	*Agrobacterium-mediated*	Sato et al. (2004)
	Overexpression of fungal diacylglycerol acetyltransferase (*DGAT2*) protein in soybean	resulted in a 1.5% increase in seed oil content without any reduction in protein content		Lardizabal et al. (2008)
	Suppression of *GmFAD3*	Resulted in soybeans with very low linolenic acid (below 2%)		Flores et al. (2008)
	SLC1 was expressed in soybean somatic embryos under the control of seed specific phaseolin promoter	resulted in a 3.2% increase in the oil content of somatic embryos and transgenic plants with a 1.5% increase in seed oil content	–	Rao et al. (2009)

TABLE 9.4 *(Continued)*

Factor/Trait	Targeted Gene	Result	Method Used	Reference
Protein and amino acid content	Aspartokinase and dihydrodipicolinate acid synthase	Increase in free lysine more than 100 times and total seed lysine 5 times	*Agrobacterium-mediated*	Falco et al. (1995)
	β-Casein	Production of bovine *β-Casein*	Particle bombardment	Maughan et al. (1999)
	Overexpression the maize zein gene	increase in 15–35% cysteine and 12–20% methionine content.	*Agrobacterium-mediated*	Dinkins et al. (2001)
	Overexpressed the O-acetylserine sulfhydrylase (*OASS*) gene	Increase in 58–74% of protein-bound cysteine and 22–32% of free cysteine in the seed		Kim et al. (2012)
	Expression of cystathionine *γ*-synthase (*AtD-CGS*)	increase in methionine content 3.8–7-fold in developing transgenic green seeds and 1.8- and 2.3-fold in mature dry seeds		Song et al. (2013)
β-Carotene	Overexpression the phytoene synthase-2A-carotene desaturase (*PAC*) gene	~62-fold higher carotenoids than non-transgenic plants (77% *β*-Carotene)		Kim et al. (2012)
	Expression of *phytoene synthase gene (crtB)*	accumulated 845 *μ*g *β*-carotene g−1 dry seed weight	Particle bombardment method	Schmidt et al. (2015)
Virus tolerance	Coat protein precursor (*CP-P*) gene	Increased resistance	*Agrobacterium*-mediated	Die et al. (1996)
	Capsid polyprotein (pCP) gene	Increased resistance	Particle bombardment method	Reddy et al. (2001)
	Coat protein gene	Highly resistant to SMV (Soybean Mosaic Virus)	*Agrobacterium*-mediated	Wang et al. (2001)
	Overexpression of *GmAKT2*	Enhances resistance to SMV		Zhou et al. (2014)
	HC-Pro gene			Gao et al. (2015)

TABLE 9.4 *(Continued)*

Factor/Trait	Targeted Gene	Result	Method Used	Reference
Insect tolerance	*Cry1A(b)*	Increased insect resistance	Microprojectile bombardment	Parrott et al. (1994)
	EPSPS	Tolerant to herbicide glyphosate	Particle bombardment	Padgette et al. (1995)
	Cry1Ac	Increased insect resistance to lepidopteran populations	Particle bombardment	Stewart et al. (1996)
	Dicamba monooxygenase	Resistance to treatment with dicamba	*Agrobacterium*-mediated	Behrens et al. (2007)
	Cry1Ac	Increased insect resistance to *A. gemmatalis*	Particle bombardment	Homrich et al. (2008)
	Cry1A	Increased insect resistance to lepidopteran populations	*Agrobacterium*-mediated	Mcpherson and MacRae, 2009
Fungal tolerance	Chitinase (*CHI*) and the barley ribosome-inactivating protein (*RIP*)	Enhances resistance against fungal disease	*Agrobacterium*-mediated and particle bombardment	Li et al. (2004)
	Expression of oxalate decarboxylase (*OXDC*) gene	Enhances white mold resistance	Particle bombardment	Cunha et al. (2010)
	Antibody gene encoding *scFv*	Enhances resistance against foliar sudden death syndrome	*Agrobacterium*-mediated	Brar and Bhattacharyya, 2012
	Non-host resistance (*NHR*) gene	Enhances resistance to rust disease		Langenbach et a. (2016)
	Introduction of *hrf2*	Enhances resistance to *Phytopthera sojae*		Niu et al. (2019)

tons of seeds were produced, accounting for 80% of all world soybean production (Voora et al., 2020). Many different types of GM soybean have been developed with various targets, such as increasing protein quality (increasing sulfur-containing amino acids (AAs) and lysine by altering their biosynthetic feedback pathways) (Falco et al., 1995), increasing oleic acid content, improving yield, delaying flowering time (Yamada et al., 2012), and enhancing disease resistance, as well as abiotic and biotic stress resistance. Hence, the demand is increasing around the globe, as producing transgenic soybean varieties with better quality and yield is the area of focus for soybean researchers and breeders.

Genome editing (GE) technologies, particularly the clustered regularly interspaced short palindromic repeats (CRISPR)-associated (Cas) technology, are recently being employed in studying the genetics of soybean and commercial trait development (reviewed in Xu et al., 2020). However, the use of these GE technologies on plants has mostly relied on efficient transformation systems and on the regeneration of plants that contain edited events (Gao, 2021; Ran et al., 2017). Thus, to use these new technologies, the key to obtaining success in soybean is an efficient and genotype-flexible transformation system. Unfortunately, soybean remains recalcitrant to routine transformations when compared with other major crops such as rice (Chen et al., 2018). Because of the low rate of transformation and inflexibility of genotype, the production of transgenic soybeans is limited. Genes responsible for biotic and abiotic stress tolerance and yield-related characteristics have already been identified; they are being targeted in soybean for crop improvement using different methods. Apart from that, whole-genome sequencing, the availability of genomic information, and new technologies have helped to increase the understanding of the connection among the genes, their functioning, and their important use in the improvement of soybean germplasm.

9.4.2 AGROBACTERIUM-MEDIATED AND PARTICLE BOMBARDMENT-MEDIATED GENETIC TRANSFORMATIONS IN SOYBEAN

Both the *Agrobacterium*-mediated and particle bombardment transformation methods have been used with relative success to produce stable transformants. In addition, transformation using *A. rhizogenes* is also being utilized as a good tool for functional studies (Homrich et al., 2012; Li et al., 2014; Wang and Xu, 2008; Yan et al., 2000). Furthermore, there are some other transformation methods for transient assays used in soybean, such as protoplast transfection, agro-infiltration, electroporation, etc.

1. **Particle Bombardment Method**: Since 1988, when the first transgenic soybean plant was produced using the particle bombardment method (Christou et al., 1988), this method has been continuously used to produce soybean transformants. Aragão et al. (2000), Christou et al. (1989), and Christou and McCabe (1992), described soybean transformation by this method, where meristems were used as the target tissues. Recently, Soto et al. (2017) showed that the transgenic soybean for herbicide tolerance (HT), through shoot regeneration, was tolerant to 1.25 kg/ha of glyphosate (Soto et al., 2017). Therefore, much emphasis should be given, and more investigation is required for a better understanding and development of transgenics for other traits of interest as well.

2. **Agrobacterium Tumefaciens-Mediated Transformation**: The transformation of soybean using this *Agrobacterium*-mediated method was first carried out by Hinchee et al. (1988) using the cotyledonary nodes. The success of this method is considerably dependent on factors like the nature of the culture medium (if it has adequate antioxidants, PGR combinations specific to genotype), the strain used, explant condition, etc. The procedure of *Agrobacterium*-mediated transformation is simple and low-cost (Hwang et al., 2017), and it has a comparatively high ratio of single-copy gene insertion. This method is used to induce transgenic roots in the soybean, which are then used to understand the functioning of genes that are needed in different processes such as symbiotic and pathogenic interactions, nutrient uptake, and transportation of hormones within the roots (Kereszt et al., 2007). Many groups have worked (Lee et al., 2011; Liu et al., 2004; Paz et al., 2004; Yan et al., 2000) and are still working to understand and unravel the potential of genetic transformation in soybean.

3. **Protoplast Transfection**: The first successful genetic transformation of soybean protoplasts was done by electroporation by Lin et al. (1987). Dhir et al. (1992) were the first to report the transformation of protoplasts derived from the immature cotyledon and the Calli of this transfected protoplast transgenic plant. However, regeneration of the

protoplast through this method has yet to be achieved in soybean, and therefore, this method is usually used to evaluate gene functions.

4. **Other Methods**: Electroporation was used by Christou et al. (1988) to deliver gene constructs into the callus of soybean and was able to successfully integrate the genes but could not successfully regenerate the plant. Later, Chowrira et al. (1995) used a different strategy by applying this method to an intact nodal meristem, but unfortunately, no transgenic plants were recovered. The agro-infiltration can be infiltrated into the plant tissue intercellular space, and from there, the gene will be delivered into the plant genome (Grimsley et al., 1986). This method was first successfully achieved for soybean by King et al. (2015), and since then, this method has been used for virus-induced gene silencing (Kim et al., 2016) and in the expression of hairpin RNA for RNAi against two-spot spider mites (Dubey et al., 2017). Other genetic engineering approaches, such as zinc finger nucleases (ZFN), transcription activator-like effector nucleases, RNA-guided CRISPR/Cas9, along with CRISPR activation and interference, should be used in developing novel varieties with economic significance (Lee et al., 2013; Ramlal et al., 2024b). General methods of in vitro propagation and transformation are shown in Figure 9.2.

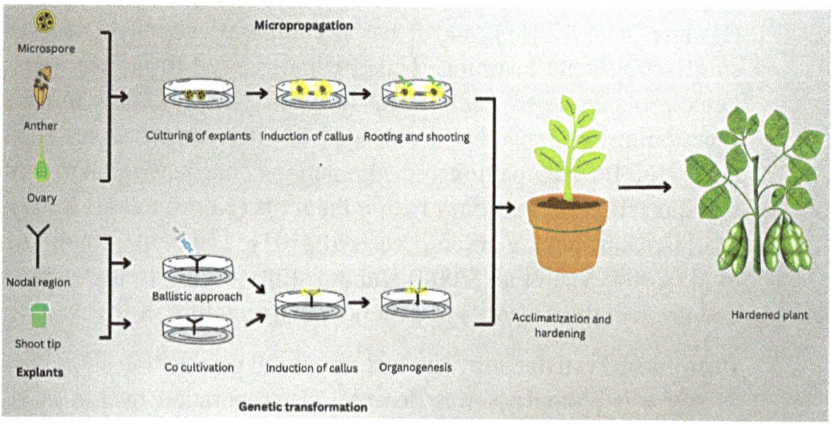

FIGURE 9.2 Schematic representation of a general process of micropropagation and genetic transformation in soybean.

9.5 CONCLUSION AND FUTURE PROSPECTS

There are several viable applications of in vitro techniques in soybean. The utilization of tissue culture techniques has significant implications in conservation efforts, offering a controlled and efficient means of generating vast numbers of plants possessing the desired qualities. The improvement of in vitro repetitive regeneration protocols from sophisticated explants like anthers, microspores, and immature embryos can provide wide-reaching impacts in soybean breeding.

KEYWORDS

- **doubled haploids**
- **genetic engineering**
- ***in vitro* techniques**
- **micropropagation**
- **plant regeneration**
- **soybean breeding**
- **soybean tissue culture**

REFERENCES

Agarwal, D. K., Billore, S. D., Sharma, A. N., Dupare, B. U., & Srivastava, S. K. (2013). Soybean: Introduction, improvement and utilization in India–problems and prospects. *Agricultural Research, 2*, 293 300.

Amarasinghe, A. A. Y., & Yuesheng, Y. (2005). Screening of high somatic embryogenic soybean varieties and innovation of culture methods. *Journal of South China Agricultural University (Hua nan Nong ye da xue xue bao), 26*(1), 84–88.

Aragão, F. J. L., Sarokin, L., Vianna, G. R., & Rech, E. L. (2000). Selection of transgenic meristematic cells utilizing a herbicidal molecule results in the recovery of fertile transgenic soybean (*Glycine max* (L.) Merrill) plants at a high frequency. *Theoretical and Applied Genetics, 101*, 1–6.

Arumuganathan, K., & Earle, E. D. (1991). Estimation of nuclear DNA content of plants by flow cytometry. *Plant Molecular Biology Reporter, 9*(3), 229–241.

Barbosa, E. G. G., Leite, J. P., Marin, S. R. R., Marinho, J. P., de Fátima Corrêa Carvalho, J., Fuganti-Pagliarini, R., et al. (2013). Overexpression of the ABA-dependent AREB1 transcription factor from *Arabidopsis thaliana* improves soybean tolerance to water deficit. *Plant Molecular Biology Reporter, 31*, 719–730.

Barwale, U. B., Kerns, H. R., & Widholm, J. M. (1986). Plant regeneration from callus cultures of several soybean genotypes via embryogenesis and organogenesis. *Planta, 167*, 473–481.

Behrens, M. R., Mutlu, N., Chakraborty, S., Dumitru, R., Jiang, W. Z., LaVallee, B. J., et al. (2007). Dicamba resistance: Enlarging and preserving biotechnology-based weed management strategies. *Science, 316*(5828), 1185–1188.

Brar, H. K., & Bhattacharyya, M. K. (2012). Expression of a single-chain variable-fragment antibody against a *Fusarium virguliforme* toxin peptide enhances tolerance to sudden death syndrome in transgenic soybean plants. *Molecular Plant-Microbe Interactions, 25*(6), 817–824.

Cardoso, M. B., Kaltchuk-Santos, E., Mundstock, E. C. D., & Bodanese-Zanettini, M. H. (2004). Initial segmentation patterns of microspores and pollen viability in soybean cultured anthers: Indication of chromosome doubling. *Brazilian Archives of Biology and Technology, 47*, 703–712.

Chen, L., Cai, Y., Liu, X., Yao, W., Guo, C., Sun, S., et al. (2018). Improvement of soybean *Agrobacterium*-mediated transformation efficiency by adding glutamine and asparagine into the culture media. *International Journal of Molecular Sciences, 19*(10), 3039.

Cheng, T. Y., Saka, H., & Voqui-Dinh, T. H. (1980). Plant regeneration from soybean cotyledonary node segments in culture. *Plant Science Letters, 19*(2), 91–99.

Chowrira, G. M., Akella, V., & Lurquin, P. F. (1995). Electroporation-mediated gene transfer into intact nodal meristems *in planta*: Generating transgenic plants without *in vitro* tissue culture. *Molecular Biotechnology, 3*, 17–23.

Christou, P., & McCabe, D. E. (1992). Prediction of germ-line transformation events in chimeric R0 transgenic soybean plantlets using tissue-specific expression patterns. *The Plant Journal, 2*(3), 283–290.

Christou, P., McCabe, D. E., & Swain, W. F. (1988). Stable transformation of soybean callus by DNA-coated gold particles. *Plant Physiology, 87*(3), 671–674.

Christou, P., Swain, W. F., Yang, N. S., & McCabe, D. E. (1989). Inheritance and expression of foreign genes in transgenic soybean plants. *Proceedings of the National Academy of Sciences, 86*(19), 7500–7504.

Cristea, V., & Cachita-Cosma, D. (1992). Studies on the reactivity of different soybean explant types (*Glycine max* L. Merr.) to *in vitro* culture conditions. *Revue Roumaine de Biologie. Série de Biologie Végétale, 37*(1), 73–81.

Cunha, W. G., Tinoco, M. L. P., Pancoti, H. L., Ribeiro, R. E., & Aragão, F. J. L. (2010). High resistance to *Sclerotinia sclerotiorum* in transgenic soybean plants transformed to express an oxalate decarboxylase gene. *Plant Pathology, 59*(4), 654–660.

Dan, Y., & Reichert, N. A. (1998). Organogenic regeneration of soybean from hypocotyl explants. *In Vitro Cellular & Developmental Biology–Plant, 34*, 14–21.

de Moraes, A. P., Bonadese-Zanettini, M. H., Callegari-Jacques, S. M., & Kaltchuk-Santos, E. (2004). Effect of temperature shock on soybean microspore embryogenesis. *Brazilian Archives of Biology and Technology, 47*, 537–544.

Deswal, K. (2018). Progress and opportunities in double haploid production in lentil (*Lens culinaris* Medik.), soybean (*Glycine max* L. Merr.), and chickpea (*Cicer arietinum* L.). *Journal of Pharmacognosy and Phytochemistry, 7*, 3105–3109.

Dinkins, R. D., Srinivasa Reddy, M. S., Meurer, C. A., Yan, B., Trick, H., Thibaud-Nissen, F., et al. (2001). Increased sulfur amino acids in soybean plants overexpressing the maize 15 kDa zein protein. *In Vitro Cellular & Developmental Biology–Plant, 37*, 742–747.

Dirks, R., Van Dun, K., De Snoo, C. B., Van Den Berg, M., Lelivelt, C. L., et al. (2009). Reverse breeding: A novel breeding approach based on engineered meiosis. *Plant Biotechnology Journal, 7*, 837–845.

Droste, A., Bodanese-Zanettini, M. H., & Ching, Y. H. (1993). Meristem culture of Brazilian soybean cultivars. *Revista Brasileira de Genetica, 16*, 135–135.

Dubey, V. K., Lee, U. G., Kwon, D. H., & Lee, S. H. (2017). Agroinfiltration-based expression of hairpin RNA in soybean plants for RNA interference against *Tetranychus urticae*. *Pesticide Biochemistry and Physiology, 142*, 53–58.

Falco, S. C., Guida, T., Locke, M., Mauvais, J., Sanders, C., Ward, R. T., & Webber, P. (1995). Transgenic canola and soybean seeds with increased lysine. *Nature Biotechnology, 13*(6), 577–582.

Ferrie, A. M., & Möllers, C. (2011). Haploids and doubled haploids in *Brassica* spp. for genetic and genomic research. *Plant Cell, Tissue and Organ Culture, 104*, 375–386.

Finer, J. J. (1988). Apical proliferation of embryogenic tissue of soybean (*Glycine max* L. Merrill). *Plant Cell Reports, 7*, 238–241.

Finer, J. J., & McMullen, M. D. (1991). Transformation of soybean via particle bombardment of embryogenic suspension culture tissue. *In Vitro Cellular & Developmental Biology–Plant, 27*, 175–182.

Flores, T., Karpova, O., Su, X., Zeng, P., Bilyeu, K., Sleper, D. A., et al. (2008). Silencing of *Gm FAD3* gene by siRNA leads to low α-linolenic acid (18:3) of *fad3*-mutant phenotype in soybean (*Glycine max* Merr.). *Transgenic Research, 17*, 839–850.

Franklin, G., Carpenter, L., Davis, E., Reddy, C. S., Al-Abed, D., Abou Alaiwi, W., & Sairam, R. V. (2004). Factors influencing regeneration of soybean from mature and immature cotyledons. *Plant Growth Regulation, 43*, 73–79.

Freytag, A. H., Rao-Arelli, A. P., Anand, S. C., Wrather, J. A., & Owens, L. D. (1989). Somaclonal variation in soybean plants regenerated from tissue culture. *Plant Cell Reports, 8*, 199–202.

Fuganti-Pagliarini, R., Marin, S. R. R., Daiane, M., Molinari, C., de Amorim Barbosa, D., Hugo Molinari, L. M. M., et al. (2020). Drought-tolerant soybean development: Evaluation of GM lines under greenhouse and field conditions. *Embrapa Soja–Capítulo Livro Científico*, 57–88.

Gai, J. Y., & Guo, Z. (1997). Efficient plant regeneration through somatic embryogenesis from germinated cotyledon of the soybean. *Soybean Genetics Newsletter, 24*, 41–46.

Gamborg, O. L., Miller, R. A., & Ojima, K. (1968). Nutrient requirements of suspension cultures of soybean root cells. *Experimental Cell Research, 50*, 151–158.

Gao, C. (2021). Genome engineering for crop improvement and future agriculture. *Cell, 184*(6), 1621–1635.

Gao, L., Ding, X., Li, K., Liao, W., Zhong, Y., Ren, R., et al. (2015). Characterization of *Soybean mosaic virus* resistance derived from inverted repeat-SMV-HC-Pro genes in multiple soybean cultivars. *Theoretical and Applied Genetics, 128*, 1489–1505.

Garda, M., Hale, B., Rao, N., Lowe, M., Bright, M., Goodling, S., & Phillips, G. C. (2020). Soybean androgenesis I: Identification of pyramidal stressors in anther cultures that sustain cell divisions and putative embryo formation from isolated microspore cultures. *In Vitro Cellular & Developmental Biology–Plant, 56*, 415–429.

Gatti, I., Guindón, F., Bermejo, C., Espósito, A., & Cointry, E. (2016). *In vitro* tissue culture in breeding programs of leguminous pulses: Use and current status. *Plant Cell, Tissue and Organ Culture, 127*, 543–559.

Germanà, M. A. (2011). Anther culture for haploid and doubled haploid production. *Plant Cell, Tissue and Organ Culture, 104*, 283–300.

Griga, M. (1993). Some factors affecting somatic embryogenesis efficiency in soybean (*Glycine max* L. Merrill). *Biologia Plantarum, 35*(2), 179–189.

Grimsley, N., Hohn, B., Hohn, T., & Walden, R. (1986). Agroinfection, an alternative route for viral infection of plants by using the *Ti* plasmid. *Proceedings of the National Academy of Sciences, 83*(10), 3282–3286.

Hale, B. M. (2020). *Induction and Characterization of Microspore Embryogenesis in Soybean (Glycine max L. Merrill)* (Doctoral dissertation, Arkansas State University, AR, US). ProQuest Dissertation & Theses. Retrieved from: https://www.proquest.com/openview/bb6ea909465affb5078c1064f5644a25/1?pqorigsite=gscholar&cbl=18750&diss=y (accessed on 25 March 2025).

Hale, B., Ferrie, A. M., Chellamma, S., Samuel, J. P., & Phillips, G. C. (2022). Androgenesis-based doubled haploidy: Past, present, and future perspectives. *Frontiers in Plant Science, 13*, 2944.

Hale, B., Lor, P., Chellamma, S., Samuel, J. P., & Phillips, G. C. (2021a). Gynoecium pubescence in soybean: A prevalent false-positive during *in vitro* androgenesis. *Plant Cell, Tissue and Organ Culture, 146*, 417–421.

Hale, B., Phipps, C., Kelley, C., & Phillips, G. C. (2019). Evaluation of soybean androgenesis by isolated microspore culture. In J. Ross (Ed.), *Arkansas Soybean Research Studies 2018* (pp. 31–37). University of Arkansas.

Hale, B., Phipps, C., Rao, N., Wijeratne, A., & Phillips, G. C. (2020b). Differential expression profiling reveals stress-induced cell fate divergence in soybean microspores. *Plants, 9*, 151.

Hassan, S. M. (2013). Soybean, nutrition, and health. In H. A. El-Shemy (Ed.), *Soybean-Bio-Active Compounds* (pp. 453–473). IntechOpen.

He, Z., Wang, X. X., Du, Y. C., Zhu, D. W., Guo, Y. M., Gao, J. C., et al. (2014). Haploid induction via *in vitro* gynogenesis in tomato (*Solanum lycopersicum* L.). *Journal of Integrative Agriculture, 13*(10), 2122–2131.

Hepher, A., Boulter, M. E., Harris, N., & Nelson, R. S. (1988). Development of a superficial meristem during somatic embryogenesis from immature cotyledons of soybean (*Glycine max* L.). *Annals of Botany, 62*(5), 513–519.

Hinchee, M. A., Connor-Ward, D. V., Newell, C. A., McDonnell, R. E., Sato, S. J., Gasser, C. S., et al. (1988). Production of transgenic soybean plants using *Agrobacterium*-mediated DNA transfer. *Nature Biotechnology, 6*(8), 915–922.

Hofmann, N. E., Raja, R., Nelson, R. L., & Korban, S. S. (2004b). Mutagenesis of embryogenic cultures of soybean and detecting polymorphisms using RAPD markers. *Biologia Plantarum, 48*, 173–177.

Hofmann, N., Nelson, R. L., & Korban, S. S. (2004a). Influence of media components and pH on somatic embryo induction in three genotypes of soybean. *Plant Cell, Tissue and Organ Culture, 77*, 157–163.

Homrich, M. S., Passaglia, L. M. P., Pereira, J. F., Bertagnolli, P. F., Pasquali, G., Zaidi, M. A., et al. (2008). Resistance to *Anticarsia gemmatalis* Hübner (Lepidoptera, Noctuidae) in transgenic soybean (*Glycine max* (L.) Merrill, Fabales, Fabaceae) cultivar IAS5 expressing a modified Cry1Ac endotoxin. *Genetics and Molecular Biology, 31*, 522–531.

Homrich, M. S., Wiebke-Strohm, B., Weber, R. L. M., & Bodanese-Zanettini, M. H. (2012). Soybean genetic transformation: A valuable tool for the functional study of genes and the production of agronomically improved plants. *Genetics and Molecular Biology, 35*, 998–1010.

Hu, C. Y., Yin, G. C., Helena, M., & Zanettini, B. (1996). Haploid of soybean. In S. M. Jain, S. K. Sopory, & R. E. Veilleux (Eds.), *Vitro Haploid Production in Higher Plants, Current Plant Science and Biotechnology in Agriculture* (Vol. *25*, pp. 377–395). Springer, Dordrecht.

Hwang, H. H., Yu, M., & Lai, E. M. (2017). *Agrobacterium*-mediated plant transformation: Biology and applications. *The Arabidopsis Book, 15*, e0186.

Isanga, J., & Zhang, G. N. (2008). Soybean bioactive components and their implications to health—A review. *Food Reviews International, 24*, 252–276.

Ivers, D. R., Palmer, R. G., & Fehr, W. R. (1974). Anther culture in soybeans. *Crop Science, 14*, 891–893.

Jian, Y. Y., Liu, D. P., Luo, X. M., & Zhao, G. L. (1986). Studies on induction of pollen plants in *Glycine max* (L.) Merr. *Chiangsu Nung Yeh Hsueh Pao/Journal of Agricultural Science, 2*, 26–30.

Joyner, E. Y., Boykin, L. S., & Lodhi, M. A. (2010). Callus induction and organogenesis in soybean (*Glycine max* (L.) Merr.) cv. Pyramid from mature cotyledons and embryos. *The Open Plant Science Journal, 4*(1).

Kadlec, M., Suchomelova, J., Smirnov, V. A., & Nikolajevna, S. L. (1991). Anther culture in soybean. *Soybean Genetics Newsletter, 18*, 121–124.

Kaltchuk-Santos, E., Bodanese-Zanettini, M. H., & Mundstock, E. (1993). Pollen dimorphism in soybean. *Protoplasma, 174*, 74–77.

Kaltchuk-Santos, E., Mariath, J. E., Mundstock, E., Hu, C. Y., & Bodanese-Zanettini, M. H. (1997). Cytological analysis of early microspore divisions and embryo formation in cultured soybean anthers. *Plant Cell, Tissue and Organ Culture, 49*, 107–115.

Kameya, T., & Widholm, J. (1981). Plant regeneration from hypocotyl sections of *Glycine* species. *Plant Science Letters, 21*(3), 289–294.

Kaneda, Y., Tabei, Y., Nishimura, S., Harada, K., Akihama, T., & Kitamura, K. (1997). Combination of thidiazuron and basal media with low salt concentrations increases the frequency of shoot organogenesis in soybeans (*Glycine max* (L.) Merr.). *Plant Cell Reports, 17*, 8–12.

Kantayos, V., & Bae, C. H. (2019). Optimized shoot induction and histological study of *in vitro* cultured Korean soybean cultivars. *Korean Journal of Plant Resources, 32*(3), 237–243.

Kartha, K. K., Pahl, K., Leung, N. L., & Mroginski, L. A. (1981). Plant regeneration from meristems of grain legumes: Soybean, cowpea, peanut, chickpea, and bean. *Canadian Journal of Botany, 59*(9), 1671–1679.

Kereszt, A., Li, D., Indrasumunar, A., Nguyen, C. D., Nontachaiyapoom, S., Kinkema, M., & Gresshoff, P. M. (2007). *Agrobacterium rhizogenes*-mediated transformation of soybean to study root biology. *Nature Protocols, 2*(4), 948–952.

Khan, S., Latif, A., Ahmad, S. Q., Ahmad, F., & Ghaffar, M. (2010). Double haploid technique: In soybean and other species. *International Journal of Applied, 5*, 649–655.

Kim, D. G., Kantayos, V., Kim, D. K., Park, H. G., Kim, H. H., Rha, E. S., et al. (2016). Plant regeneration by *in vitro* tissue culture in Korean soybean (*Glycine max* L.). *Korean Journal of Plant Research, 29*(1), 143–153. https://doi.org/10.7732/KJPR.2016.29.1.143.

Kim, K. H., Lim, S., Kang, Y. J., Yoon, M. Y., Nam, M., Jun, T. H., et al. (2016). Optimization of a virus-induced gene silencing system with *Soybean yellow common mosaic virus* for gene function studies in soybeans. *Plant Pathology Journal, 32*(2), 112.

Kim, K. H., Park, H. K., Park, M. S., & Yeo, U. D. (2001). Effects of auxins and cytokinin's on organogenesis of soybean *Glycine max* L. *Journal of Plant Biotechnology, 3*(2), 95–100.

Kim, M. J., Kim, J. K., Kim, H. J., Pak, J. H., Lee, J. H., Kim, D. H., et al. (2012). Genetic modification of the soybean to enhance the β-carotene content through seed-specific expression. *PLoS One, 7*(10), e48287.

Kim, W. S., Chronis, D., Juergens, M., Schroeder, A. C., Hyun, S. W., Jez, J. M., & Krishnan, H. B. (2012). Transgenic soybean plants overexpressing O-acetylserine sulfhydrylase accumulate enhanced levels of cysteine and Bowman–Birk protease inhibitor in seeds. *Planta, 235*, 13–23.

Kim, Y. J., Park, T. I., Kim, H. S., Park, H. K., Chon, S. U., & Yun, S. J. (2004). Factors affecting organogenesis from mature cotyledon explants and regeneration in soybean. *Journal of Plant Biotechnology, 6*(1), 39–43.

Kimball, S. L., & Bingham, E. T. (1973). Adventitious bud development of soybean hypocotyl sections in culture. *Crop Science, 13*(6), 758–760.

King, J. L., Finer, J. J., & McHale, L. K. (2015). Development and optimization of agroinfiltration for soybean. *Plant Cell Reports, 34*, 133–140.

Komatsuda, T. (1989). Somatic embryogenesis and plant regeneration in soybean. *Research Journal of Food and Agriculture, 12*, 8, 9.

Komatsuda, T. (1990). Ability of soybean (*Glycine max* L. Merrill) genotypes to produce somatic embryos on a medium containing a low concentration of sucrose. *Japanese Journal of Breeding, 40*, 371–375.

Komatsuda, T. (1992). Research on somatic embryogenesis and plant regeneration in soybean. *Bulletin of the National Institute of Agrobiological Resources, 1*, 78–113.

Komatsuda, T., & Kao, S. W. (1990). Screening of soybean (*Glycine max* L. Merrill) genotypes for somatic embryo production from immature embryo. *Japanese Journal of Breeding, 40*, 249–251.

Komatsuda, T., & Ohyama, K. (1986). Callus and somatic embryo formation by immature embryo culture in soybean. *Japanese Journal of Breeding, 36*, 230–231.

Komatsuda, T., & Ohyama, K. (1988). Genotypes of high competence for somatic embryogenesis and plant regeneration in soybean (*Glycine max*). *Theoretical and Applied Genetics, 75*, 695–700.

Komatsuda, T., Kaneko, K., & Oka, S. (1991). Genotype × sucrose interactions for somatic embryogenesis on soybean. *Crop Science, 31*, 333–337.

Kothari, S. L., Chandra, N., & Sharma, V. K. (1991). Multiple shoot formation from cotyledonary node and shoot apex cultures of *Glycine max*. *Soybean Genetics Newsletter, 18*, 148, 149.

Langenbach, C., Schultheiss, H., Rosendahl, M., Tresch, N., Conrath, U., & Goellner, K. (2016). Interspecies gene transfer provides soybean resistance to a fungal pathogen. *Plant Biotechnology Journal, 14*(2), 699–708.

Lardizabal, K., Effertz, R., Levering, C., Mai, J., Pedroso, M. C., & Jury, T., et al. (2008). Expression of *Umbelopsis ramanniana* DGAT2A in seed increases oil in soybean. *Plant Physiology, 148*(1), 89–96.

Lauxen, M. D. S., Kaltchuk-Santos, E., Hu, C. Y., Callegari-Jacques, S. M., & Bodanese-Zanettini, M. H. (2003). Association between floral bud size and developmental stage in soybean microspores. *Brazilian Archives of Biology and Technology, 46*, 515–520.

Lazzeri, P. A., Hildebrand, D. F., & Collins, G. B. (1987a). Soybean somatic embryogenesis: Effects of hormones and culture manipulations. *Plant Cell, Tissue and Organ Culture, 10*, 197–208.

Lazzeri, P. A., Hildebrand, D. F., & Collins, G. B. (1987b). Soybean somatic embryogenesis: Effects of nutritional, physical, and chemical factors. *Plant Cell, Tissue and Organ Culture, 10*, 209–220.

Lazzeri, P. A., Hildebrand, D. F., & Collins, G. B. (1987c). Soybean somatic embryogenesis: Effects of nutritional, physical, and chemical factors. *Plant Cell, Tissue and Organ Culture, 10*, 209–220.

Lazzeri, P. A., Hildebrand, D. F., Sunega, J., Williams, E. G., & Collins, G. B. (1988). Soybean somatic embryogenesis: Interactions between sucrose and auxin. *Plant Cell Reports, 7*, 517–520.

Lee, H., Park, S. Y., & Zhang, Z. J. (2013). An overview of genetic transformation of soybean. In *A Comprehensive Survey of International Soybean Research–Genetics, Physiology, Agronomy and Nitrogen Relationships*. IntechOpen.

Lee, K., Yi, B. Y., Kim, K. H., Kim, J. B., Suh, S. C., Woo, H. J., et al. (2011). Development of efficient transformation protocol for soybean (*Glycine max* L.) and characterization of transgene expression after *Agrobacterium*-mediated gene transfer. *Journal of the Korean Society for Applied Biological Chemistry, 54*, 37–45.

Li, C., Zhang, H., Wang, X., & Liao, H. (2014). A comparison study of *Agrobacterium*-mediated transformation methods for root-specific promoter analysis in soybean. *Plant Cell Reports, 33*, 1921–1932.

Li, D. W., Schmid, J., & Keller, E. R. (1989). Application of growth regulators in the induction and maintenance of calluses of soybean. *Hereditas, 11*, 1–4.

Li, H. Y., Zhu, Y. M., Chen, Q., Conner, R. L., Ding, X. D., Li, J., & Zhang, B. B. (2004). Production of transgenic soybean plants with two anti-fungal protein genes via *Agrobacterium* and particle bombardment. *Biologia Plantarum, 48*, 367–374.

Li, Y., Zhang, J., Hao, L., Hua, J., Duan, L., et al. (2013). Expression of an *Arabidopsis* molybdenum cofactor sulfurase gene in soybean enhances drought tolerance and increases yield under field conditions. *Plant Biotechnology Journal, 11*(6), 747–758.

Lin, W., Odell, J. T., & Schreiner, R. M. (1987). Soybean protoplast culture and direct gene uptake and expression by cultured soybean protoplasts. *Plant Physiology, 84*(3), 856–861.

Lippmann, B., & Lippmann, G. (1993). Soybean embryo culture: Factors influencing plant recovery from isolated embryos. *Plant Cell, Tissue and Organ Culture, 32*, 83–90.

Liu, D. P., & Zhao, G. L. (1986). Callus attained from soybean pollen culture *in vitro*. *Soybean Science, 1*, 17.

Liu, H. K., Yang, C., & Wei, Z. M. (2004). Efficient *Agrobacterium tumefaciens*-mediated transformation of soybeans using an embryonic tip regeneration system. *Planta, 219*(6), 1042–1049.

Loganathan, M., Maruthasalam, S., Shiu, L. Y., et al. (2010). Regeneration of soybean (*Glycine max* L. Merrill) through direct somatic embryogenesis from the immature embryonic shoot tip. *In Vitro Cellular & Developmental Biology–Plant, 46*, 265–273. https://doi.org/10.1007/s11627-009-9263-1.

Ma, X. H., & Wu, T. L. (2008). Rapid and efficient regeneration in soybean (*Glycine max* L. Merrill) from whole cotyledonary node explants. *Acta Physiologiae Plantarum, 30*, 209–216.

Maranna, S., Nataraj, V., Kumawat, G., Chandra, S., Rajesh, V., Ramteke, R., Patel, R. M., Ratnaparkhe, M. B., Husain, S. M., Gupta, S., & Khandekar, N. (2021). Breeding for higher yield, early maturity, wider adaptability and waterlogging tolerance in soybean (*Glycine max* L.): A case study. *Scientific Reports, 11*, 22853.

Maughan, P. J., Philip, R., Cho, M. J., Widholm, J. M., & Vodkin, L. O. (1999). Biolistic transformation, expression, and inheritance of bovine β-casein in soybean (*Glycine max*). *In Vitro Cellular & Developmental Biology–Plant, 35*, 344–349.

McCabe, D. E., Swain, W. F., Martinell, B. J., & Christou, P. (1988). Stable transformation of soybean (*Glycine max*) by particle acceleration. *Nature Biotechnology, 6*(8), 923–926.

McPherson, R. M., & MacRae, T. C. (2009). Evaluation of transgenic soybean exhibiting high expression of a synthetic *Bacillus thuringiensis cry1A* transgene for suppressing lepidopteran population densities and crop injury. *Journal of Economic Entomology, 102*(4), 1640–1648.

Medic, J., Atkinson, C., & Hurburgh, C. R. (2014). Current knowledge in soybean composition. *Journal of the American Oil Chemists' Society, 91*, 363–384.

Muthu, R. K., Shanmugasundaram, G., Karuppan, T., Balakrishnan, J., Doss, G., & Shanmugam, K. (2023). Combined supplementation of 2,4-D and kinetin elevates callus induction from cotyledon segments of different soybean cultivars (*Glycine max* (L.) Merrill). *Plant Archives, 23*(2), 25–31.

Nawracala, J., & Konieczny, G. (1996). Regeneration of soybean plants via somatic embryogenesis. *Biuletyn Instytutu Hodowli i Aklimatyzacji Roślin, 198*, 21–29.

Nitsch, J. P., & Nitsch, C. (1969). Haploid plants from pollen grains. *Science, 163*, 85–87.

Niu, L., Yang, J., Zhang, J., He, H., Xing, G., Zhao, Q., et al. (2019). Introduction of the *harpin Xooc*-encoding gene *hrf2* in soybean enhances resistance against the oomycete pathogen *Phytophthora sojae*. *Transgenic Research, 28*, 257–266.

Padgette, S. R., Kolacz, K. H., Delannay, X., Re, D. B., LaVallee, B. J., Tinius, C. N., et al. (1995). Development, identification, and characterization of a glyphosate-tolerant soybean line. *Crop Science, 35*(5), 1451–1461.

Pagano, M. C., & Miransari, M. (2016). The importance of soybean production worldwide. In M. Miransari (Ed.), *Abiotic and Biotic Stresses in Soybean Production* (pp. 1–26). Academic Press.

Parrott, W. A., All, J. N., Adang, M. J., Bailey, M. A., Boerma, H. R., & Stewart, C. N. (1994). Recovery and evaluation of soybean plants transgenic for a *Bacillus thuringiensis var. kurstaki* insecticidal gene. *In Vitro Cellular & Developmental Biology–Plant, 30*, 144–149.

Paz, M. M., Shou, H., Guo, Z., Zhang, Z., Banerjee, A. K., & Wang, K. (2004). Assessment of conditions affecting *Agrobacterium*-mediated soybean transformation using the cotyledonary node explant. *Euphytica, 136*, 167–179.

Pratap, A., & Gupta, S. K. (2007). Advances in doubled haploid technology of oilseed rape. *Indian Journal of Crop Science, 2*, 267–271.

Pratap, A., & Kumar, J. (2011). *Biology and Breeding of Food Legumes*. CABI.

Pratap, A., Gupta, S. K., Kumar, J., Mehandi, S., & Pandey, V. R. (2016). Soybean. In S. K. Gupta (Ed.), *Breeding Oilseed Crops for Sustainable Production: Opportunities and Constraints* (pp. 293–315). Elsevier.

Radhakrishnan, R., & Ranjithakumari, B. D. (2007). Callus induction and plant regeneration of Indian soybean (*Glycine max* (L.) Merr. cv. CO_3) via half-seed explant culture. *Journal of Agricultural Technology, 3*(2), 287–297.

Radhakrishnan, R., Ramachandran, A., & Ranjitha Kumari, B. D. (2009). Rooting and shooting: Dual function of thidiazuron in *in vitro* regeneration of soybean (*Glycine max* L.). *Acta Physiologiae Plantarum, 31*, 1213–1217. https://doi.org/10.1007/s11738-009-0356-6.

Rajendran, A., & Lal, S. K. (2020). Assessing the need of pre-germination anaerobic stress-tolerant varieties in Indian soybean (*Glycine max* (L.) Merrill). *National Academy Science Letters, 43*, 593–597.

Ramlal, A., Bhat, I., Nautiyal, A., Baweja, P., Mehta, S., Kumar, V., et al. (2023a). *In silico* analysis of angiotensin-converting enzyme (ACE) inhibitory compounds obtained from soybean (*Glycine max* (L.) Merr.). *Frontiers in Physiology, 14*, 1172684. https://doi.org/10.3389/fphys.2023.1172684.

Ramlal, A., Dey, N., Sharma, U., & Rajendran, A. (2022c). *In silico* studies to reveal the potential inhibitory capacity of soy isoflavonoids against angiotensin-converting enzyme. In H. Chopra, V. Ghuriani, P. M. Arora, S. Babbar, & P. Baweja (Eds.), *Sustainable Future for Humanity: The New Learning Curve* (pp. 155–167). Imperial Publications.

Ramlal, A., Mehta, S., Nautiyal, A., Baweja, P., Sharma, D., Lal, S. K., et al. (2024a). Androgenesis in soybean (*Glycine max* (L.) Merr.): A critical revisit. *In Vitro Cellular & Developmental Biology–Plant, 60*(8). https://doi.org/10.1007/s11627-023-10402-z.

Ramlal, A., Nautiyal, A., Baweja, P., Kumar, V., Mehta, S., Mahto, R. K., et al. (2022a). Angiotensin-converting enzyme (ACE) inhibitory peptides and isoflavonoids from soybean (*Glycine max* (L.) Merr.). *Frontiers in Nutrition, 9*, 1068388. https://doi.org/10.3389/fnut.2022.1068388.

Ramlal, A., Nautiyal, A., Baweja, P., Mahto, R. K., Mehta, S., Mallikarjun, B. P., et al. (2022b). Harnessing heterosis and male sterility in soybean (*Glycine max* (L.) Merr.): A critical revisit. *Frontiers in Plant Science, 13*, 981768. https://doi.org/10.3389/fpls.2022.981768.

Ramlal, A., Nautiyal, A., Lal, S. K., & Chigeza, G. (2023c). Editorial: A wonder legume, soybean: Prospects for improvements. *Frontiers in Plant Science, 14*, 1294185. https://doi.org/10.3389/fpls.2023.1294185.

Ramlal, A., Sharma, D., Lal, S. K., Raju, D., Shivam, & Rajendran, A. (2023b). First report of ovary-derived calli induction in soybean (*Glycine max* (L.) Merr.). *Plant Cell, Tissue and Organ Culture, 153*, 439–445.

Ramlal, A., Singhal, H., Kumar, L., Pandey, V., & Mehta, S. (2024). Transcriptional activation and suppression using CRISPR/Cas9 in plants. In A. Alok, J. Kumar, &

M. S. Shekhawat (Eds.), *Basics of CRISPR/Cas Mediated Plant Genome Editing* (pp. 190–203). CRC Press, Taylor & Francis, UK. (In production).

Ran, Y., Liang, Z., & Gao, C. (2017). Current and future editing reagent delivery systems for plant genome editing. *Science China Life Sciences, 60*, 490–505.

Ranch, J. P., & Buchheim, J. A. (1991). Propagating multiple whole fertile plants from immature leguminous. *U.S. Patent No. 5,008,200*. United Agri Seeds Inc.

Raza, G., Singh, M. B., & Bhalla, P. L. (2017). *In vitro* plant regeneration from commercial cultivars of soybean. *BioMed Research International, 2017,* 7379693.

Reddy, M. S., Ghabrial, S. A., Redmond, C. T., Dinkins, R. D., & Collins, G. B. (2001). Resistance to *Bean pod mottle virus* in transgenic soybean lines expressing the capsid polyprotein. *Phytopathology, 91*(9), 831–838.

Rizzo, G., & Baroni, L. (2018). Soy, soy foods and their role in vegetarian diets. *Nutrients, 10*, 43.

Rodrigues, L. R., Forte, B. D. C., & Bodanese-Zanettini, M. H. (2006). Isolation and culture of soybean (*Glycine max* L. Merrill) microspores and pollen grains. *Brazilian Archives of Biology and Technology, 49*, 537–545.

Rodrigues, L. R., Forte, B. D. C., Oliveira, J. M. S., Mariath, J. E. A., & Bodanese-Zanettini, M. H. (2004). Effects of light conditions and 2,4-D concentration in soybean anther culture. *Plant Growth Regulation, 44*, 125–131.

Rodrigues, L. R., Oliveira, J. M. S., Mariath, J. E. A., Iranco, L. B., & Bodanese-Zanettini, M. H. (2005). Anther culture and cold treatment of floral buds increased symmetrical and extranuclei frequencies in soybean pollen grains. *Plant Cell, Tissue and Organ Culture, 8*, 101–104.

Rodrigues, L. R., Terra, T. D. F., Bered, F., & Bodanese-Zanettini, M. H. (2004). Origin of embryo-like structures in soybean anther culture investigated using SSR marker. *Plant Cell, Tissue and Organ Culture, 77*, 287–289.

Rudolf-Pilih, K., Petkovšek, M., Jakše, J., Štajner, N., Murovec, J., & Bohanec, B. (2019). Proposal of a new hybrid breeding method based on genotyping, inter-pollination, phenotyping, and paternity testing of selected elite F1 hybrids. *Frontiers in Plant Science, 10*, 1111.

Sairam, R. V., Franklin, G., Hassel, R., Smith, B., Meeker, K., Kashikar, N., Parani, M., Abed, D. A., Ismail, S., Berry, K., & Goldman, S. L. (2003). A study on the effect of genotypes, plant growth regulators, and sugars in promoting plant regeneration via organogenesis from soybean cotyledonary nodal callus. *Plant Cell, Tissue and Organ Culture, 75*, 79–85.

Santarem, E. R., Pelissier, B., & Finer, J. J. (1997). Effect of explant orientation, pH, solidifying agent, and wounding on initiation of soybean somatic embryos. *In Vitro Cellular & Developmental Biology–Plant, 33*, 13–19.

Santos, K. G. B. D., Mundstock, E., & Bodanese-Zanettini, M. H. (1997). Genotype-specific normalization of soybean somatic embryogenesis through the use of an ethylene inhibitor. *Plant Cell Reports, 16*, 859–864.

Sato, S., Xing, A., Ye, X., Schweiger, B., Kinney, A., Graef, G., & Clemente, T. (2004). Production of γ-linolenic acid and stearidonic acid in seeds of marker-free transgenic soybean. *Crop Science, 44*(2), 646–652.

Schmidt, M. A., Parrott, W. A., Hildebrand, D. F., Berg, R. H., Cooksey, A., Pendarvis, K., et al. (2015). Transgenic soybean seeds accumulating β-carotene exhibit the collateral enhancements of oleate and protein content traits. *Plant Biotechnology Journal, 13*(4), 590–600.

Seeja, G., & Sreekumar, S. (2020). Doubled haploids in genetic improvement: A review. *International Journal of Recent Scientific Research, 11*, 36941–36949.

Sellars, R. M., Southward, G. M., & Phillips, G. C. (1990). Adventitious somatic embryogenesis from cultured immature zygotic embryos of peanut and soybean. *Crop Science, 30*(2), 408–414.

Seo, J. S., Sohn, H. B., Noh, K., Jung, C., An, J. H., Donovan, C. M., et al. (2012). Expression of the *Arabidopsis* AtMYB44 gene confers drought/salt-stress tolerance in transgenic soybean. *Molecular Breeding, 29*, 601–608.

Shan, Z., Raemakers, K., Tzitzikas, E. N., Ma, Z., & Visser, R. G. (2005). Development of a highly efficient, repetitive system of organogenesis in soybean (*Glycine max* (L.) Merr). *Plant Cell Reports, 24*, 507–512.

Singh, D., Swapnil, A. S., & Kumar, M. (2020). Androgenesis: A valuable technique to shorten the breeding cycle. *Food Science Reports, 1*, 18–20.

Skinner, D. Z., & Liang, G. H. (1996). Haploidy in alfalfa. In S. M. Jain, S. K. Sopory, & R. E. Veilleux (Eds.), *In Vitro Haploid Production in Higher Plants. Current Plant Science and Biotechnology in Agriculture* (Vol. 25, pp. 365–375). Springer, Dordrecht. https://doi.org/10.1007/978-94-017-1858-5_19.

Song, S., Hou, W., Godo, I., Wu, C., Yu, Y., Matityahu, I., et al. (2013). Soybean seeds expressing feedback-insensitive cystathionine γ-synthase exhibit a higher content of methionine. *Journal of Experimental Botany, 64*(7), 1917–1926.

Soto, N., Delgado, C., Hernández, Y., Rosabal, Y., Ferreira, A., Pujol, M., et al. (2017). Efficient particle bombardment-mediated transformation of Cuban soybean (INCASoy-36) using glyphosate as a selective agent. *Plant Cell, Tissue and Organ Culture, 128*, 187–196.

Stewart Jr, C. N., Adang, M. J., All, J. N., Boerma, H. R., Cardineau, G., Tucker, D., & Parrott, W. A. (1996). Genetic transformation, recovery, and characterization of fertile soybean transgenic for a synthetic *Bacillus thuringiensis* cryIAc gene. *Plant Physiology, 112*(1), 121–129.

Sussex, I. M. (1989). Developmental programming of the shoot meristem. *Cell, 56*(2), 225–229.

Swallah, M. S., Yang, X., Li, J., Korese, J. K., Wang, S., Fan, H., Yu, H., & Huang, Q. (2023). The pros and cons of soybean bioactive compounds: An overview. *Food Reviews International, 39*(8), 5104–5131.

Tang, W. T., Ling, T. S., & Chang, C. S. (1973). Effects of kinetin and auxin on callus formation in anther tissue cultures of soya bean. *Journal of Agricultural Association China (Chung-hua Nung-hsueh Hui Pao), 83*, 1–7.

Texeira, L. R., Churata, B. G. M., Vieira, E. S. N., Martins, P. K., & Schuster, I. (2011). Evaluation of soybean cultivars on the embryogenic and organogenic potential. *Acta Scientiarum. Agronomy, 33*, 67–74.

Tian, L. N., Brown, D. C. W., Voldeng, H., & Webb, J. (1994). *In vitro* response and pedigree analysis for somatic embryogenesis of long-day photoperiod adapted soybean. *Plant Cell, Tissue and Organ Culture, 36*, 269–273.

Tiwari, S., Shanker, P., & Tripathi, M. (2004). Effects of genotype and culture medium on *in vitro* androgenesis in soybean (*Glycine max* Merr.). *Indian Journal of Biotechnology, 3*, 441–444.

Tiwari, S., Tripathi, M. K., & Kumar, S. (2021). Improvement of soybean through plant tissue culture and genetic transformation: A review. *JNKVV Research Journal, 45*(1), 1–18.

Tripathi, M., & Tiwari, S. (2003). Epigenesis and high frequency plant regeneration from soybean (*Glycine max* (L.) Merr.) hypocotyls. *Plant Cell, Tissue and Organ Culture, 13*(1), 61–73.

Tripathi, M., & Tiwari, S. (2004). Morphogenesis and plantlet regeneration from soybean (*Glycine max* L. Merrill) leaf discs influenced by genotypes and plant growth regulators. *Legume Research, 27*(2), 88–93.

Vasil, I. K., & Thorpe, T. A. (2013). *Plant Cell and Tissue Culture*. Springer Science & Business Media.

Wang, G., & Xu, Y. (2008). Hypocotyl-based *Agrobacterium*-mediated transformation of soybean (*Glycine max*) and application for RNA interference. *Plant Cell Reports, 27*, 1177–1184.

Wang, X., Eggenberger, A. L., Nutter, F. W., & Hill, J. H. (2001). Pathogen-derived transgenic resistance to soybean mosaic virus in soybean. *Molecular Breeding, 8*, 119–127.

Wang, Y., Jiang, L., Chen, J., Tao, L., An, Y., Cai, H., & Guo, C. (2018). Overexpression of the alfalfa WRKY11 gene enhances salt tolerance in soybean. *PLOS One, 13*(2), e0192382.

Wijnker, E., van Dun, K., de Snoo, C. B., Lelivelt, C. L., Keurentjes, J. J., Naharudin, N. S., Ravi, M., Chan, S. W. L., de Jong, H., & Dirks, R. (2012). Reverse breeding in *Arabidopsis thaliana* generates homozygous parental lines from a heterozygous plant. *Nature Genetics, 44*, 467–470.

Wright, M. S., Williams, M. H., Pierson, P. E., & Carnes, M. G. (1987). Initiation and propagation of *Glycine max* L. Merr.: Plants from tissue-cultured epicotyls. *Plant Cell, Tissue and Organ Culture, 8*, 83–90.

Xu, H., Zhang, L., Zhang, K., & Ran, Y. (2020). Progresses, challenges, and prospects of genome editing in soybean (*Glycine max*). *Frontiers in Plant Science, 11*, 571138.

Yamada, T., Takagi, K., & Ishimoto, M. (2012). Recent advances in soybean transformation and their application to molecular breeding and genomic analysis. *Breeding Science, 61*(5), 480–494.

Yan, B., Srinivasa Reddy, M. S., Collins, G. B., & Dinkins, R. D. (2000). *Agrobacterium tumefaciens*-mediated transformation of soybean (*Glycine max* (L.) Merrill.) using immature zygotic cotyledon explants. *Plant Cell Reports, 19*, 1090–1097.

Ye, X. G., Fu, Y. Q., & Wang, L. Z. (1994). Study on several problems of soybean anther culture. *Soybean Science, 13*, 193–199.

Yeh, M. S. (1989). *In vitro* culture of immature soybean embryos. I. The abilities of somatic embryogenesis and regeneration from different aged immature embryos in *Glycine max*. *Journal of Agriculture and Forestry, 38*, 139–150.

Yeh, M. S., & Chyuan, J. H. (1991). *In vitro* culture of immature soybean embryos: Organogenesis studies on soybean embryonic axes culture. *Journal of Agriculture and Forestry, 40*, 77–89.

Yeh, M. S., & Chyuan, J. H. (1992). *In vitro* culture of immature soybean embryos V. Effects of agar, Gelrite and casein hydrolysate on immature soybean embryo culture. *Journal of Agriculture and Forestry, 41*, 13–22.

Yeung, E. C., & Sussex, I. M. (1979). Embryogeny of *Phaseolus coccineus*: The suspensor and the growth of the embryo-proper *in vitro*. *Zeitschrift für Pflanzenphysiologie, 91*, 423–433.

Yin, G. C. (1981). Study on pollen culture of soybean. *Heilongjiang Agricultural Science, 1*, 12–14.

Yin, G. C., Zhu, Z. Y., Xu, Z., Chen, L., Li, X. Z., & Bi, F. Y. (1982). Studies on induction of pollen plant and their androgenesis in *Glycine max* (L.) Merr. *Soybean Science, 1*, 69–76.

Zhao, G., Liu, Y., Yin, A., & Li, J. (1998). Germination of embryo in soybean anther culture. *Chinese Science Bulletin, 43*, 1991–1995.

Zhou, L., He, H., Liu, R., Han, Q., Shou, H., & Liu, B. (2014). Overexpression of *GmAKT2* potassium channel enhances resistance to soybean mosaic virus. *BMC Plant Biology, 14*(1), 1–11.

Zhuang, X. J., Hu, C. Y., Chen, Y., & Yin, G. C. (1991). Embryoids from soybean anther culture. *Soybean Genetics Newsletter, 18*, 265.

Zulkarnain, Z. Z. Z. (2014). The effect of different levels and sources of auxin and cytokinin to callus formation on soybean anther culture. *Journal of Agroteknologi Tropika, 3*, 58.

PART III

Soybean Diseases and Biotechnological Interventions

CHAPTER 10

Pathogens of Soybean: Symptoms, Crop Losses, and Biotechnological Strategies to Control

MEGHA KUMARI,[1] SHRISHTI NAIDU,[1] and ARCHANA SINGH[1,2,*]

[1]*Department of Botany, Hansraj College, University of Delhi, Delhi, India*

[2]*Delhi School of Climate Change and Sustainability, Institution of Eminence, Maharishi Karnad Bhawan, University of Delhi, Delhi, India*

[*]Corresponding author

ABSTRACT

Soybean (*Glycine max* (L.) Merr.) is one of the most important leguminous crops and has a wide range of agricultural significance because of its high protein and oil content. It is principally utilized for commercial food products, dietary supplements, fodder, the oil industry, and medicinal values. Soybean is affected by hundreds of pathogens, including fungi, bacteria, and viruses. Pathogen attacks have already been documented to afflict soybean production all around the globe, with 30 of them deemed to be a danger to soybean production. Infections by pathogens have a negative impact on the soybean crop, leading to lower yields, huge economic losses, and financial crises. A couple of control measures have been taken into consideration to reduce the crop losses caused by pathogen infection. Out of these, biotechnological interventions such as molecular breeding and genetic engineering have played a crucial role in controlling disease symptoms. This chapter contains an overview of illnesses caused by bacterial, fungal, and viral pathogens in *G. max*/soybean, with a specific emphasis on disease symptoms and alternative means of control.

Soybeans, Volume 2: Genome, Biotechnology, Stress, and Disease.
R. Ambika Rajendran, Ayyagari Ramlal, Sanjay Kumar Lal, and Godfree Chigeza (Eds.)

DOI: 10.1201/9781998511730-13

10.1 INTRODUCTION

Glycine max (L.) Merr./soybean is by far the most commonly cultivated legume crop, supplying 25% of total edible oil and two-thirds of all protein requirements for cattle and human feed (Krishnamurthy, 1975; Rajendran et al., 2022). Soybeans have been used as sources of oil, bread, milk, and cheese by the people of China, Korea, Japan, Manchuria, Indonesia, and the Philippines for decades (Horvath, 1926). Soybeans have long been used as a major ingredient in formulated poultry feed and fish meals. Soybeans are highly valued around the world for their high-quality protein and oil, which account for roughly 40% and 20% of the soybean, respectively (Clemente and Cahoon, 2009). Soybean oil is often employed in food and cooking, but it is also used as a source of biodiesel as part of current attempts to reduce carbon emissions and dependency on non-renewable energy (Mofijur et al., 2013). Soybean meal (SM) accounts for approximately 70% of the value of soybeans, while seed accounts for the remaining 30% (Raghuvanshi and Bisht, 2010). Soybeans have industrial applications, such as core ingredients in adhesives and sealants (Raghuvanshi and Bisht, 2010). Soybean protein is referred to as a complete protein because of its composition of amino acids (AAs). It is widely recognized for its nutritional benefits in the treatment of heart disease and diabetes (Isanga and Zhang, 2008; Ramlal et al., 2022). Currently, the "Big-3" producers on the globe are the United States, Brazil, and Argentina (Tiwari, 2017). In the United States, yearly production losses owing to soybean diseases are estimated to be over 11% (Hartman et al., 2015). Soybean rust recently reduced output by 60% on around 500 acres of soybeans in the US (Sikora et al., 2014a, b). From 1996 to 2016, the overall projected economic loss in the United States due to soybean diseases was $95.48 billion, with $80.89 billion from the northern and $14.59 billion from the southern parts of the US, respectively (Bandara et al., 2020).

This leguminous plant is cultivated worldwide, and its production is limited by the number of insects and diseases. There are indeed a number of diseases that target and harm legume crops, resulting in lower quality and yield. Over 135 diseases of this crop have been identified, with roughly 30 species thought to be capable of causing major economic harm (Roy et al., 2001). Pests, pathogens, and weeds all have negative impacts on soybean yield, with weeds being the most damaging culprit (Powles, 2010). Nematodes, fungi, viruses, bacteria, and oomycetes are among the microorganisms that cause economic harm each year. Aphids, beetles, mites, and stinkbugs are all soybean pests that target the entire plant (Hartman and Hill, 2010; O'Neal and Johnson, 2010). As a result of pathogen infection, both the production and quality of the soybean crop decrease. Several infections appear frequently in large concentrations in an agro-ecological zone, whereas others are either absent or appear irregularly.

The aforementioned soybean infections can cause severe harm, putting the viability of soybean farming in jeopardy. As a result, all control strategies specified under integrated control of the infection must be used in the suppression of diseases. Since chemical interventions are rarely used, emphasis should be directed toward improving crop management practices by breeding less vulnerable genotypes (Miloš et al., 2013). In most breeding programs, a great deal of emphasis is placed on introducing the highest possible levels of parasite resistance. The disease and pest resistance in soybean has proven to be a valuable strategy in reducing potential losses. Therefore, this challenge can be solved by increasing the utilization of genetic resources via molecular breeding and/or bioengineering. We will address the fundamental overview of soybean infections caused by bacteria, fungi, viruses, and nematodes in this chapter, as well as the soybean's molecular defense mechanisms, biotechnological approaches, and genome editing (GE) to help avoid these infections.

10.2 SOYBEAN PATHOGENS AND THEIR ASSOCIATED DISEASES

A pathogen is an organism that infects its host and causes illness, with virulence suggesting the severity of the illness. A vast number of pathogens affect soybean fields, causing bacterial, fungal, viral, and nematode infections on the leaves, stems, and roots. In certain cases, these illnesses can be quite harmful. As a result, a full evaluation of soybean disease symptoms, epidemiology, and management is necessary.

Environmental factors also influence soybean growth and development, with noteworthy research indicating that each 1° increment in global mean temperature results in a 3.1% loss in soybean production (Zhao et al., 2017). However, this estimate does not take into account yield depletion caused by diseases, which are equally affected by environmental conditions (Rajendran et al., 2022, 2023; Roth et al., 2020). For plant disease to arise, all three parts of the disease triangle have to be present, that is, a favorable environment, a virulent pathogen, and a susceptible host. There may be genetic or other agronomic factors that alter susceptibility to disease or resistance to it. The virulence and availability of pathogens can influence disease load. Many elements influence the disease environment, the most important of which are temperature, soil type, light amount and quality, and moisture (Roth et al., 2020).

Since soybeans are such an important part of the world's agricultural economy, protecting the crop against persistent, intermittent, and developing diseases is critical. Even though possible hazards to soybean yield differ by growing location (Allen et al., 2017), stem and root rot caused by *Phytophthora*, sudden death syndrome (SDS), and stem rot caused by *Sclerotinia*, on the other hand, are economically important soybean diseases that are seasonal. Soybean interaction with *Phytophthora sojae* has demonstrated a close gene-for-gene linkage (Flor, 1955), which has sparked a lot of studies and publications (Dorrance, 2018). Reniform and root-knot nematodes (RKNs), *Diaporthe*, and frog eye leaf rot infections are all becoming more common (Roth et al., 2020). Downy mildew and bacterial blight were found to be the most common soybean pathogens, followed by stem canker, white mold on stems, charcoal rot of roots, and seed decay (Jasnić and Vidić, 2008). Other soybean infections appear on a rare basis, while epiphytotic attacks have been well documented.

These diseases are burdensome economically, which has led to the study of their underlying epidemiology, biology, and management. Illnesses affect yield by around 10% on average, but losses can be substantially greater in individual areas and with various diseases. Because soybean pathogens are ubiquitous across the states, precise disease detection and understanding the prospect of disease losses are critical for the soybean industry to thrive. To effectively control soybean diseases, good diagnosis and early detection are essential. Knowing a field's disease history and staying up-to-date on reported disease development during the growing season will assist in making key management decisions that will help us safeguard yield potential.

Crop rotation, genomic resistance, nematicide and fungicide seed treatments, mid-season spraying of fungicides, and biocontrol are all important disease management tactics in row-crop agriculture (Roth et al., 2020). Plant population, row spacing, tillage, and irrigation are all examples of other cultural practices that might be beneficial (Roth et al., 2020). For successful illness reduction, every disease requires a particular and often distinctive combination of care measures. Because environmental conditions range from one place to another, some disease control measures may be more successful in specific regions than others. As a result, disease prevention and control need a method that is both regulated and effective.

10.2.1 BACTERIAL DISEASES

There are billions of bacterial species in nature. Many of them carry out biochemical activities that are necessary for life to continue, but others generate diseases in many organisms, such as humans, plants, and animals. Every year, millions of acres of cropland are planted with soybeans. Several ailments, including the two most prominent bacterial infections in soybeans, bacterial blight and bacterial pustule (Giesler and Weissling, 2004), can affect their development and health. Both of these foliar diseases can originate in the lower canopy and are sometimes mistaken for *Septoria* leaf spot or soybean rust. Wind and rain, as well as natural breaches or wounds produced by high winds, hailstorms, insects, and other factors, are common ways for these diseases to propagate from plant to plant. Leaf spots, water-soaked sores, blights, and withering are some of the symptoms of bacterial infections. The two crucial bacterial diseases for the soybean are:

10.2.1.1 BACTERIAL BLIGHT

The most common bacterial soybean disease is bacterial blight (Johnson and Chamberlain, 1953). The bacterium *Pseudomonas amygdali* pv. *glycinea* causes bacterial blight in soybeans. Yield losses have been observed to range between 4% and 40%, depending on the severity of the situation (Jagtap et al., 2012). It is prevalent throughout the Midwest and common in cool, wet weather (Wrather and Koenning, 2009). Although it is mostly a foliar disease, it can also affect parts of the plant, such as petioles, pods, and stems (Kirkpatrick et al., 2014). Under high disease load, bacterial blight has been documented to produce considerable output losses on vulnerable cultivars. Economic losses might occur when soybeans are cultivated for seed (Giesler and Weissling, 2004). Early in the season, the disease is most frequent. It overwinters in plant leftovers in the field. The first infection of soybeans happens when bacterial cells are carried from plant waste on the soil surface to the leaves by wind-driven water droplets. Disease outbreaks are often preceded by rainstorms with strong winds. The leaf surface must be damp for infection to take place. Infested seeds can cause infection in seedlings.

The major symptoms are the formation of tiny, pointed, transparent, water-drenched light brown-yellow spots (Kirkpatrick et al., 2014). The spots' cores deepen to a reddish-brown color, become sunken, and thus are bordered by a waterlogged perimeter with a yellowish-green halo as they mature. On the top of the leaf surface, the halo is much more visible. Spots may expand and unite to produce vast, irregular, necrotic patches when environmental factors encourage disease development. Strong winds and pounding rains cause these large dead sections of the leaves to fall off or tear away, giving the leaves a ragged look. Premature defoliation can occur when the condition is severe (Johnson and Chamberlain, 1953). The indicators of bacterial blight on leaves are similar to those of the fungus brown spot (Kirkpatrick et al., 2014). However, bacterial blight spreads quickly through the canopy, whereas brown spot spreads slowly. By holding infested foliage up to the light, bacterial blight patches may be seen, which seem transparent. Lesions form on the cotyledons of contaminated seeds and can spread to the leaves, causing a secondary infection. Stomata or wounds allow the disease to penetrate the plant. Cultivation or spraying when the leaves are not dry and wind-blown rain transmits the bacteria from affected tissue. During the growing season, seeds can become contaminated by the pods. The growth of bacterial blight is favored by windy, chilly, wet weather with temperatures between 75°F and 79°F, but it is suppressed by hot and dry weather (Kirkpatrick et al., 2014).

The control of the disease is mostly based on cultivars that are immune to bacterial leaf blight. Raising high-quality, disease-free seeds and utilizing tillage practices that promote quick decomposition of crop residue are examples of cultural management approaches. In fields with a history of bacterial blight, narrow row widths and large plant numbers should be avoided (Kirkpatrick et al., 2014).

10.2.1.2 BACTERIAL PUSTULE

Bacterial pustules were discovered in some regions of soybean fields (Johnson and Chamberlain, 1953). *Xanthomonas axonopodis* (syn. *campestris*) pv *glycines*, which causes bacterial pustules, can induce premature defoliation and decrease seed size and quantity. This illness has been recorded in almost every country where soybeans are produced when the weather is hot and rainy. It is most common during early growth stages. Bacterial leaf spot significantly reduces soybean output (Zhao et al., 2022). In 1998, for example, 88% of soybeans grown in South Korean fields were seriously affected (Zhao et al., 2022). Similarly, it was responsible for a 20.7–34.9% loss of Thailand's soybean production in 1983 (Kim et al., 2010; Prathuangwong and Amnuaykit, 1987).

The emergence of small pale-green spots on leaves is the first symptom. Such spots have raised centers and can appear on either the lower or upper sides of the leaf, with the lower surface being more common (Giesler and Weissling, 2004). The primary leaf veins are frequently connected with lesions. Tiny, light-colored pustules will emerge in the center of the spots as the condition worsens. Many spots can combine to produce irregular lesions. This disease can be readily mistaken for soybean rust. A tiny, circular aperture at the top of soybean rust pustules allows spores to escape. Bacterial pustule lesions do not generally open or generate spores. If a bacterial pustule has an opening, it is usually a linear fracture on the surface of the pustule (Giesler and Weissling, 2004).

Bacterial pustule, like bacterial blight, overwinters in crop waste and is transferred to the plant by air-driven rain or water drops dripping from the ground. Furthermore, the disease can spread while the foliage is damp during cultivation. The bacteria might penetrate the plant through wounds and natural openings. The development of this condition is aided by warm weather and frequent rain. Warm temperatures, unlike bacterial blight, have no impact on the formation of bacterial pustules. It grows best at temperatures ranging from 86°F to 92°F (Giesler and Weissling, 2004).

The adoption of genotypes that are immune to bacterial pustule is crucial in the treatment of this disease. Planting high-quality, disease-free seeds and utilizing tillage procedures that expedite crop residue decomposition are examples of cultural practices (Kirkpatrick et al., 2014).

10.2.2 FUNGAL DISEASES

Soybean fields have a variety of foliar, stem, and root fungal infections. Fungal infections can cause significant losses in certain circumstances. A plethora of fungal diseases can have a significant influence on seedlings, seed decay, purple seed stain, frog eye leaf spot on seed, anthracnose, downy mildew, and numerous other supplementary fungal invaders of injured pods, such as *Fusarium* and *Cladosporium*.

10.2.2.1 ANTHRACNOSE

Anthracnose is a plant disease that reduces plant height (PH), seed quality, and production by 16 to 26% (Kirkpatrick et al., 2014). Anthracnose is a significant worldwide disease, with pod infections having a bigger effect on yield reduction than stem and petiole infections. Soybeans are vulnerable at every stage of growth. When contaminated seeds are sown, damping-off occurs both before and after they develop. The cotyledons of growing seedlings acquire dark brown, sunken blemishes. When circumstances encourage disease growth, these lesions can spread down the stem, prompting one or perhaps all cotyledons to become moist, wither, and abscise from the stem (Kirkpatrick et al., 2014). Several tiny lesions can destroy young plants if the illness progresses rapidly.

Foliar symptoms appear as irregularly sized brown lesions on stems, petioles, and pods during early reproductive development stages (Johnson and Chamberlain, 1953). Whenever anthracnose lesions encircle the petiole of a leaf, resulting in a shepherd's crook sign, premature defoliation can occur across the canopy on maturing plants (Kirkpatrick et al., 2014). Infestation of pods and pedicels early in the season might result in fewer, shorter seeds or even no seed development (Johnson and Chamberlain, 1953). Black fungal fruiting forms acervuli that generate microscopic black spines (setae) that are plentiful and randomly dispersed on infected tissue at later stages of the disease progression, near soybean maturity (R7, R8). Setae can be observed with an effective hand lens or a dissecting microscope and are indicative of anthracnose. The fruiting bodies of *Diaporthe phaseolorum*, a common disease on pods and stems that causes pod as well as stem blight, do not have setae and are commonly grouped in rows. *Colletotrichum truncatum* and other similar species produce anthracnose, which can be found as mycelium on crop leftovers or in infected seeds. Seedlings could become infected and colonized by the fungus without symptom development until the early reproductive phase, or seedlings may be attacked and colonized by the fungus without symptoms (Kirkpatrick et al., 2014). Warm, wet weather encourages stem and pod infection, but dry weather reduces the severity of the illness.

The use of high-quality, disease-free seedlings, plus tillage or rotation measures that decrease soybean residue, are all part of anthracnose management. Anthracnose can be controlled by applying a fungicide between the start of pod development (R3) and the start of seed production (R5).

Seed treatments with fungicides are also effective at mitigating the severity of anthracnose in seedlings (Kirkpatrick et al., 2014).

10.2.2.3 BROWN SPOT

Septoria leaf spot, otherwise called brown spot, has been documented in various soybean-growing regions all over the world. It is one of the most common foliar soybean diseases known globally. Brown spot is extremely widespread in the Northern Plains and Midwestern U.S. soybean fields; however, the illness seldom causes economic damage (Maitra, 2018). When there is a lot of defoliation, yield loss of up to 2.5% to 9.5% might occur (Cruz et al., 2010). When vulnerable cultivars are grown and disease conditions are favorable, this disease can induce early defoliation, which leads to production losses (Kirkpatrick et al., 2014). Irregular, small brown leaf spots appear on the top and lower leaf surfaces and range in size from just a speck to a fifth of an inch in diameter (Johnson and Chamberlain, 1953). Nearby leaf spots may clump together, forming uneven blotches. The lesions intensify to a blackish-brown color, and a yellow halo frequently forms all around the leaf spot. On the stem, petioles, and pods, irregularly sized brown lesions with indistinct edges can also develop (Kirkpatrick et al., 2014).

Although leaf spots typically affect the lower canopy, this disease can be transmitted to the higher canopy under favorable conditions. Severely diseased leaves turn rusty brown as the plant matures and may fall prematurely. Holding sick leaves up to the light exposes dense and opaque brown spot leaf lesions. Bacterial blight lesions, on the other hand, are transparent and may look identical to the untrained eye. *Septoria glycines*, a fungus that overwinters on crop leftovers and infected seed, causes a brown spot (Kirkpatrick et al., 2014). Conidia (spores) released by pycnidia, flask-shaped fruiting structures that occur on crop detritus, cause initial infections on cotyledons and leaves (Johnson and Chamberlain, 1953). Conidia grow on the surface of the leaf and pass through stomata to enter the plant. Conidia are spread upwards in the canopy by the action of wind or dripping rainfall on leaves, petioles, stems, and pods, causing secondary infection (Johnson and Chamberlain, 1953). Infection and sickness can strike at any point during the season. Warm (79°F to 83°F) and humid conditions are ideal for disease growth. On the other hand, hot, dry weather inhibits the spread of illness. As a result, the brown spot is much more serious when soybeans are planted early, especially after lengthy periods of rain, when soybeans are cultivated consistently in the same field, or if the crop is planted in fields with inadequate drainage (Kirkpatrick et al., 2014).

Brown spot susceptibility varies among soybean cultivars; therefore, sowing a less susceptible, suitable cultivar will reduce the disease. Planting good quality seeds, cycling with non-host crop plants (corn, cotton, rice, or sorghum) for two years, and employing tillage measures that limit crop residue on the top of the field are all cultural practices that may help eliminate brown spots. A fungicide application scheduled between the commencement of pod fill (R3) and the first seed production (R5) can be beneficial in avoiding yield loss in rare cases where the brown spot is extensive enough to threaten yield, yield potential is high, and conditions encourage prolonged disease (Kirkpatrick et al., 2014).

10.2.3 VIRAL DISEASES

Viruses infecting soybean are considered a major economic constraint to soybean production. Soybean is infected by almost 46 viruses (Hill and Whitham, 2014), and only a few of them are economically important, such as *soybean mosaic virus* (SMV), *peanut stunt virus* (PSV), *tobacco ringspot virus* (TRSV), *alfalfa mosaic virus* (AMV), *soybean dwarf virus* (SbDV), *bean pod mottle virus* (BPMV), and *soybean vein necrosis virus* (SVNV) (Allen et al., 2017).

10.2.3.1 SMV DISEASE

Among the most deadly and universal diseases, soybean mosaic disease is a major concern. It is caused by a virus called *SMV*. Low-lying areas and hot soybean-growing regions are particularly prone to this problem. The SMV is included in the *Potyvirus* genus in the *Potyviridae* family of plant viruses (Hartman et al., 1999). It is an RNA virus with only one strand containing a genome of 9.5 kb in length that encodes for 11 proteins. The first case of this virus was found in Connecticut in 1915, and it was described in 1921 (Gardner and Kendrick, 1921). SMV can be found in every soybean-cultivating region. Around 21 strains of SMV have been found (Xiuqiang et al., 2003) that have been categorized into seven major groups (G1-G7) (Cho and Goodman, 1979). It is the most predominant viral pathogen of soybean, responsible for causing soybean mosaic disease. The virus causes a yield loss of 35% to 94% (Chen et al., 2015; Liao et al., 2002). It is commonly transmitted by *Aphis glycines* (soybean aphid), and infected seeds can potentially spread the disease. Aphids of many species are involved in the transmission of SMV in healthy plants, and among them, *Acyrthosiphon pisum*, *Myzus persicae*, *A. fabae*, *A. glycines*, and *Rhophalosiphum maidis* are the chief aphid species (Edwardson and Christie, 1991). Infected seeds are smaller in size, less vigorous, and have less oil but more protein and AAs (El-Amretz et al., 1987). Infected leaves are puckered, lightly or darkly mottled, and curled downward. Plants have stunted growth and can also be asymptomatic (Brand et al., 1993), affecting the reproductive parts of the plant. Early plant infection includes reduced seed size and weight because of seed coat mottling, while late infection slightly affects yield and seed quality (Hill, 1987). Symptoms may vary depending on the strain, cultivar, plant age, and environment (Hartman et al., 1999). SMV-infected seeds are also susceptible to *Phomopsis* species (Hepperly et al., 1979). The severity of symptoms is influenced by temperature; symptoms can be severe at a temperature of more than 30°C (Hartman et al., 1999). Besides *Fabaceae*, SMV also causes infection in the members of Chenopodiaceae, Amaranthaceae, Solanaceae, Scrophulariaceae, and Passifloraceae (Hill and Whitham, 2014).

10.2.3.2 BEAN POD MOTTLE VIRUS (BPMV) DISEASE

BPMV is a member of the *Comovirus* genus and is a member of the family Secoviridae. It contains single-stranded RNA as its genetic material and is divided into two isometric particles, each of which is necessary for the disease to manifest (Gergerich, 1999). It was reported in common beans in South Carolina for the first time (Zaumeyer and Thomas, 1948), and it was first found in soybean in Arkansas (Walters, 1958). BPVM is usually passed on by bean beetles and various leaf-eating beetles such as *Diabrotica undecimpunctata howardi*, *Cerotoma trifurcata*, *Epicauta vittata*, *Popillia japonica*, and *Colaspis brunnea*, *Epilachna varivestis* (Gergerich, 1999; Wickizer and Gergerich, 2007). Its reach is limited to members of Fabaceae (Bradshaw et al., 2007). It is ubiquitous in soybean-growing land, including the southern and southeastern United States, and the north-central US

(Giesler et al., 2002). The yield loss caused by BPMV can be as high as 52.6%, depending on the time of infection and the variety of soybean (Hopkins and Mueller, 1984). When BPMV and SMV are combined, yield loss might reach 85% (Ross, 1968). U.S. strains have been categorized into two subcategories and one hybrid group (I/II) depending on nucleic acid hybridization studies. The severity of symptoms varies among isolates of BPMV (Gu et al., 2002), with symptoms being more apparent during rapid growth or in the early season. Symptoms become distinct during low temperatures (Gergerich, 1999). BPMV infection causes a reduction in seed quality and increased susceptibility to *Phomopsis* spp. (Stuckey et al., 1982), discoloration in plants, acute leaf mottling, mild chlorosis, extreme mosaic (Ross, 1968), terminal necrosis, delay in the maturity of soybean stems (Schwenk and Nickell, 1980), and seed coat mottling. The hilum appears to be bleeding due to mottling (Quiniones et al., 1971). BPMV interacts with SMV, thereby reducing seed quality and yield substantially (Calvert and Ghabrial, 1983; Ross, 1968).

10.2.4 NEMATODE DISEASES

Nematodes are one of the disastrous parasites of soybean, as they collectively cause yield loss of 30–40% (Rebois et al., 1970). Yield loss mainly depends on the susceptibility of the cultivar, time of planting, temperature, soil quality, and species of nematode (Lima et al., 2017). While several nematode species are involved in causing disease in soybean, only a few are mentioned as a serious impediment to soybean planting worldwide. These include soybean cyst nematode (SCN) (*Heterodera glycines*), RKN (*Meloidogyne incognita*), root-lesion nematode (*Pratylenchus brachyurus*), and reniform nematode (*Rotylenchulus reniformis*) (Lima et al., 2017).

10.2.4.1 SOYBEAN CYST NEMATODE DISEASE (SCN)

SCN is among the most devastating root parasites of soybean, causing a yield loss of 80%. Yield loss due to this pathogen mainly depends on the type of cultivar, the genetics of the cultivar, and the region (Dias et al., 2010). Around 16 races have been identified to date (Riggs and Schmitt, 1988). It was first found in Japan in 1915. Worldwide, it has spread to many regions and continues to spread to new regions of North America (Tylka and Marett, 2021). It is prevalent in the U.S. (Yu, 2011). It often causes a high level of damage since aboveground symptoms are not evident, but also because the infection is usually discovered at a later stage when a significant amount of damage has already been done (Davis and Tylka, 2000). The life cycle of SCN has three stages: egg, juvenile (j1-j4), and adult. Only the juvenile (j2) stage is responsible for causing infection in a soybean plant (Davis and Tylka, 2000). Symptoms of SCN are similar to those of other stresses such as drought stress, nutrient deficiency, and herbicide toxicity, which can make it difficult to identify the causative agent (Browde et al., 1994). The leaves of soybeans with SCN infection become chlorotic, darker, and faded. Other symptoms include less developed roots, stunting, and reduced nodulation with nitrogen-fixing bacteria. Infected roots become susceptible to soil-borne pathogenic fungi and bacteria. The type of soil also affects how severe a symptom is (Davis and Tylka, 2000). Nematodes living in soybean cysts might persist there for years. They are difficult to eradicate once they are introduced to soybean fields (Dias et al., 2010).

10.2.4.2 ROOT-KNOT DISEASE (RKN)

Root-knot nematode (RKN) commonly parasitizes crop plants due to their predominant nature. They pose grave constraints to crop production worldwide as they cause significant yield loss. Almost 90 species of RKN have been involved in affecting a large number of plant species (Moens et al., 2009). They are endo-sedentary pathogens of soybean as they exploit the root's physiological and molecular pathways for their growth and feeding (Lima et al., 2017). RKN can cause a yield loss of up to 75% when infected by *Meloidogyne incognita* (Emerson et al., 2018). The root-knot disease is an economically important disease caused by several species of Meloidogyne. The most significant RKNs, however, are *M. incognita* and *M. javanica* (Dias et al., 2010). *M. incognita* has widely spread in soybean-producing regions of Brazil, while *M. javanica* has spread globally due to its aggressiveness and favorable conditions. After surveying hundreds of crop fields in the U.S., *M. incognita* was found to be more pervasive (70%) than other species of *Meloidogyne* (Rich, 1985). RKN has three life stages: egg, juvenile, and adult. J2 (second juvenile) is the one that infects soybean plants. By leveraging special enzymes and stylets, penetration is forced inside the vascular bundle of roots, where they set their location for feeding by prompting hyperplasia and hypertrophy. As a result, giant cells form, and swelling occurs (Moens et al., 2009). RKN infection symptoms include irregular stunting, yellowing, necrosis in the leaf vein, production of galls in infected tissue due to hyperplasia and hypertrophy, and altered nutrient and water uptake (Moens et al., 2009).

10.3 MOLECULAR DEFENSE MECHANISM

The effects of disease-causing pathogens are rapid, occurring on molecular and microscopic levels, reducing the chance of plants succeeding. However, every plant is capable of shielding itself from pathogens. Soybean plants rely on a combination of constitutive and inducible defense mechanisms that include oxidative burst, calcium burst, mitogen-activated protein kinase (MAPK) activation, hypersensitivity response (HR), the release of phenolic compounds and phytoalexins, callose production, and expression of pathogenesis-related proteins (Arora et al., 2020; El Rabey et al., 2011; Kumar et al., 2021; Singh and Singh, 2018; Singh et al., 2017b; Suman et al., 2021). The PAMP-triggered immunity system (PTI) causes such a response (Zhang and Zhou, 2010). The pathogenic infection led to significant changes in both metabolomic and transcriptomic levels. Signaling pathways involve the induction of systemic acquired resistance (SAR) that is instigated by salicylic acid (SA) (Fu and Dong, 2013; Singh et al., 2017a). Bacteria interacting with soybean cells trigger a unique recognition event, and a program of cell death is initiated as a result, known as the HR. In order for this process to occur, SA is required (Tenhaken and Rubel, 1997). During bacterial infection, *Glycine max* receptor FLAGELLIN SENSING 2 (GmFLS2) recognizes the bacterial elicitor flg22 protein and initiates defense signaling through GmMAPK3 and GmMAPK6 (Figure 10.1) (Tian et al., 2020). *Glycine max* antimicrobial peptides are found to induce a defense response against nematodes (RKN and SCN) by producing and accumulating reactive oxygen species (ROS) at the apoplast, which can spread throughout the plant, inducing SAR (Lee et al., 2018). Bacterial hairpin induces the expression of other defensive genes such as non-race specific disease resistance 1 (NDR1), rhg4 serine hydroxymethyltransferase, enhanced disease susceptibility (EDS1-2), and rhg1 during effector-triggered immunity (Aljaafri et al., 2017).

A defense response is then elicited by NDR1 via the MAPK cascade (Lee et al., 2001). In contrast, when there is a viral infection, the process appears to have evolved differentially: by disintegrating vesicles, releasing enzymes and cellular content, and inducing programmed cell death (PCD). Upon SMV infection, soybean induces ER stress-induced autophagy by preventing apoptosis on a larger scale through the alleviation of cellular dysfunctions (Díaz-Cruz and Cassone, 2018). WRKY and MYB transcription factors (TFs) play a critical role in soybean immunity to invaders (Liu et al., 2015). During fungal infection, leucine-rich repeat (LRR) serine-threonine protein kinases are activated. Induction of other proteins such as nodulins, chitinases, SNF-1-related protein kinases, and brassinosteroid-insensitive 1-associated receptor kinase 1 also takes place during fungal infection. As noted in soybean, the host defense mechanism is found to be highly active (Figure 10.1) (Lanubile et al., 2015).

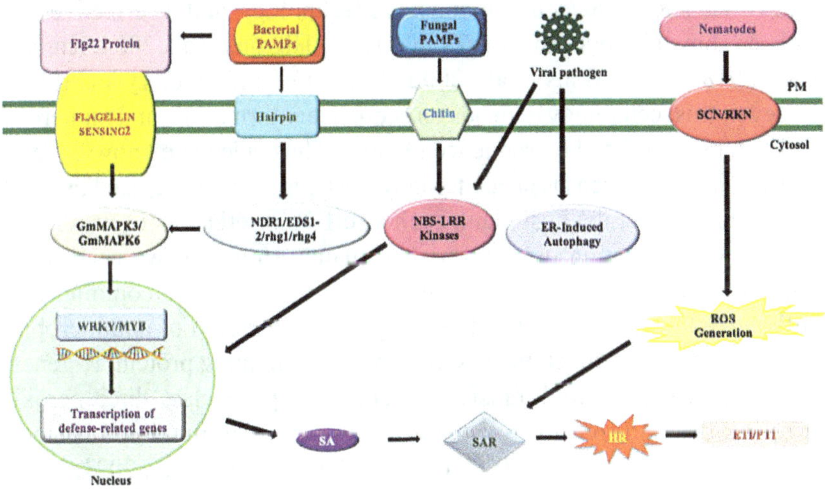

FIGURE 10.1 Soybean defense signaling against its pathogens.

Abbreviations: PM: plasma membrane; PAMPs: pathogen-associated molecular patterns; MAPK: mitogen-activated protein kinase; SCN: soybean cyst nematode; NBS-LRR: nucleotide-binding site–leucine-rich repeat; RKN: root-knot nematode; NDR1: non-race-specific disease resistance 1; ROS: reactive oxygen species; EDS1: enhanced disease susceptibility 1; HR: hypersensitive response; SA: salicylic acid; PTI: PAMP-triggered immunity; SAR: systemic acquired resistance; and ETI: effector-triggered immunity.

10.4 BIOTECHNOLOGICAL INTERVENTIONS AND GENOME EDITING (GE) TO PREVENT DISEASES

Many challenges impede the sustainable generation of soybeans, such as pests, pathogens, weeds, and even nutrient deficiency. Hence, different approaches are being taken to cultivate soybeans sustainably. Transgenesis, for instance, has proven to be effective in eliminating such challenges and ensuring the food supply. By using genetic engineering techniques such as gene silencing and transgenics, soybeans can be significantly more adaptable (Dunwell, 2000). With genetic engineering, it is now possible to express and incorporate desirable genes from one organism into another. Genetic engineering offers the possibility of incorporating a disease-resistance trait in plants without harming the environment. For example, transgenic plants based on RNA silencing technology for viral diseases have already been constructed in other plants, where mostly the coat protein of viruses has been targeted via siRNA silencing or antisense technology, and its effectiveness has been shown in soybean crops (Table 10.1) (Higgins et al., 2004; Kumari et al., 2018; Ozias-Akins et al., 2002; Tougou et al., 2006; Tylka, 1994). However, this technology depends on sequence information, and numerous viral strains are related by sequence identity. Thus, adequate knowledge of viral genome sequences is needed. Many host genes are being used by viral pathogens to infect the host plant, such as eIF4E or eIF(iso)4E, which can be utilized for creating disease-resistant plants. However, we need more significant information on how soybean and viral proteins communicate (Hill and Whitham, 2014). Contrastingly, overexpression of antimicrobial genes such as defensin, chitinases, ribosome-inactivating protein, R-genes, beta-1,3 glucanases, lipid-transfer proteins, and plant class three peroxidases has been utilized for resistance to microbes like fungi and bacteria (Table 10.1) (Jaiswal et al., 2012; Li et al., 2004; Mehta et al., 2021; Mohr and Cahill, 2007; Punja, 2001; Sarowar et al., 2009; Singh et al., 2018; Soto et al., 2020). For creating nematode-resistant crops, various protease inhibitors (PIs), namely cowpea trypsin inhibitors, serine proteases, and cystatins, have been overexpressed (Lilley et al., 1999). Targeting of nematode effectors via RNA interference (RNAi) technology has also been done (Klink et al., 2009). In soybean, conventional breeding along with transgenic techniques has been applied for decades. However, transgenic soybeans are viewed with disapproval by the public because of social concerns like transgenes. Hence, people are shifting to GE or new plant breeding technologies (NPBT) such as CRISPR/Cas, zinc finger nucleases (ZFN), and transcription activators like effector nucleases (TALENs) because they are cost-effective and transgene-free (Rahman et al., 2022). There have been many improvements made to soybean traits with TALEN and CRISPR/Cas for fatty acid (FA) and carotenoid biosynthesis (Haun et al., 2014). GE enables us to create desirable alterations in an organism's genomic DNA. NPBT operates by creating double-strand breaks (DSBs) in a targeted sequence of DNA by using artificial nucleases that are specific to a certain sequence. Non-homologous end-joining (NHEJ) and homologous recombination repair are the two possible routes for restoring DSBs (Pacher and Puchta, 2017). ZFN depends on the interaction of Cys2-His2 zinc finger protein with FokI nucleases for producing DSBs (Urnov et al., 2005). TALEN is similar to ZFN as it also depends on DNA-protein interaction; it is formed through the fusion of TALE repeats and the domain of FokI nuclease cleavage (Christian et al., 2010). The CRISPR/Cas system has been derived from prokaryotic adaptive immunity and depends on RNA-DNA interaction, where CRISPR RNA guides Cas nucleases to the targeted sequence. The type II system of CRISPR/Cas, which has three different types, is the most frequently employed for GE (Barrangou et al., 2007; Deltcheva et al., 2011). NPBT predicts genes' functions either by knocking them out or by knocking them in Mohanta et al. (2017). Due to recent advancements in CRISPR/Cas, gene knock-in is possible, while this system is still lacking when it comes to soybean (Soda et al., 2018). More research on the CRISPR/Cas system is needed for creating transgene-free disease-resistant soybean plants. It is important to find disease-susceptible genes that can be modified via NPBT. Other plants such as rice, tomato, and *Arabidopsis* have successfully used these technologies (Nekrasov et al., 2017; Pyott et al., 2016). Assembly of multiple *R*-genes, such as the nucleotide-binding site-leucine-rich repeat (*NBS-LRR*) mediated by CRISPR/Cas, may facilitate broad-spectrum disease resistance (Giannakopoulou et al., 2015). Newly evolved CRISPR interference (CRISPRi) controls specific genes and enables multiplex gene editing and can also regulate similar genes that can ensure extensive resistance. By modifying the sequence of target genes with an open reading frame, it is possible to regulate their translation (Zhang et al., 2018). Even though there are many obstacles to GE in soybean production, it is also an area of opportunity. However, as other NPBT systems evolve, GE for soybeans will have a higher degree of flexibility.

TABLE 10.1 Genetic Engineering Methods, Targeted Genes/Proteins for Disease-Resistant Transgenic Soybean Plants

Disease Category	Causative Agent	Symptoms	Prevalence	Targeted Genes/ Protein	Expression Pattern	Genetic Engineering Method/Other Control Method	References
Bacterial Disease							
Wildfire	*Pseudomonas syringae* pv. *tabaci*	Brown necrotic spots, yellowish haloes on leaf	—	*nptII* gene and *ttr* (tabtoxin resistance) gene	Normal expression	*Agrobacterium*-mediated transformation (Tobacco)	Batchvarova et al. (1998); Myung et al. (2009)
Fungal Disease							
Anthracnose	*Colletotrichum truncatum*	Depressed spots, pod rot, premature pod opening, smooth necrotic veins	Worldwide	*NmDef02* (defensin)	Normal expression	Particle bombardment, foliar fungicides, seed treatment	Boufleur et al. (2021); Soto et al. (2020)
Soybean rust disease	*Phakopsora pachyrhizi*	Rust pustules, small lesions on leaf veins	Worldwide	*Rpp1–6* genes	—	*Rpp* gene pyramiding via genetic crossings	Langenbach et al. (2016)
Brown spot	*Septoria glycines*	Brown necrotic lesions of uneven size	Southern USA	Tobacco osmotin (*Tbosm*)	Overexpression	*Agrobacterium*-mediated transformation (Tobacco), foliar fungicides	Singh et al. (2020); Subramanyam et al. (2012)
Charcoal rot	*Macrophomina phaseolina*	Yellow pods, chlorosis, blackleg, vascular blockages	Worldwide	Chitinase (*chi11*)	Normal expression	*Agrobacterium*-mediated transformation (Jute), strip/ no-till, irrigation, crop rotation	Khalili et al. (2020); Majumder et al. (2018)
Damping-off	*Phytophthora sojae*	Seedling rot, root-stem rot, vascular blockages	Worldwide	Isoflavone synthase (*IFS*)/ Chalcone reductase (*CHR*)	Suppression	*Agrobacterium rhizogenes*-mediated RNAi	Graham et al. (2007); Tyler (2007)
Fusarium root rot	*Fusarium* spp.	Chlorosis, necrosis, wilting, reduced pod count	Worldwide	*rVrD1*	Overexpression	*Agrobacterium*-mediated transformation, fungicides, stress reduction	Jasnić et al. (2005); Li et al. (2004)
Powdery mildew	*Microsphaera diffusa*	White colonies, chlorosis, necrotic specks	South China, Brazil, other soybean regions	*Tbosm*	Overexpression	*Agrobacterium*-mediated transformation (Tobacco), fungicides	Subramanyam et al. (2012); Zhou et al. (2022)

TABLE 10.1 *(Continued)*

Disease Category	Causative Agent	Symptoms	Prevalence	Targeted Genes/ Protein	Expression Pattern	Genetic Engineering Method/Other Control Method	References
Sclerotium blight	*Sclerotium rolfsii*	Yellowing, wilting, interveinal chlorosis	Worldwide	Rice chitinase gene	Normal expression	Particle bombardment, *Agrobacterium*-mediated transformation (Taro), rotation, weed control	He et al. (2010); Lin et al. (2022)
Sclerotinia stem rot	*Sclerotinia minor*, *S. sclerotiorum*	Watery lesions, wilting, pod infection	North Central USA, China, Canada	Oxalate oxidase gene	Normal expression	Transformation (Barley), spacing, rotation, fungicides	Lin et al. (2022); Livingstone et al. (2005)
Foliar blight/ RAB	*Rhizoctonia solani*	White specks on leaves, non-regular brown spots on leaves upon maturity, water-soaked lesions, dark brown streaks on petiole of leaf, blight in leaves and floral parts, and yellow-lemon ring surrounding spots	Tropical regions – Brazil, India, Japan, China, Philippines, Puerto Rico, Taiwan, and United State	*rVrD1*	Overexpression	*Agrobacterium*-mediated transformation, tillage, less dense planting, fungicides	Li et al. (2004); Yang (1989)
				Viral Disease			
Alfalfa mosaic virus disease	*Alfalfa mosaic virus*	Dwarfing, vein browning, leaf distortion, mosaic/ mottled leaf pattern	Worldwide	*dsABS* (short RNA hairpins)	Normal expression	*Agrobacterium*-mediated transformation, RNA silencing	Mueller and Grau (2007); Zhang et al. (2011)
Bean pod mottle viral disease	*Bean pod mottle virus*	Pod/seed coat mottling, chlorotic leaf mottling, terminal necrosis	USA	*PAC1* (RNase)	Overexpression	*Agrobacterium*-mediated transformation, leaf beetle management	Giesler et al. (2002); Yang et al. (2019)
Soybean mosaic disease	*Potyvirus soybean mosaic virus*	Curled, puckered leaves, small mottled seeds	Worldwide	*PAC1* (RNase)	Overexpression	*Agrobacterium*-mediated transformation, early planting	Usovsky et al. (2022); Yang et al. (2019)

TABLE 10.1 *(Continued)*

Disease Category	Causative Agent	Symptoms	Prevalence	Targeted Genes/ Protein	Expression Pattern	Genetic Engineering Method/Other Control Method	References
Dwarf disease	*Luteovirus soybean dwarf virus*	Stunting, yellowing, rugosity, puckered leaves	USA	*SbDV-CP* gene	Suppression	RNA silencing, particle bombardment, insecticide-treated resistant lines	Hill and Whitham (2014); Tougou et al. (2006)
Soybean vein necrosis	*Orthotospovirus soybean vein necrosis virus*	Chlorotic patches, necrosis, leaf death, darkened veins	North America	–	–	Thrips management	Zhou and Tzanetakis (2013)
Bud blight	*Nepovirus tobacco ringspot virus*	Shepherd's crook, bud necrosis, floral bud proliferation, dwarf leaflets	Worldwide	*PDS* (phytoene desaturase)	Suppression	Virus-induced gene silencing, *Agrobacterium*-mediated transformation, dagger nematode control	Hartman et al. (2015); Zhao et al. (2016)
Peanut stripe	*Peanut stripe virus*	Leaf curling, viral mosaic, necrotic lesions	Africa, Asia, North America	Coat protein	Suppression	Particle bombardment (peanut plant)	Higgins et al. (2004); Hill and Whitham (2014)
Bean yellow mosaic disease	*Mung bean yellow mosaic virus*	Yellow blotches, crinkled leaves, rugosity, stunted growth	South Asia	Coat protein	Suppression	*Agrobacterium*-mediated transformation, RNA silencing, weed removal	Kumari et al. (2018); Usharani et al. (2005)
Spotted wilt disease	*Orthotospovirus tomato spotted wilt virus*	Atypical or absent TSWV symptoms	Iran, Georgia, Korea	Coat protein	Suppression	Particle bombardment (peanut plant), antisense technology	Ozias-Akins et al. (2002); Yoon et al. (2018)
				Nematode Disease			
Soybean cyst nematode disease	*Heterodera glycines*	Symptoms mimic abiotic stress	USA	*Snb-1* (synaptobrevin), *cry14AB*	Suppression, expression	RNA interference, aerosol beam microinjector, crop rotation	Kahn et al. (2021); Klink et al. (2009)

TABLE 10.1 *(Continued)*

Disease Category	Causative Agent	Symptoms	Prevalence	Targeted Genes/ Protein	Expression Pattern	Genetic Engineering Method/Other Control Method	References
Root-knot disease	*Meloidogyne* spp.	Stunting, yellowing, vein necrosis, gall formation	Worldwide	*L-lactate dehydrogenase, tyrosine phosphatase, stress-70 protein precursor, ATP synthase subunit beta*	Suppression	RNA interference, resistant cultivars, crop rotation	Ibrahim et al. (2011)
Reniform nematode disease	*Rotylenchulus reniformis*	Necrotic, stunted roots; chlorotic plants	USA	Recombinant proteinase inhibitors	Normal expression	*Agrobacterium*-mediated transformation (*Arabidopsis thaliana*), tolerant cultivars	Niblack et al. (2004); Urwin et al. (2000)
Lesion nematode disease	*Pratylenchus* spp.	Chlorosis, root lesions, sunken necrotic regions	Africa, Brazil, Southern USA	Troponin, calponin	Suppression	RNAi (soaking method), carrot plant, tolerant cultivars	Niblack et al. (2004); Tan et al. (2013)

10.5 CONCLUSION AND FUTURE PERSPECTIVE

Since soybeans are such an important part of the global agricultural economy, crop protection against persistent, intermittent, and new diseases is critical. This analysis looks at all of the known yield-limiting diseases that influence soybean cultivation across the world. Increased pressure from agricultural pests, such as weeds and insects, along with pathogens and nematodes, is expected when temperatures and precipitation rise, as in the case of soybean illness or disease.

The development of disease prediction systems that can be used to predict within-season risks and forthcoming epidemics should receive more attention. Technologies like these might be used to make long-lasting or interminable decisions in order to clearly predict the developing infection concerns and modify treatment approaches accordingly. In conclusion, future soybean output is assumed to rise in proportion to growing demand. Two stages will be critical for fulfilling the rising demand for agricultural goods: (i) Predicting major disease outbreaks and efficiently protecting crops against infections; and (ii) prioritizing yield improvements via genetic resistance. Given the time and money savings, as well as the fair yield boost, soybean growers have begun to turn to NBPT and biotechnology for crop enhancement. Future technological advancement assures enhanced grain value and high yield through gene modification. Understanding the primary soybean diseases, especially those that are developing, is critical in order to be prepared for future challenges. An increase in the usage of hereditary materials or genetic resources, via bioengineering and molecular breeding, might provide the answers required to combat future pests, illnesses, and food-related issues.

KEYWORDS

- **biotechnological interventions**
- **crop protection**
- **disease prediction systems**
- **genetic engineering**
- **molecular breeding**
- **pathogen resistance**
- **soybean diseases**

REFERENCES

Aljaafri, W. A. R., McNeece, B. T., Lawaju, B. R., Sharma, K., Niruala, P. M., Pant, S. R., et al. (2017). A harpin elicitor induces the expression of a coiled-coil nucleotide binding leucine rich repeat (CC-NB-LRR) defense signaling gene and others functioning during defense to parasitic nematodes. *Plant Physiology and Biochemistry, 121*, 161–175.

Allen, T. W., Bradley, C. A., Sisson, A. J., Byamukama, E., Chilvers, M. I., Coker, C. M., et al. (2017). Soybean yield loss estimates due to diseases in the United States and Ontario, Canada, from 2010 to 2014. *Plant Health Progress, 18*(1), 19–27.

Arora, R., Kumar, A., Singh, I. K., & Singh, A. (2020). Pathogenesis-related proteins: A defensin for plants but an allergen for humans. *International Journal of Biological Macromolecules, 157*, 659–672.

Bandara, A. Y., Weerasooriya, D. K., Bradley, C. A., Allen, T. W., & Esker, P. D. (2020). Dissecting the economic impact of soybean diseases in the United States over two decades. *PLoS ONE, 15*(4), e0231141.

Barrangou, R., Fremaux, C., Deveau, H., Richards, M., Boyaval, P., Moineau, S., et al. (2007). CRISPR provides acquired resistance against viruses in prokaryotes. *Science, 315*(5819), 1709–1712.

Batchvarova, R., Nikolaeva, V., Slavov, S., Bossolova, S., Valkov, V., Atanassova, S., et al. (1998). Transgenic tobacco cultivars resistant to *Pseudomonas syringae* pv. *tabaci*. *Theoretical and Applied Genetics, 97*, 986–989.

Boufleur, T. R., Ciampi-Guillardi, M., Tikami, Í., Rogério, F., Thon, M. R., Sukno, S. A., et al. (2021). Soybean anthracnose caused by *Colletotrichum* species: Current status and future prospects. *Molecular Plant Pathology, 22*(4), 393–409.

Bradshaw, J. D., Rice, M. E., & Hill, J. H. (2007). No-choice preference of *Cerotoma trifurcata* (Coleoptera: Chrysomelidae) to potential host plants of *Bean pod mottle virus* (Comoviridae) in Iowa. *Journal of Economic Entomology, 100*(3), 808–814.

Brand, R. J., Burger, J. T., & Rybicki, E. P. (1993). Cloning, sequencing, and expression in *Escherichia coli* of the coat protein gene of a new potyvirus infecting South African *Passiflora*. *Archives of Virology, 128*(1), 29–41.

Browde, J. A., Pedigo, L. P., Owen, M. D. K., & Tylka, G. L. (1994). Growth of soybean stressed by nematodes, herbicides, and simulated insect defoliation. *Agronomy Journal, 86*(6), 968–974.

Calvert, L. A., & Ghabrial, S. A. (1983). Enhancement by *Soybean mosaic virus* of *Bean pod mottle virus* titer in doubly infected soybean. *Phytopathology, 73*(7), 992–997.

Chen, H., Zhang, L., Yu, K., & Wang, A. (2015). Pathogenesis of *Soybean mosaic virus* in soybean carrying *Rsv1* gene is associated with miRNA and siRNA pathways, and breakdown of *AGO1* homeostasis. *Virology, 476*, 395–404.

Cho, E. K., & Goodman, R. M. (1979). Strains of *Soybean mosaic virus*: Classification based on virulence in resistant soybean cultivars. *Phytopathology, 69*(5), 467–470.

Christian, M., Cermak, T., Doyle, E. L., Schmidt, C., Zhang, F., Hummel, A., et al. (2010). Targeting DNA double-strand breaks with TAL effector nucleases. *Genetics, 186*(2), 757–761.

Clemente, T. E., & Cahoon, E. B. (2009). Soybean oil: Genetic approaches for modification of functionality and total content. *Plant Physiology, 151*(3), 1030–1040.

Cruz, C. D., Mills, D., Paul, P. A., & Dorrance, A. E. (2010). Impact of brown spot caused by *Septoria glycines* on soybean in Ohio. *Plant Disease, 94*(7), 820–826.

Davis, E. L., & Tylka, G. L. (2000). Soybean cyst nematode disease. *The Plant Health Instructor.*

Deltcheva, E., Chylinski, K., Sharma, C. M., Gonzales, K., Chao, Y., Pirzada, Z. A., et al. (2011). CRISPR RNA maturation by trans-encoded small RNA and host factor RNase III. *Nature, 471*(7340), 602–607.

Dias, W. P., Garcia, A., Silva, J. F. V., & Carneiro, G. E. de S. (2010). *Nematóides em soja: Identificação e Controle*. Londrina.

Díaz-Cruz, G. A., & Cassone, B. J. (2018). A tale of survival: Molecular defense mechanisms of soybean to overcome *Soybean mosaic virus* infection. *Physiological and Molecular Plant Pathology, 102*, 79–87.

Dorrance, A. E. (2018). Management of *Phytophthora sojae* of soybean: A review and future perspectives. *Canadian Journal of Plant Pathology, 40*(2), 210–219.

Dunwell, J. M. (2000). Transgenic approaches to crop improvement. *Journal of Experimental Botany, 51*(suppl_1), 487–496.

Edwardson, J. R., & Christie, R. G. (1991). *The Potyvirus Group* (Vols. *I–IV*). Florida Agricultural Experiment Station Monograph.

El Rabey, H., Abdallah, S., & Ahmed, A. (2011). Biochemical and molecular investigations on prime mechanisms of induced resistance in French bean (*Phaseolus vulgaris* L.) and soybean (*Glycine max* L.). *African Journal of Agricultural Research, 6*(28), 6132–6146.

El-Amretz, A. A., El-Said, H. M., & Salem, D. E. (1987). Effect of *Soybean mosaic virus* infection on quality of soybean seed. *Agricultural Research Review, 63*, 155–164.

Emerson, M., Brown, K., Faske, T. R., & Kirkpatrick, T. L. (2018). Field performance of several glyphosate-resistant maturity group 4 and 5 soybean cultivars in a root-knot nematode-infested field. *Arkansas Soybean Research Studies, 648*, 40–42.

Flor, H. H. (1955). Host-parasite interaction in flax rust—Its genetics and other implications. *Phytopathology, 45*, 680–685.

Fu, Z. Q., & Dong, X. (2013). Systemic acquired resistance: Turning local infection into global defense. *Annual Review of Plant Biology, 64*, 839–863.

Gardner, N. W., & Kendrick, J. B. (1921). Soybean mosaic. *Journal of Agricultural Research, 22*, 111–114.

Gergerich, R. C. (1999). Comoviruses: *Bean pod mottle comovirus*. *Compendium of Soybean Diseases* (4th ed., pp. 61–62). American Phytopathological Society.

Giannakopoulou, A., Steele, J. F. C., Segretin, M. E., Bozkurt, T. O., Zhou, J., Robatzek, S., et al. (2015). Tomato *I2* immune receptor can be engineered to confer partial resistance to the oomycete *Phytophthora infestans* in addition to the fungus *Fusarium oxysporum*. *Molecular Plant-Microbe Interactions, 28*(12), 1316–1329.

Giesler, L. J., & Weissling, T. J. (2004). *Bacterial Diseases of Soybean.* Cooperative Extension, Institute of Agriculture and Natural Resources.

Giesler, L. J., Ghabrial, S. A., Hunt, T. E., & Hill, J. H. (2002). *Bean pod mottle virus*: A threat to U.S. soybean production. *Plant Disease, 86*(12), 1280–1289.

Graham, T. L., Graham, M. Y., Subramanian, S., & Yu, O. (2007). RNAi silencing of genes for elicitation or biosynthesis of 5-deoxyisoflavonoids suppresses race-specific resistance and hypersensitive cell death in *Phytophthora sojae* infected tissues. *Plant Physiology, 144*(2), 728–740.

Gu, H., Clark, A. J., De Sa, P. B., Pfeiffer, T. W., Tolin, S., & Ghabrial, S. A. (2002). Diversity among isolates of *Bean pod mottle virus*. *Phytopathology, 92*(4), 446–452.

Hartman, G. L., & Hill, C. B. (2010). Diseases of soybean and their management. In G. Singh (Ed.), *The Soybean: Botany, Production and Uses* (pp. 276–299). CABI.

Hartman, G. L., Rupe, J. C., Sikora, E. J., Domier, L. L., Davis, J. A., & Steffey, K. L. (2015). *Compendium of Soybean Diseases and Pests.* American Phytopathological Society.

Hartman, G. L., Sinclair, J. B., & Rupe, J. C. (1999). *Compendium of Soybean Diseases* (pp. 37–39). American Phytopathological Society.

Haun, W., Coffman, A., Clasen, B. M., Demorest, Z. L., Lowy, A., Ray, E., et al. (2014). Improved soybean oil quality by targeted mutagenesis of the fatty acid desaturase 2 gene family. *Plant Biotechnology Journal, 12*(7), 934–940.

He, X., Miyasaka, S. C., Zou, Y., Fitch, M. M. M., & Zhu, Y. J. (2010). Regeneration and transformation of taro (*Colocasia esculenta*) with a rice chitinase gene enhances resistance to *Sclerotium rolfsii*. *HortScience, 45*(7), 1014–1020.

Hepperly, P. R., Bowers, G. R., Jr., Sinclair, J. B., & Goodman, R. M. (1979). Predisposition to seed infection by *Phomopsis sojae* in soybean plants infected by *Soybean mosaic virus*. *Phytopathology, 69*(8), 846–848.

Higgins, C. M., Hall, R. M., Mitter, N., Cruickshank, A., & Dietzgen, R. G. (2004). *Peanut stripe potyvirus* resistance in peanut (*Arachis hypogaea* L.) plants carrying viral coat protein gene sequences. *Transgenic Research, 13*, 59–67.

Hill, J. H. (1987). *Soybean mosaic virus*: Effects of primary disease incidence on yield and seed quality. *Plant Disease, 71*(3), 237–239.

Hill, J. H., & Whitham, S. A. (2014). Control of virus diseases in soybeans. *Advances in Virus Research* (pp. 355–390). Elsevier.

Hopkins, J. D., & Mueller, A. J. (1984). Effect of *Bean pod mottle virus* on soybean yield. *Journal of Economic Entomology, 77*(4), 943–947.

Horvath, A. A. (1926). Changes in the blood composition of rabbits fed on raw soybeans. *Journal of Biological Chemistry, 68*(2), 343–355.

Ibrahim, H. M. M., Alkharouf, N. W., Meyer, S. L. F., Aly, M. A. M., Abd El Kader, Y., Hussein, E. H. A., et al. (2011). Post-transcriptional gene silencing of root-knot nematode in transformed soybean roots. *Experimental Parasitology, 127*(1), 90–99.

Isanga, J., & Zhang, G. N. (2008). Soybean bioactive components and their implications to health—A review. *Food Reviews International, 24*(2), 252–276.

Jagtap, G. P., Dhopte, S. B., & Dey, U. (2012). Bio-efficacy of different antibacterial antibiotic, plant extracts and bioagents against bacterial blight of soybean caused by *Pseudomonas syringae* pv. *glycinea*. *Scientific Journal of Microbiology, 1*(1), 1–9.

Jaiswal, P., Cheruku, J. R., Kumar, K., Yadav, S., Singh, A., Kumari, P., et al. (2012). Differential transcript accumulation in chickpea during early phases of compatible interaction with a necrotrophic fungus *Ascochyta rabiei*. *Molecular Biology Reports, 39*, 4635–4646.

Jasnić, S. M., Vidić, M. B., Bagi, F. F., & Đorđević, V. B. (2005). Pathogenicity of *Fusarium* species in soybean. *Zbornik Matice Srpske za Prirodne Nauke, 109*, 113–121.

Jasnić, S., & Vidić, M. (2008). Diseases of the soybean root and basal stem. *Biljni Lekar, 36*(3, 4), 214–221.

Johnson, H. W., & Chamberlain, D. W. (1953). Bacteria, fungi and viruses on soybeans. *USDA Yearbook, 238*, 247.

Kahn, T. W., Duck, N. B., McCarville, M. T., Schouten, L. C., Schweri, K., Zaitseva, J., et al. (2021). A *Bacillus thuringiensis* Cry protein controls soybean cyst nematode in transgenic soybean plants. *Nature Communications, 12*(1), 3380.

Khalili, E., Kouchaki, S., Ramazi, S., & Ghanati, F. (2020). Machine learning techniques for soybean charcoal rot disease prediction. *Frontiers in Plant Science, 11*, 590529.

Kim, D. H., Kim, K. H., Van, K., & Lee, S. H. (2010). Fine mapping of a resistance gene to bacterial leaf pustule in soybean. *Theoretical and Applied Genetics, 120*, 1443–1450.

Kirkpatrick, T. L., Faske, T., & Robbins, R. T. (2014). *Nematode Management in Arkansas Soybean Production Handbook* (Vol. 197). Division of Agriculture, Cooperative Extension Service.

Klink, V. P., Kim, K. H., Martins, V., MacDonald, M. H., Beard, H. S., Alkharouf, N. W., et al. (2009). A correlation between host-mediated expression of parasite genes as tandem inverted repeats and abrogation of development of female *Heterodera glycines* cyst formation during infection of *Glycine max*. *Planta, 230*, 53–71.

Krishnamurthy, K. (1975). *Soybean Production in Karnataka*. University of Agricultural Sciences.

Kumar, J., Ramlal, A., Kumar, K., Rani, A., & Mishra, V. (2021). Signaling pathways and downstream effectors of host innate immunity in plants. *International Journal of Molecular Sciences, 22*(16), 9022.

Kumari, A., Hada, A., Subramanyam, K., Theboral, J., Misra, S., Ganapathi, A., et al. (2018). RNAi-mediated resistance to yellow mosaic viruses in soybean targeting coat protein gene. *Acta Physiologiae Plantarum, 40*, 1–12.

Langenbach, C., Campe, R., Beyer, S. F., Mueller, A. N., & Conrath, U. (2016). Fighting Asian soybean rust. *Frontiers in Plant Science, 7*, 797.

Lanubile, A., Muppirala, U. K., Severin, A. J., Marocco, A., & Munkvold, G. P. (2015). Transcriptome profiling of soybean (*Glycine max*) roots challenged with pathogenic and non-pathogenic isolates of *Fusarium oxysporum*. *BMC Genomics, 16*, 1–14.

Lee, J., Klessig, D. F., & Nürnberger, T. (2001). A harpin binding site in tobacco plasma membranes mediates activation of the pathogenesis-related gene HIN1 independent of extracellular calcium but dependent on mitogen-activated protein kinase activity. *The Plant Cell, 13*(5), 1079–1093.

Lee, M. W., Huffaker, A., Crippen, D., Robbins, R. T., & Goggin, F. L. (2018). Plant elicitor peptides promote plant defenses against nematodes in soybean. *Molecular Plant Pathology, 19*(4), 858–869.

Li, H. Y., Zhu, Y. M., Chen, Q., Conner, R. L., Ding, X. D., Li, J., et al. (2004). Production of transgenic soybean plants with two antifungal protein genes via *Agrobacterium* and particle bombardment. *Biologic Plantarum, 48*, 367–374.

Liao, L., Chen, P., Buss, G. R., Yang, Q., & Tolin, S. A. (2002). Inheritance and allelism of resistance to soybean mosaic virus in Zao18 soybean from China. *Journal of Heredity, 93*(6), 447–452.

Lilley, C. J., Devlin, P., Urwin, P. E., & Atkinson, H. J. (1999). Parasitic nematodes, proteinases, and transgenic plants. *Parasitology Today, 15*(10), 414–417.

Lima, F. S. O., Correa, V. R., Nogueira, S. R., & Santos, P. R. R. (2017). Nematodes affecting soybean and sustainable practices for their management. In *Soybean—Basis, Yield, Biomass Production* (pp. 95–110). InTech.

Lin, F., Chhapekar, S. S., Vieira, C. C., Da Silva, M. P., Rojas, A., Lee, D., et al. (2022). Breeding for disease resistance in soybean: A global perspective. *Theoretical and Applied Genetics, 135*(11), 3773–3872.

Liu, J. Z., Graham, M. A., Pedley, K. F., & Whitham, S. A. (2015). Gaining insight into soybean defense responses using functional genomics approaches. *Briefings in Functional Genomics, 14*(4), 283–290.

Livingstone, D. M., Hampton, J. L., Phipps, P. M., & Grabau, E. A. (2005). Enhancing resistance to *Sclerotinia minor* in peanut by expressing a barley oxalate oxidase gene. *Plant Physiology, 137*(4), 1354–1362.

Maitra, S. (2018). Soybean (Bhatman). In *Forage Crop World, Vol. II: Minor Forage Crops* (pp. 121–140). Apple Academic Press.

Majumder, S., Datta, K., Sarkar, C., Saha, S. C., & Datta, S. K. (2018). The development of *Macrophomina phaseolina* (fungus)-resistant and glufosinate (herbicide)-tolerant transgenic jute. *Frontiers in Plant Science, 9*, 920.

Mehta, S., Chakraborty, A., Roy, A., Singh, I. K., & Singh, A. (2021). Fight hard or die trying: Current status of lipid signaling during plant–pathogen interaction. *Plants, 10*(6), 1098.

Miloš, V., Vuk, Đ., Kristina, P., & Jegor, M. (2013). Review of soybean resistance to pathogens. *Institute of Field and Vegetable Crops, Novi Sad, 50*(2), 52–61.

Moens, M., Perry, R. N., & Starr, J. L. (2009). *Meloidogyne* species—A diverse group of novel and important plant parasites. In *Root-Knot Nematodes* (Vol. 1, pp. 483). CAB International.

Mofijur, M., Atabani, A. E., Masjuki, H. H., Kalam, M. A., & Masum, B. M. (2013). A study on the effects of promising edible and non-edible biodiesel feedstocks on engine performance and emissions production: A comparative evaluation. *Renewable and Sustainable Energy Reviews, 23*, 391–404.

Mohanta, T. K., Bashir, T., Hashem, A., Abd_Allah, E. F., & Bae, H. (2017). Genome editing tools in plants. *Genes, 8*(12), 399.

Mohr, P. G., & Cahill, D. M. (2007). Suppression by ABA of salicylic acid and lignin accumulation and the expression of multiple genes in *Arabidopsis* infected with *Pseudomonas syringae* pv. *tomato*. *Functional & Integrative Genomics, 7*, 181–191.

Mueller, E. E., & Grau, C. R. (2007). Seasonal progression, symptom development, and yield effects of *Alfalfa mosaic virus* epidemics on soybean in Wisconsin. *Plant Disease, 91*(3), 266–272.

Myung, I. S., Kim, J. W., An, S. H., Lee, J. H., Kim, S. K., Lee, Y. K., et al. (2009). Wildfire of soybean caused by *Pseudomonas syringae* pv. *tabaci*, a new disease in Korea. *Plant Disease, 93*(11), 1214.

Nekrasov, V., Wang, C., Win, J., Lanz, C., Weigel, D., & Kamoun, S. (2017). Rapid generation of a transgene-free powdery mildew-resistant tomato by genome deletion. *Scientific Reports, 7*(1), 1–6.

Niblack, T. L., Tylka, G. L., & Riggs, R. D. (2004). Nematode pathogens of soybean. In *Soybeans: Improvement, Production, and Uses* (Vol. 16, pp. 821–851). Wiley Online Library.

O'Neal, M. E., & Johnson, K. D. (2010). Insect pests of soybean and their management. In *The Soybean* (pp. 300).

Ozias-Akins, P., Yang, P., Culbreath, A. K., Gorbet, D. W., & Weeks, J. R. (2002). Field resistance to *Tomato spotted wilt virus* in a transgenic peanut (*Arachis hypogaea* L.). *Proceedings of the American Peanut Research and Education Society, 70.*

Pacher, M., & Puchta, H. (2017). From classical mutagenesis to nuclease-based breeding—Directing natural DNA repair for a natural end-product. *The Plant Journal*, 819–833.

Powles, S. B. (2010). Gene amplification delivers glyphosate-resistant weed evolution. *Proceedings of the National Academy of Sciences, 107*(3), 955, 956.

Prathuangwong, S., & Amnuaykit, K. (1987). Studies on tolerance and rate-reducing bacterial pustule of soybean cultivars/lines. *Agriculture and Natural Resources, 21*(4), 408–426.

Punja, Z. K. (2001). Genetic engineering of plants to enhance resistance to fungal pathogens—A review of progress and future prospects. *Canadian Journal of Plant Pathology, 23*(3), 216–235.

Pyott, D. E., Sheehan, E., & Molnar, A. (2016). Engineering of CRISPR/Cas9-mediated potyvirus resistance in transgene-free *Arabidopsis* plants. *Molecular Plant Pathology, 17*(8), 1276–1288.

Quiniones, S. S., Dunleavy, J. M., & Fisher, J. W. (1971). Performance of three soybean varieties inoculated with soybean mosaic virus and bean pod mottle virus. *Crop Science, 11*(5), 662–664.

Raghuvanshi, R. S., & Bisht, K. (2010). *18 Uses of Soybean: Products and Preparation.* The Soybean.

Rahman, M., Zulfiqar, S., Raza, M. A., Ahmad, N., & Zhang, B. (2022). Engineering abiotic stress tolerance in crop plants through CRISPR genome editing. *Cells, 11*(22), 3590.

Rajendran, A., Lal, S. K., Raju, D., & Ramlal, A. (2022). Associations of direct and indirect selection for pregermination anaerobic stress tolerance in soybean (*Glycine max*). *Plant Breeding, 141*(5), 634–643.

Rajendran, A., Lal, S. K., Raju, D., Mallikarjun, B. P., Ramlal, A., & Sharma, D. (2023). Waterlogging tolerance evaluation methods for soybean (*Glycine max* (L.) Merr.) at the pregermination stage. *Genetic Resources and Crop Evolution, 70*, 1–11.

Ramlal, A., Nautiyal, A., Baweja, P., Kumar, V., Mehta, S., Mahto, R. K., et al. (2022). Angiotensin-converting enzyme inhibitory peptides and isoflavonoids from soybean (*Glycine max* (L.) Merr.). *Frontiers in Nutrition, 9*, 1068388.

Rebois, R. V., Epps, J. M., & Hartwig, E. E. (1970). Correlation of resistance in soybeans to *Heterodera glycines* and *Rotylenchulus reniformis*. *Phytopathology, 60*(4), 695–700.

Rich, J. R. (1985). Root-knot nematodes in north-central Florida soybean fields. *Nematropica, 15*(1), 43–48.

Riggs, R. D., & Schmitt, D. P. (1988). Complete characterization of the race scheme for *Heterodera glycines*. *Journal of Nematology, 20*(3), 392.

Ross, J. P. (1968). Effect of single and double infections of soybean mosaic and bean pod mottle viruses on soybean yield and seed characters. *Plant Disease Reporter, 52*(5), 344–348.

Roth, M. G., Webster, R. W., Mueller, D. S., Chilvers, M. I., Faske, T. R., Mathew, F. M., et al. (2020). Integrated management of important soybean pathogens of the United States in changing climate. *Journal of Integrated Pest Management, 11*(1), 17.

Roy, K. W., Baird, R. E., & Abney, T. S. (2001). A review of soybean (*Glycine max*) seed, pod, and flower mycofloras in North America, with methods and a key for identification of selected fungi. *Mycopathologia, 150*, 15–27.

Sarowar, S., Kim, Y. J., Kim, K. D., Hwang, B. K., Ok, S. H., & Shin, J. S. (2009). Overexpression of lipid transfer protein (LTP) genes enhances resistance to plant pathogens and LTP functions in long-distance systemic signaling in tobacco. *Plant Cell Reports, 28*, 419–427.

Schwenk, F. W., & Nickell, C. D. (1980). Soybean green stem caused by bean pod mottle virus. *Plant Disease, 64*, 863–865.

Sikora, E. J., Allen, T. W., Wise, K. A., Bergstrom, G., Bradley, C. A., Bond, J., et al. (2014a). A coordinated effort to manage soybean rust in North America: A success story in soybean disease monitoring. *Plant Disease, 98*(7), 864–875.

Sikora, E. J., Conner, K., Delaney, D., Zhang, L., & Delaney, M. (2014b). Observations on soybean rust and soybean vein necrosis virus in Alabama in 2013. *Proceedings of the Southern Soybean Disease Workers 41st Annual Meeting, Pensacola Beach, FL, USA.*

Singh, A., & Singh, I. K. (2018). *Molecular Aspects of Plant-Pathogen Interaction.* Springer.

Singh, A., Lim, G., & Kachroo, P. (2017a). Transport of chemical signals in systemic acquired resistance. *Journal of Integrative Plant Biology, 59*(5), 336–344.

Singh, A., Tyagi, C., Nath, O., & Singh, I. K. (2018). *Helicoverpa*-inducible thioredoxin h from *Cicer arietinum*: Structural modeling and potential targets. *International Journal of Biological Macromolecules, 109*, 231–243.

Singh, G., Dukariya, G., & Kumar, A. (2020). Distribution, importance, and diseases of soybean and common bean: A review. *Biotechnology Journal International, 24*(6), 86–98.

Singh, I. K., Kumar, S., Singh, S., & Singh, A. (2017b). Expression profiling of mitogen-activated protein kinase genes from chickpea (*Cicer arietinum* L.) in response to *Helicoverpa armigera*, wounding, and signaling compounds. *Journal of Asia-Pacific Entomology, 20*(3), 942–948.

Soda, N., Verma, L., & Giri, J. (2018). CRISPR-Cas9-based plant genome editing: Significance, opportunities, and recent advances. *Plant Physiology and Biochemistry, 131*, 2–11.

Soto, N., Hernández, Y., Delgado, C., Rosabal, Y., Ortiz, R., Valencia, L., et al. (2020). Field resistance to *Phakopsora pachyrhizi* and *Colletotrichum truncatum* of transgenic soybean expressing the *NmDef02* plant defensin gene. *Frontiers in Plant Science, 11*, 562.

Stuckey, R. E., Ghabrial, S. A., & Reicosky, D. A. (1982). Increased incidence of *Phomopsis* sp. in seeds from soybeans infected with bean pod mottle virus. *Plant Disease, 66*(9), 826–829.

Subramanyam, K., Arun, M., Mariashibu, T. S., Theboral, J., Rajesh, M., Singh, N. K., et al. (2012). Overexpression of tobacco osmotin (*Tbosm*) in soybean conferred resistance to salinity stress and fungal infections. *Planta, 236*, 1909–1925.

Suman, S., Bagal, D., Jain, D., Singh, R., Singh, I. K., & Singh, A. (2021). Biotic stresses on plants: Reactive oxygen species generation and antioxidant mechanism. *Frontiers in Plant-Soil Interactions, 2021*, 381–411.

Tan, J. A. C., Jones, M. G. K., & Fosu-Nyarko, J. (2013). Gene silencing in root lesion nematodes (*Pratylenchus* spp.) significantly reduces reproduction in a plant host. *Experimental Parasitology, 133*(2), 166–178.

Tenhaken, R., & Rubel, C. (1997). Salicylic acid is needed in hypersensitive cell death in soybean but does not act as a catalase inhibitor. *Plant Physiology, 115*(1), 291–298.

Tian, S. N., Liu, D. D., Zhong, C. L., Xu, H. Y., Yang, S., Fang, Y., et al. (2020). Silencing *GmFLS2* enhances the susceptibility of soybean to bacterial pathogen through attenuating the activation of *GmMAPK* signaling pathway. *Plant Science, 292*, 110386.

Tiwari, S. P. (2017). *Emerging Trends in Soybean Industry*.

Tougou, M., Furutani, N., Yamagishi, N., Shizukawa, Y., Takahata, Y., & Hidaka, S. (2006). Development of resistant transgenic soybeans with inverted repeat-coat protein genes of soybean dwarf virus. *Plant Cell Reports, 25*, 1213–1218.

Tyler, B. M. (2007). *Phytophthora sojae*: Root rot pathogen of soybean and model oomycete. *Molecular Plant Pathology, 8*(1), 1–8.

Tylka, G. (1994). *Soybean Cyst Nematode*. Iowa State University Cooperative Extension Service.

Tylka, G. L., & Marett, C. C. (2021). Known distribution of the soybean cyst nematode, *Heterodera glycines*, in the United States and Canada in 2020. *Plant Health Progress, 22*(1), 72–74.

Urnov, F. D., Miller, J. C., Lee, Y. L., Beausejour, C. M., Rock, J. M., Augustus, S., et al. (2005). Highly efficient endogenous human gene correction using designed zinc-finger nucleases. *Nature, 435*(7042), 646–651.

Urwin, P. E., Levesley, A., McPherson, M. J., & Atkinson, H. J. (2000). Transgenic resistance to the nematode *Rotylenchulus reniformis* conferred by *Arabidopsis thaliana* plants expressing proteinase inhibitors. *Molecular Breeding, 6*, 257–264.

Usharani, K. S., Surendranath, B., Haq, Q. M. R., & Malathi, V. G. (2005). Infectivity analysis of a soybean isolate of *Mungbean yellow mosaic India virus* by agroinoculation. *Journal of General Plant Pathology, 71*, 230–237.

Usovsky, M., Chen, P., Li, D., Wang, A., Shi, A., Zheng, C., et al. (2022). Decades of genetic research on *Soybean mosaic virus* resistance in soybean. *Viruses, 14*(6), 1122.

Walters, H. J. (1958). A virus disease complex in Arkansas. *Phytopathology, 48*, 346.

Wickizer, S. L., & Gergerich, R. C. (2007). First report of Japanese beetle (*Popillia japonica*) as a vector of *Southern bean mosaic virus* and *Bean pod mottle virus*. *Plant Disease, 91*(5), 637.

Wrather, A., & Koenning, S. (2009). Effects of diseases on soybean yields in the United States 1996 to 2007. *Plant Health Progress, 10*(1), 24.

Xiuqiang, W., Junyi, G., & Zuqin, P. (2003). Classification and distribution of strain groups of *Soybean mosaic virus* in middle and lower Huang-Huai and Changjiang valleys. *Soybean Science, 22*(2), 102–107.

Yang, X. B. (1989). *Ecology and Epidemiology of Rhizoctonia foliar Blights of Soybean*. Louisiana State University and Agricultural & Mechanical College.

Yang, X., Niu, L., Zhang, W., He, H., Yang, J., Xing, G., et al. (2019). Increased multiple virus resistance in transgenic soybean overexpressing the double-strand RNA-specific ribonuclease gene *PAC1*. *Transgenic Research, 28*, 129–140.

Yoon, Y. N., Jo, Y., Cho, W. K., Choi, H., Jang, Y., Lee, Y. H., et al. (2018). First report of *Tomato spotted wilt virus* infecting soybean in Korea. *Plant Disease, 102*(2), 461.

Yu, Q. (2011). Soybean cyst nematode (*Heterodera glycines* Ichinohe). In H. El-Shemy (Ed.), *Soybean Physiology and Biochemistry* (pp. 461–474).

Zaumeyer, W. J., & Thomas, H. R. (1948). Pod mottle, a virus disease of beans. *Journal of Agricultural Research, 77*, 81–96.

Zhang, H., Si, X., Ji, X., Fan, R., Liu, J., Chen, K., et al. (2018). Genome editing of upstream open reading frames enables translational control in plants. *Nature Biotechnology, 36*(9), 894–898.

Zhang, J., & Zhou, J. M. (2010). Plant immunity triggered by microbial molecular signatures. *Molecular Plant, 3*(5), 783–793.

Zhang, X., Sato, S., Ye, X., Dorrance, A. E., Morris, T. J., Clemente, T. E., et al. (2011). Robust RNAi-based resistance to mixed infection of three viruses in soybean plants expressing separate short hairpins from a single transgene. *Phytopathology, 101*(11), 1264–1269.

Zhao, C., Liu, B., Piao, S., Wang, X., Lobell, D. B., Huang, Y., et al. (2017). Temperature increase reduces global yields of major crops in four independent estimates. *Proceedings of the National Academy of Sciences, 114*(35), 9326–9331.

Zhao, F., Cheng, W., Wang, Y., Gao, X., Huang, D., Kong, J., et al. (2022). Identification of novel genomic regions for bacterial leaf pustule (BLP) resistance in soybean (*Glycine max* L.) via integrating linkage mapping and association analysis. *International Journal of Molecular Sciences, 23*(4), 2113.

Zhao, F., Lim, S., Igori, D., Yoo, R. H., Kwon, S. Y., & Moon, J. S. (2016). Development of tobacco ringspot virus-based vectors for foreign gene expression and virus-induced gene silencing in a variety of plants. *Virology, 492*, 166–178.

Zhou, J., & Tzanetakis, I. E. (2013). Epidemiology of soybean vein necrosis-associated virus. *Phytopathology, 103*(9), 966–971.

Zhou, Q., Jiang, B., Cheng, Y., Ma, Q., Xia, Q., Jiang, Z., et al. (2022). Fine mapping of an adult-plant resistance gene to powdery mildew in soybean cultivar Zhonghuang 24. *Crop Journal, 10*(4), 1103–1110.

CHAPTER 11

Enhancing Soybean Production Through Genetic Engineering and Molecular Breeding: A Review of Soybean Pests and Their Management Strategies

MANISHA YADAV,[1] INDRAKANT K. SINGH,[2] and ARCHANA SINGH[1,3,*]

[1]*Department of Botany, Hansraj College, University of Delhi, Delhi, India*

[2]*Department of Zoology, Molecular Biology Research Lab, Deshbandhu College, University of Delhi, Kalkaji, New Delhi, India*

[3]*Delhi School of Climate Change and Sustainability, Institution of Eminence, Maharishi Karnad Bhawan, University of Delhi, Delhi, India*

*Corresponding author

ABSTRACT

Glycine max (L.) Merr. is an economically important leguminous crop native to Southeastern Asia and now cultivated worldwide. Soybean is often known as the "miracle crop" due to its notable therapeutic and nutritional potential, along with its high oilseed content, which contributes to strong production demand. Being a luxuriant crop and a rich protein source, soybean attracts a large number of insect pests nearly 380 species globally—at various growth stages, causing an estimated 25% reduction in yield annually. Beyond conventional pest control strategies, biotechnology is rapidly advancing, offering cost-effective pest management and improved productivity. Molecular breeding uses genomic tools

Soybeans, Volume 2: Genome, Biotechnology, Stress, and Disease.
R. Ambika Rajendran, Ayyagari Ramlal, Sanjay Kumar Lal, and Godfree Chigeza (Eds.)

DOI: 10.1201/9781998511730-14

to identify quantitative trait loci (QTLs) associated with pest resistance and applies molecular markers to tag QTLs or target genes for developing resistant soybean varieties. Genetic engineering enables gene expression or silencing to create pest-resistant transgenic soybeans with minimal risk and greater adaptability. Emerging genome-editing tools such as transcription activator-like effector nucleases, zinc finger nucleases (ZFNs), and CRISPR/Cas9 have further refined soybean genome modification. A few pest-resistant soybean varieties developed through these approaches have been adopted by farmers, resulting in yield gains. However, transgenic soybeans face challenges related to cost, environmental acceptance, and consumer testing. This chapter reviews major soybean pests, their biology and damage, and advances in molecular breeding and genetic engineering.

11.1 INTRODUCTION

Soybean, a member of the Fabaceae family, is an oilseed and protein-rich legume crop. Soybean is commonly known as the "Crop of the Planet," "Golden Bean," and "Miracle Bean" due to its high nutritional value. Soybean provides high-quality protein (40%) with well-proportioned essential amino acids (AAs), oil (18–20%), minerals (6–7%), crude fibers (5–6%), and is rich in Omega-5 and Omega-6 FAs, as well as 17–19% carbohydrates (Chauhan and Joshi, 2005; Naresh et al., 2019; Ramlal et al., 2022). The demand for soybean and its products increased rapidly due to their wide range of uses in food, animal feed, and industrial products (Wilson, 2008). Soybean contributes to more than 60% of global oil production. Despite all this, non-edible products of soybean include FAs, paints, soaps, varnishes, lubricants, resins, surfactants, and biodiesel (Marking, 1998; Longhini, 1996). Soybean biodiesel is a lower-cost and flexible alternative fuel, whereas soybean-derived plastic has recycling alternatives. Soybean enriches soil fertility through the fixation of atmospheric nitrogen, up to 65–100 kg/hectare, into the soil with nitrogen-fixing bacteria (Patil et al., 2018). It is a native crop of Southeastern Asia, first domesticated in the eastern half of China and now cultivated worldwide (Badole and Bodhankar, 2012). World soybean production in 2020–21 was 367.76 million tons, with the largest contributions from Brazil, the United States, Argentina, China, India, and Canada. The soybean genome is fully sequenced, and the processes of nodulation, disease resistance, and gene regulation are well established (Gresshoff, 2017).

Sustainable soybean production has been challenged by many factors such as insect infestation, viral and bacterial diseases, weeds, and the anti-nutritional impacts of grain. However, soybean insect pests are major constraints in meeting world soybean demand and are responsible for increasing soybean production costs. Luxuriant growth, succulent and soft foliage, nutrient richness, shelter, and space provided by the soybean crop captivate various insect pests (Brahman et al., 2018; Seagraves et al., 2004). It has been revealed that nearly 380 insect species across the globe are known to infest soybean. The soybean yield loss caused by insect pests is more than 25% (Harish et al., 2009; Oerke, 2006). Soybean insect pests like stem borer, pod borer, sucking insects, and defoliators are reported as economically important pests infesting soybean from the seedling phase to maturity (Turnipseed and Kogan, 1976). Different approaches are being implemented for sustainable soybean production to curb insect infestation, viral diseases, weeds, and anti-nutritional effects. Among all of them, molecular breeding and genetic engineering have persisted as modes of choice for improving soybean productivity. In this article, we focus on insect pests of soybean that result in significant yield loss, plant defense mechanisms, and the application of molecular breeding and genetic engineering for enhancing sustainable soybean production

11.2 MAJOR INSECT PESTS OF SOYBEAN

Soybean crop yield is challenged by various biotic constraints such as insects and pests. Defoliating insects, nematodes, and weeds account for nearly 85% of total economic loss due to biotic stress in soybean (Hammond et al., 1991) A large number of insects belonging to different insect classes like Hemiptera, Diptera, Lepidoptera, Coleoptera, Orthoptera, Collembola, and Thysanoptera feed upon soybean from the seedling stage until maturity, affecting soybean quality and quantity (Gaur and Mogalapu, 2018; Kogan and Turnipseed, 1987). Insect pests of soybean pose a severe threat to soybean yield. Aphids and whiteflies are also known for transmitting viral diseases. Identification, life cycle, and damage caused by five major pests of soybean are described below:

11.2.1 SPODOPTERA LITURA (LEAF DEFOLIATOR)

Spodoptera litura (Fabricius) (Noctuidae: Lepidoptera), commonly called the tobacco caterpillar and common cutworm, is an economically important pest of soybean, widespread across temperate and tropical Asia, Australia, and also in the Pacific Islands. It is a polyphagous pest infesting nearly 112 cultivated plant species, including economic crops like tobacco, cotton, groundnut, soybean, castor, and rice, and so forth (Maree et al., 1999; Pogue, 2006). It has great potential to invade and adapt to a vast ecological range, with enormous migratory and reproductive ability (Armes et al., 1997; Qin and Ye, 2007). The moths show sexual dimorphism. Forewings in males are 14–17 mm long, while females have 15–18 mm long greenish-brown forewings (Arslan et al., 2015). Neonate larvae are green in color with a black head, while fully grown caterpillars are black with yellowish-green stripes on the dorsolateral side and a lateral whitish band along with a ring-like dark band on both the anterior and posterior ends of the body (Bhattacharya et al., 1977). Females lay eggs in groups of 3–4 on leaves. The egg incubation period varies from 3 to 6 days. Neonates pass through six larval stages (L1-L6), spending 3 to 4 days in each stage (Arslan et al., 2015). Pupation occurs in soil for 7–15 days. Pupae measure 15 × 20 × 5 mm in size and are lustrous reddish-brown. Adult moths have a lifespan of 7–8 days. The life cycle of the common cutworm is completed in 30–37 days, thus it completes 8 generations in a year.

The damaging stage for a crop is its larvae. Neonate larvae gregariously feed on the lower surface of leaves by scraping chlorophyll from the leaf lamina, while grown caterpillars feed voraciously by making holes and ultimately leaving only petioles and veins (Bughio et al., 1994). The pest is responsible for a significant loss of 26–29% as it remains active from the end of July or August to late October, synchronizing with soybean peak reproductive stages and a warm, humid climate (Punithavalli et al., 2014).

11.2.2 APROAEREMA MODICELLA (LEAF MINER PEST)

Aproaerema modicella (Deventer) (Gelechiidae: Lepidoptera) – although the leaf miner insect is a critical pest of soybean and groundnut, it also infests mung bean, pigeon pea, clover, hairy indigo, babchi seeds, hyacinth bean, and mung bean (Shanower et al., 1993). It is widespread in Asia and has extended to India, China, Pakistan, Sri Lanka, Bangladesh, Thailand, Vietnam, the Philippines, Laos, Indonesia, and Malaysia (Mohammad, 1981). The moths are small in size, having a 6–8 mm wingspan. High mobility, small-sized bodies, and scales over the body hamper the identification of female and male moths. The presence of pink gonads that are seen through cuticles and the position of the genital aperture in males are helpful characteristics in distinguishing male and female larvae and pupae, respectively (Shanower et al., 1993; Walt et al., 2008). Larvae are 6–8 mm long and yellowish-green in color. The female lays 87–186 eggs singly on the petioles and leaves, which have an egg period of 3–4 days. Newly hatched larvae, which are 1.5 mm long, mine into leaves and exit as brownish-grey adults. Larvae progress through three instar stages that last for 8–13 days, and pupation is completed in webbed leaves in 6–8 days (Badgujar et al., 2009).

Neonate larvae result in the formation of leaf mines by feeding on the mesophyll between the lower and upper epidermis. Later instar larvae roll and spin a web among leaflets, forming an enclosure for pupation (Du Plessis, 2003). Severe infestation causes defoliation and sound yield loss varying from 37.3% to 67.5%, depending on soybean varieties (Crop, 2019; Kenis et al., 2006;).

11.2.3 *MELANAGROMYZA SOJAE (STEM FLY)*

Melanagromyza sojae (Zehnter) (Diptera: Agromyzidae)—*M. sojae* is one of the crucial pests of soybean on the Asian continent (India, China, Taiwan, Nepal, Japan, Indonesia), parts of Russia, Northeast Africa (Saudi Arabia and Egypt), and Australia (Spencer, 1973). Stem fly is an oligophagous pest infesting several legume crops like *Cajanus indicus, Phaseolus radiatus, Medicago sativa, Vigna radiata, Dolichos biflorus*, etc. However, soybean is its main host (Spencer, 1973; Thapa, 2012). It is brown when it emerges from the pupal case. Hyaline wings have a definite notch in the coastal region. Maggots are generally yellowish (Spencer, 1973). Female stem flies have the potential to lay 60–85 eggs singly on the abaxial surface of young leaves, and the eggs hatch after 2–3 days. The maggot bores through leaf tissue, reaches the main stem, and feeds on plant pith. The maggot phase takes 8–12 days (Singh, 1982; Spencer, 1973). The maggot, before pupation, makes an exit hole for the ease of adult emergence through the stem. Pupation occurs inside the stem tunnel and lasts for nearly 7–10 days. The stem fly's lifespan varies from 16–26 days. The stem fly's activity is positively favored by warm temperatures, high humidity, and high rainfall (Brier and Charleston, 2020; Gain et al., 1986).

Soybean is prone to stem fly pest incidence throughout its growth and development, from seedling to maturity stages, whereas infestation is severe on crops sown in July-August (Adimani, 1976). Newly emerged stem fly maggots enter the stem by boring into leaf tissue and making zigzag stem tunnels by mining in both upward and downward directions. Infestation during the early growth stages results in higher seedling mortality; however, infestation in late growth phases affects plant yield-attributing characteristics (Savajji, 2006). Stem fly infestation in soybean has been reported to cause an average yield loss of 20–30% in various parts of the world (Jayappa, 2000; Talekar, 1990).

11.2.4 HELICOVERPA ARMIGERA (POD BORER)

Helicoverpa armigera Hubner (Noctuidae: Lepidoptera), generally known as the corn earworm and cotton bollworm, is reported as one of the most devastating insect pests distributed throughout Southeastern and Central Asia, Southern Europe, the Middle East, Africa, and Australasia (Karim, 2000; Tay et al., 2017). Being a polyphagous pest, *H. armigera* infests over 181 vegetable, grain, and fiber crop species across the world counting economically beneficial crops viz maize, rice, cotton, chickpea, sunflower, sorghum, and tomato and so forth (Cunningham and Zalucki, 2014; Gulzar et al., 2017; Yucel and Hanife, 2018).

Males of the cotton bollworm are generally yellowish-brown, while females are orange to brown-colored (Gaur and Mogalapu, 2018; Patankar et al., 2001). Larvae exhibit color polymorphism. Final larvae range from 3.5 to 4.2 cm in size and are brownish or pale green (King, 1994). Females usually lay 500–2000 spherical eggs singly on plant parts. Newly laid eggs are yellowish and change their color to dark grey/black (Naseri et al., 2011). The incubation period for eggs varies from 2 to 4 days. Newly emerged larvae are yellowish-colored and translucent, pass through six larval stages, and take 20 days. Pupae are brown-colored, with a length and width of 14–22 mm and 4.5–6.5 mm. Six instar larvae pupate in soil for 13–14 days (Reigada et al., 2016). The lifespan for this insect is 38–45 days. It completes a total of eight generations in a year.

H. armigera is a destructive pest with high fecundity and reproductive rates, having good migratory abilities (Feng et al., 2009). The larvae aggressively feed on grains as they bore into pods during the reproductive phase, whereas they feed on leaves during the vegetative phase (Azambuja et al., 2015; Biswas, 2013). *H. armigera* causes a global loss of $3 billion per annum and a significant loss of 30–40% in infested crops (Haile et al., 2021).

11.2.5 APHIS GLYCINES (SUCKING INSECT APHID)

Aphis glycines Matsumura (Hemiptera: Aphididae) – though soybean aphid (SBA) is less globally distributed, it is widespread in Asian countries, which include Japan, Malaysia, Korea, China, Taiwan, the Philippines, and Thailand, as well as southern Canada and the north-central US (Blackman, 2000; Wang et al., 1994). Matsumura, 1917, first described the SBA. It has both wingless and winged adult forms. Aphids range from yellow to light green, having pale antennae, red eyes, and dark-tipped tail pipes (cornicles) at the posterior end of the abdomen. These cornicles help differentiate SBAs from green aphids. Nymphs with broad shoulders will generate wing pads and ultimately develop into winged aphids (Hodgson and Koch, 2018; Voegtlin et al., 2004). SBAs have a composite life cycle, that is, holocyclic (producing sexual morphs from an egg), with heteroccious pests (Ragsdale et al., 2004). In the spring and summer seasons, the aphids undergo asexual reproduction on soybean (the secondary host) to produce all wingless female aphids. Aphids complete 15–16 generations on soybean with a doubling time of 6–7 days (Ragsdale et al., 2007). In September or October (the end of the soybean growing season), gynoparae (winged females) migrate back to *Rhamnus cathartica* L. (common buckthorn), the primary host plant, where they complete 3–4 generations. SBAs progress through four nymphal instars (Ghosh, 2008). SBAs possess piercing and sucking mouthparts that they use to feed on plant sap from the phloem of leaves and tender plant parts. Their distribution on the host plant varies according to the growing season (McCornack et al., 2008). Heavy infestation by aphids results in distorted or yellow leaves, aborted leaves, and stunted plants (Lin et al., 1993). Aphids excrete sugar-rich, sticky honeydew that deposits on various plant parts and supports black sooty mold growth while reducing photosynthesis by hindering the amount of light reaching the plant (Beckendorf et al., 2008). Heavy infestation also affects pod number, seed quality, and size, resulting in a significant loss of 40% yield (Ragsdale et al., 2007). SBAs have been reported to transmit AMV and SMV in soybean (Clark et al., 2002; Hill et al., 2001).

11.3 PLANT DEFENSE MECHANISM AGAINST HERBIVORY

Plants are challenged by numerous biotic and abiotic stresses in their life cycle; among them, insect pests are the most crucial threats to plant productivity. In the course of time, plants have evolved sophisticated defense mechanisms to cope with herbivory (Rajendran et al., 2023; Schuman and Baldwin, 2016). Plants confer resistance via both constitutive and inducible defensive strategies (Kumar et al., 2021; Singh and Singh, 2018;). Constitutive defense responses involve traits like cuticles, spines, trichomes, and specialized defense metabolites, which are expressed regardless of herbivory, while induced defenses mount only in insect-challenged plants, thus considered a cost-effective way (Karban et al., 1997). Elaborate defense signaling cascades, cell wall enforcement, expression of plant defense genes, and synthesis of defense responses are events under induced defense responses with lower resource allocation (Bostock, 2005; Singh et al., 2018, 2021). Secondary metabolites, viz., glucosinolates and flavanols, are characterized as regulators of defense deployment in plants.

For the successful implementation of plant defense responses, pattern recognition receptors (PRRs) in plants recognize herbivore-associated molecular patterns and damage-associated molecular patterns associated with insects to perceive herbivory (Erb and Reymond, 2019). Insect-derived elicitors or insect attacks lead to early signaling events, which include depolarization of plasma membrane (PM) potential (V_m) and enhanced cytosolic calcium concentration (Dodd et al., 2010; Tian et al., 2020). The calcium signals are decoded by calcium sensor proteins, which act as both relay and responder proteins (Aldon et al., 2018; Kumar et al., 2020; Yadav et al., 2022). V_m depolarization also leads to the production of ROS in the apoplast and cell organelles, depending on NADPH oxidase, peroxidase, amine-oxidase, and superoxide dismutase (SOD). These ROS are reported as both positive and negative modulators of plant-herbivore defense (Miller et al., 2009). Mitogen-activated protein kinases (MAPKs) are also immediately activated to regulate the synthesis of jasmonate and ethylene phytohormones for defense gene activation upon insect attack (Keshan et al., 2021). MAPKs are also known to activate WRKY transcription factors (TFs) and NADPH oxidase (Li et al., 2015). TFs, WRKYs, are known for their anti-herbivore effect, whereas MYC TFs regulate plant defense by mediating phytohormone signaling (Onkokesung et al., 2014; Schweizer et al., 2013). Multiple phytohormones induce fine-tuned defense responses through antagonistic and synergistic crosstalk against herbivory, but the jasmonate pathway is the foremost signaling cassette interceding the information delivered at the interface of plant-insect interaction and initiating defense responses in plants (Major et al., 2006; Singh et al., 2016). Upon insect damage, α-linolenic acid in plastids undergoes oxygenation and cyclization reactions to produce OPDA, which is transported into peroxisomes (Stintzi et al., 2001). (+)-7-iso-JA is produced through β-oxidation of reduced OPDA (Taki et al., 2005). (+)-7-iso-JA is conjugated to L-isoleucine in the cytosol by the JAR1 enzyme. JA-Ile, a bioactive form of JA, signals through binding to a nuclear receptor complex consisting of JAZ and COI1 (reported as an F-box component of the E3 ubiquitin ligase complex, SCFCOI1). JA-Ile on interaction triggers degradation of JAZ repressors through the 26S proteasome, which leads to the release of repressed MYC TFs and ultimately the expression of plant defense genes against herbivores (Devoto and Turner, 2005; Thines et al., 2007). Salicylic acid (SA), abscisic acid (ABA), and ethylene are modulators of resistance to herbivores and are induced upon feeding by herbivory guilds. However, SA phytohormones antagonize the jasmonate signaling pathway under herbivory (Pieterse et al., 2012; Singh et al., 2008).

11.4 GENETIC ENGINEERING FOR PEST RESISTANCE

Various strategies are being employed for insect pest control and sustainable soybean production. The usage of insecticides has been limited as they may kill the natural enemies of pests, degrade soil fertility, cause environmental pollution, and affect human health. Since the 1990s, the use of transgenics coupled with conventional breeding has dramatically enhanced soybean improvement. Techniques of genetic engineering are employed for editing the genomes of crops to improve a trait or insert a novel trait. These crops are known as transgenic/ genetically modified organisms (GMO) crops. In comparison to the traditional method of plant selection and breeding strategies, transgenic production is a more precise method of genetic alteration and lessens the time for identifying desired traits (Figure 11.1). Insect resistance and herbicide tolerance (HT) are two forefront traits in early transgenic crops. In 1988, the first genetically modified (GM) soybean was produced by employing Agrobacterium-mediated and particle bombardment transformation methods (Hinchee et al., 1988). Insect-resistant Bt-crops were engineered with a gene from soil bacteria *Bacillus thuringiensis* (Bt) that codes for Cry protein. The Cry protein, when ingested by insect pests, exhibits insecticidal activity upon activation by the alkaline pH of the insect gut and ultimately leads to insect death (McPherson et al., 2003). Bt-crops are efficient in controlling insect infestation and lessening insecticide pulverization and environmental contamination. RNA interference (RNAi) is also used for silencing a gene to produce desired traits in GMO crops. Transgenic breeding is performed with engineered transgenic crops. However, widespread acceptance of GMO crops faces many challenges, including potential toxicity, environmental risks, and allergenicity to living organisms. Table 11.1 lists a number of transgenic soybean plants engineered for insect resistance.

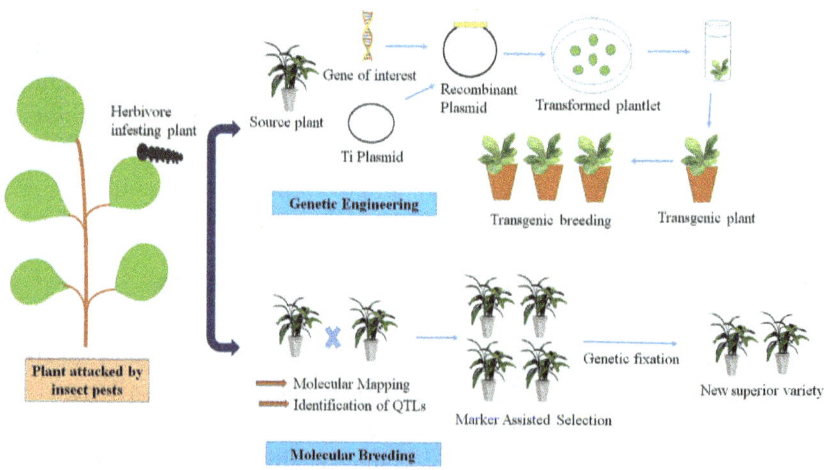

FIGURE 11.1 Illustrating generation of genetically engineered plants using transgenic technology and molecular breeding.

TABLE 11.1 Transgenic Soybean with Improved Insect Resistance

Serial No.	Resistance	Gene	Method	Reference
1.	*Anticarsia gemmatalis* (Velvetbean caterpillar)	*Cry 1Ab*	Particle bombardment	Parrot et al. (1994)
2.	*Helicoverpa Zea* (Corn earworm)	*Cry 1Ac*	Particle bombardment	Stewart et al. (1996)
3.	*Anticarsia gemmatalis* (Velvetbean caterpillar)	*Cry1Ac*	Particle bombardment	Stewart et al. (1996)
4.	*Anticarsia gemmatalis* (Velvetbean caterpillar)	*Cry1Ab*	Particle bombardment	Dufourmantel et al. (2005)
5.	*Helicoverpa armigera* (Cotton bollworm)	*Cry1Ac*	Particle bombardment	Dang et al. (2007)
6.	*Helicoverpa zea* (Soybean podworm)	*Cry1Ac (tic107)*	*Agrobacterium* Transformation	Miklos et al. (2007)
7.	*Anticarsia gemmatalis* (Velvetbean caterpillar)	*Cry1Ac*	Particle bombardment	Homrich et al. (2008)
8.	*Heliothis virescens* (Tobacco budworm)	*Cry1Ac*	*Agrobacterium* transformation	Monsanto (2012)
9.	*Helicoverpa armigera* (Corn earworm)	*Cry1Ac*	*Agrobacterium* transformation	Monsanto (2012)
10.	*Spodoptera litura* (Common cutworm)	*GmAOC3*	*Agrobacterium* transformation	Wu et al. (2015)
11.	*Leguminivora glycinivorella* (Soybean podborer)	*Ribosomal protein P0*	*Agrobacterium* transformation	Meng et al. (2017)
12.	*Spodoptera litura* (Common Cutworm)	*GmVQ58* (VQ motif)	*Agrobacterium* transformation	Li et al. (2020)
13.	*Aphis glycine* (Soybean aphid)	*TREH, ATPD, ATPE, CHSI*	RNA-interference	Yan et al. (2020)
14.	*Holotrichia parallela* (Dark Black Chafer)	*Cry8-like*	*Agrobacterium* transformation	Qin et al. (2019)
15.	*Tetranychus urticae* (Spider mites)	*GmDR1*	*Agrobacterium* Transformation	Ngaki et al. (2021)
16.	*Aphis glycines* (Soybean aphid)	*GmDR1*	*Agrobacterium* transformation	Ngaki et al. (2021)
17.	*Leguminivora glycinivorella* (Soybean podborer)	*Sp[illegible]try1* Serine protease	RNA-interference	Meng et al. (2017)

11.5 MOLECULAR BREEDING FOR PEST RESISTANCE

Plant breeding is a process of crossing plants with particular traits to produce a plant with desired traits that are inherited by the genes of the next generation. Crop breeding aims to produce crops with desirable traits, such as higher grain yield, better nutritional quality, and greater tolerance to biotic and abiotic stresses (Rajendran et al., 2022; Ramlal et al., 2022). Indeed, crop breeding is one of the most symbolic innovations in human history that has facilitated human survival. In the last 60–65 years, conventional breeding has resulted in a cogent contribution to enhancing soybean yield; however, progress is passive due to the time-consuming process of selection, difficulties in the selection of appropriate genotypes, and handling quantitative traits. The limitations of conventional breeding were broadly overcome by techniques of molecular breeding (Tanksley, 1993; Tuberosa et al., 2002). Marker-assisted breeding facilitates the identification of relative contribution, location, and number of quantitative trait loci (QTL) corresponding with soybean pest resistance and also enhances the efficiency of conventional breeding based on phenotype (Kulamarva et al., 2009; Kumar et al., 2023). Marker-assisted selection has numerous advantages, such as being independent of seasonal and temporal variation, allowing early selection using seed DNA gene pyramiding, characterizing new valuable alleles from wild germplasm, and enabling parental line selection from a broad genetic base (Collard et al., 2008). New plant breeding technologies (NPBT), along with advances in molecular biology, provide a fast, efficient, and promising method for crop improvement (Liu et al., 2017; Zaidi et al., 2019). NPBTs can be used to obtain desired products either by gene knockout or gene knock-in and also allow for selection segregation without transgenes in the next generation (Mohanta et al., 2017). CRISPR/Cas9, TALENs, and ZFNs are three methods of NPBT used for creating the desired mutation in crops, which introduce a double-stranded break in the genomic DNA that is repaired by homologous or non-homologous end joining (Boch et al., 2009; Curtin et al., 2018; Deveau et al., 2010; Joung et al., 2013). Soybean is a paleopolyploid plant where the haploid genome has 20 chromosomes (Chr) (i.e., 2n=40) and 1100 million base pairs. A soybean integrated genetic linkage map has been constructed with 2989 markers, including 1141 SNPs (Choi et al., 2007). Plant resistance is one of the most potential and sustainable methods for insect pest control. In soybean, using marker-assisted selection and genome-wide association studies (GWAS), various cultivars with pest resistance have been developed.

11.6 MOLECULAR BREEDING IN SOYBEAN FOR INSECT PEST RESISTANCE

The breeding for insect resistance started with the discovery of resistance to insect pests in plant introductions (PIs) in soybean. The three primary modes of plant resistance to insects (PRI) are antixenosis, antibiosis, and tolerance, which can be inferred individually by designing bioassays (Kogan and Ortman, 1978). The three primary sources of insect resistance in soybean are from the soybean lines introduced from Japan, namely PI 171451, PI 227687, and PI 22958 (Van Duyn et al., 1971, 1972). These PIs were found to possess resistance (both antibiosis and antixenosis) against a number of insects. These three plant introductions were low-yielding, poor agronomic types and susceptible to diseases and nematodes compared to cultivars (Kilen and Lambert, 1986). Clark (1972) studies inferred that genetic variation present in soybean cultivars provides opportunities to develop insect-resistant cultivars against multiple insect species. PI 165989, PI 227687, PI 229321, PI 22935, and PI 171451 were among the cultivars examined that exhibited significant resistance levels against one or multiple insect pests of soybean, that is, bean leaf beetle, striped blister beetle, and bollworm moth. Out of the 21 experimental cultivars studied, PI 227687 emerged as the most intriguing cultivar, which exhibited very little pod damage throughout the growing season, indicating the potential presence of an antibiosis defense mechanism against bollworm. Inheritance studies on the nature of resistance showed that resistance against insects is quantitative in nature (Clark et al., 1972; Kilen and Lambert, 1986; Rector et al., 1998, 1999, 2000).

A laboratory experiment was conducted to assess 22 soybean genotypes to characterize antibiosis or antixenosis resistance. The genotypes PI 227687, PI 171451, PI 274453, PI 229358, PI 274454, and "IAC 17" were identified as soybean introductions possessing resistance to *H. armigera* that could be useful in breeding lepidopteran-resistant soybean cultivars (Coelho et al., 2020). Inheritance of insect pest resistance in *G. max* is quantitative, and its incorporation into cultivars with high yield has been challenging. The quantitative nature of the resistance and the poor agronomic performance of these primary sources of resistance were responsible for slow progress in breeding for insect resistance. Later on, molecular markers were used for transferring the QTLs. Around 139 RFLP markers were employed to characterize QTLs linked with antixenosis resistance against corn earworm in soybean. Two minor and one major QTL for resistance were characterized, and an allele from PI229358 provided insect resistance at all three QTLs (Rector, 1997). RFLP was also used to characterize QTLs for antixenosis resistance in *Helicoverpa zea* in soybean PI227687 and PI171451, compared to QTLs in PI229358, and also identified their location on the soybean linkage map. The experimental study observed a QTL shared among PI171451, PI229358, and PI227687 on the "H" linkage group (LG), while PI229358 and PI171451 had a QTL in common on the "M" LG. Minor QTLs on the "Dl" LG and LG "C2" were characterized, which were unique to PI229358 and PI227687, respectively (Rector, 1999). Rector (2000) also characterized QTLs linked with antibiosis resistance against corn earworm.

In the third phase of breeding, a combination of QTLs from the primary source of insect resistance, along with *Bacillus thuringiensis* gene, was used to achieve better resistance against insect pests (Walker et al., 2002, 2004). Walker et al. (2002; 2004) first reported complementary and additive effects between plant QTLs with insect resistance and the Bt transgene, which is introgressed by MAS. Soybean lines with a factorial combination of QTLs from PI 229358 and the Bt cry1Ac synthetic gene were developed with SSR markers. Through larval weight gain analysis and field defoliation on detached leaves, it was observed that lines containing both the PI 229358 strain at the quantitative trait locus (QTL) on "M" LG exhibited significantly more resistance to Bt-resistant pests, that is, TBW, and to lepidopteran pests compared to other lines. The synthetic Bt cry1Ac transgene from "Jack-Bt" and a QTL associated with corn earworm resistance in the PI 229358 introduction of soybean were pyramided by MAS into soybean lines. These lines with pyramided genes are viable strategies for pest resistance with the employment of SSR markers (Walker, 2002).

The sources of resistance to numerous insect pests in soybean germplasm were identified with the help of GWAS. GWAS analysis performed using SNP markers identified regions on Chr 6 and 7 having QTLs for resistance against corn earworm and common cutworm, respectively. *JAM2-like* and *MYC2-like* are putative genes identified on

Chr 7, associated with common cutworm (Liu et al., 2016). qSCCW18-3, a QTL mapped in the promoter sequence of the Rubisco activase gene (*GmRCAb*) of soybean, exhibits resistance to common cutworm (Wang et al., 2015). These studies laid the foundation for MAS breeding for resistance against common cutworms. Isoflavones, specialized secondary metabolites in soybean seeds, act as defense factors against plant pests and diseases. QTLs associated with soybean pod borer resistance were investigated by crossing "Zhongdou 27" and "Jiunong 20," which have higher and lower content of isoflavone, respectively. Four QTLs with phenotypic variation for resistance to SPB ranged from 2–4% on soybean Chr 7, 10, 13, and 17, and 11 QTLs associated with isoflavone content in similar marker intervals were identified. Finding these QTLs would be beneficial in MAS in breeding soybean with high isoflavone content and SPB resistance (Zhao et al., 2015). Phenotypic data for damage caused by beet armyworm, potato leafhopper (PLH), Mexican beetle (MBB), soybean looper (SBL), and velvet bean caterpillar (VBC) were integrated from USDA soybean germplasm. Genes with leucine-rich regions and MYC factors are identified around the SNP markers for VBC, PLH, SBL, and MBB. This information provides insight into resistance to these four insect pests in soybean germplasm (Chang et al., 2017). Four molecular markers associated with stink bug resistance were identified on Chr 15 and 6 through association mapping in soybean germplasm. It was reported that out of 112 genes, 31 genes would be translated to proteins involved in the herbivore-triggered defense pathway, most probably in the jasmonic acid pathway. These molecular markers will be a valuable tool in MAS for breeding stink bug-resistant soybean (Ghione et al., 2021).

11.7 MOLECULAR BREEDING FOR IMPROVED APHID RESISTANCE

SBA biotypes have been promptly evolving, indicating the need for the characterization of new aphid resistance genes in soybean germplasm. Aphid resistance in soybean cultivars "Jackson" and "Dowling" is controlled by a dominant gene. In the cultivar "Dowling," this gene is known as *Rag1*. The Jackson gene and *Rag1* were both mapped at LG M in soybean germplasm between SSR markers Satt4463 and Satt435 (Li et al., 2007). Soybean PI 243540 exhibits antibiosis resistance to the

Ohio biotype of aphid, which is also controlled by a single dominant gene (Kang et al., 2008). Inheritance and mapping of genes responsible for SBA resistance in soybean PI 200538 were determined using DNA markers by crossing with susceptible genotypes, that is, "Ina," "Williams 82," and LD02-4485. The genetic ratio of the F2 population shows the dominance of the resistance gene, which is mapped to the LG of soybean flanked by SSR marker loci Sct033, Satt510, Satt114, and Soyhsp176. Since these loci are present in the same region as *Rag2* and also provide resistance to aphid biotypes 1 and 2, PI 200538 might be considered a source of *Rag2* (Hill et al., 2009). A study was conducted both in the greenhouse and field to validate and map the SBA resistance gene in early maturing PI567541B germplasm by employing molecular markers. Two QTLs controlling SBA resistance were found to be located in the F and M LGs of soybean introduction 567541B. Multiple interval methods of mapping reported additive interaction between these QTLs and the resulting phenotypic variation. These QTLs were also validated in breeding lines derived from soybean introduction 567541B, which could be beneficial in breeding practices for aphid resistance through marker assisted selection (MAS) (Zhang et al., 2009). A gene controlling aphid resistance was mapped on Chr 16 between the Satt339 and Satt414 loci with the help of molecular markers in soybean PI567543C. This gene was found to be additive at this locus in heterozygous lines for aphid resistance and mapped on a separate Chr in germplasm compared to previously known resistance genes. Thus, *Rag3* is a designated new gene for aphid resistance, which will be beneficial for breeding SBA-resistant soybean cultivars (Zhang et al., 2010). *Rag1* is an aphid resistance gene in soybean. For the ease of MAS during breeding, the single nucleotide polymorphism (SNP) marker associated with *Rag1* is fine-mapped through genome shotgun sequencing of the population with a recombinant event near the *Rag1* gene. The result showed that the *Rag1* gene is located between 21A and 46169.7 SNP markers, with a physical distance of 115 kb (Kim et al., 2010). PI 567301B was known to have anti-xenosis resistance to SBA biotypes 1 and 2 (Jun et al., 2012). In soybean PI 567598B, aphid resistance was controlled by recessive genes. QTL located on Chr 16 is responsible for aphid resistance, and the marker associated with it could be used in MAS breeding (Bales et al., 2013). QTL_13_1 was mapped on Chr 13 near the *Rag2* gene, whereas QTL_13_2 was mapped close to Rag4 in PI 567324. The oligogenic resistance in this soybean introduction provides more durable and broader resistance to SBAs (Jun et al., 2013). Xiao et al. (2014) conducted a study to find the location of a dominant R_P746 gene controlling antibiosis resistance to soybean P746. The gene was mapped on Chr 13 and flanked by Satt334 and Satt335 on either side. Results of chromosome walking identified its location between BARCSOYSSR_13_1278 and 13_1363, with distances of 4.2 cM and 2.6 cM, respectively. In soybean PI366121, a novel QTL *Raso2* was characterized using an SNP microarray on chromosome 7, responsible for resistance to the Korean biotype Fox-glove aphid (Lee et al., 2015). *Rag6* and *Rag3c* QTLs in the E08934 soybean breeding line (wild *soybean* × *G. Soja 85–32*) were identified to provide antibiosis resistance against aphids and were mapped on chromosomes 8 and 16 (Zhang et al., 2017). PI 606390A possesses resistance to all four biotypes of aphids. qRag-G is the new locus mapped on chromosome 18 at the 14 cM region, conferring resistance to biotypes 3 and 4 of aphids (LaMantia et al., 2019).

11.8 CONCLUSION AND FUTURE PROSPECTIVE

Soybean is a leguminous crop with high nutritional content, and it is used for food, fodder, and fuel globally. However, insect pests are a major threat to soybean production, resulting in significant yield loss. Conventional breeding is a time-consuming and laborious process still used alongside GMO crops. Nowadays, biotechnological interventions like plant molecular breeding and genetic engineering, along with advancements in genetic tools, are employed for pest resistance in soybean. When genes are known, transgenic soybean provides a fast and durable method of pest resistance. However, transgenic soybean faces low acceptance among people due to ethical issues and restrictions. Molecular breeding with the assistance of molecular markers has been used to increase the efficiency and generation of soybean pest-resistant varieties for the last few years. NPBT are a cost-effective, fast, and transgene-free strategy for improving soybean pest-resistant varieties. However, there are many challenges to NPBT in soybean; it also provides opportunities for superior soybean varieties. These genetic engineering and fast breeding methods play a pivotal role in controlling soybean pests. Among them, CRISPR/Cas9 is the most efficient and cost-effective system for crop improvement by knocking down/knocking in genes for desired traits and can also target multiple genes at once. Although CRISPR/Cas9 has not been implemented in soybean for insect resistance, it provides us with future endeavors in the field of soybean improvement.

KEYWORDS

- **CRISPR/Cas9**
- **genetic engineering**
- **genome editing**
- **molecular breeding**
- **next-generation breeding technologies**
- **soybean pest resistance**
- **transgenic soybean**

REFERENCES

Adimani, B. D. (1976). *Studies on the Insects of Soybean [Glycine max (L.) Merrill] with Special Reference to Bionomics and Control of Pod Borer, Cydia ptychora Meyrick (Lepidoptera: Torticidae)* (M.Sc. thesis). University of Agricultural Sciences, Bangalore.

Aldon, D., Mbengue, M., Mazars, C., & Galaud, J. P. (2018). Calcium signalling in plant biotic interactions. *International Journal of Molecular Sciences, 19*, 665.

Armes, N. J., Wightman, J. A., Jadhav, D. R., & Ranga Rao, G. V. (1997). Status of insecticide resistance in *Spodoptera litura* in Andhra Pradesh, India. *Pesticide Science, 50*, 240–248.

Arslan, Z. F., Uludag, A., & Uremis, I. (2015). Status of invasive alien plants included in EPPO Lists in Turkey. *EPPO Bulletin, 45*, 66–72.

Azambuja, R., Degrande, P. E., dos Santos, R. O., de Souza, E. P., & Gomes, C. E. C. (2015). Effect of Bt soybean on larvae of *Helicoverpa armigera* (Hübner) (Lepidoptera: Noctuidae). *Journal of Agricultural Science, 7*, 90.

Badgujar, A. G., Shetgar, S. S., Patait, D. D., Subhan, S., & Dhurgude, S. S. (2009). Life-fecundity tables of leaf miner, *Aproaerema modicella* (Deventer) on three different host plants. *Journal of Entomological Research, 33*, 257–260.

Badole, S. L., & Bodhankar, S. L. (2012). *Glycine max* (soybean) treatment for diabetes. In *Bioactive Food as Dietary Interventions for Diabetes: Bioactive Foods in Chronic Disease States* (p. 77).

Bales, C., Zhang, G., Liu, M., Mensah, C., & Gu, C. (2013). Mapping soybean aphid resistance genes in PI 567598B. *Theoretical and Applied Genetics, 126*, 2081–2091.

Beckendorf, E. A., Catangui, M. A., & Riedell, W. E. (2008). Soybean aphid feeding injury and soybean yield, yield components, and seed composition. *Agronomy Journal, 100*, 237–246.

Bhattacharya, A. K., & Rathore, Y. S. (1977). *Survey and Study of the Bionomics of Major Soybean Insects and Their Chemical Control*. G. B. Pant University of Agriculture and Technology.

Biswas, G. C. (2013). Insect pests of soybean (*Glycine max* L.), their nature of damage and succession with the crop stages. *Journal of the Asiatic Society of Bangladesh, Science, 39*, 1–8.

Blackman, R. L., & Eastop, V. F. (2000). *Aphids on the World's Crops: An Identification and Information Guide* (2nd ed.).

Boch, J., Scholze, H., Schornack, S., Landgraf, A., Hahn, S., & Kay, S. (2009). Breaking the code of DNA binding specificity of TAL-type III effectors. *Science, 326*, 1509–1512.

Bostock, R. M. (2005). Signal crosstalk and induced resistance: Straddling the line between cost and benefit. *Annual Review of Phytopathology, 43*, 545.

Brahman, S. K., Awasthi, A. K., & Singh, S. (2018). Studies on insect pests of soybean (*Glycine max*) with special reference to seasonal incidence of Lepidopteran defoliators. *Journal of Pharmacognosy and Phytochemistry, 7*, 1808–1811.

Brier, H., & Charleston, K. (2020). Soybean stem fly outbreak in soybean crops.

Bughio, A. R., Hussain, T., Qureshi, Z. A., Ahmad, M., & Shakoori, A. R. (1994). Quantitative studies of food consumption and growth of *Spodoptera litura* (F.) on soybean. In *Proceedings of the Pakistan Congress of Zoology* (pp. 99–104).

Chang, H. X., & Hartman, G. L. (2017). Characterization of insect resistance loci in the USDA soybean germplasm collection using genome-wide association studies. *Frontiers in Plant Science, 8*, May, 1–12.

Chauhan, G. S., & Joshi, O. P. (2005). Soybean (*Glycine max*): The 21st-century crop. *Indian Journal of Agricultural Sciences, 75*, 461–469.

Choi, I. Y., Hyten, D. L., Matukumalli, L. K., Song, Q., Chaky, J. M., & Quigley, C. V. (2007). A soybean transcript map: Gene distribution, haplotype and single-nucleotide polymorphism analysis. *Genetics, 176*, 685–696.

Clark, A. J., & Perry, K. L. (2002). Transmissibility of field isolates of soybean viruses by *Aphis glycines*. *Plant Disease, 86*, 1219–1222.

Clark, W. J., Harris, F. A., Maxwell, F. G., & Hartwig, E. E. (1972). Resistance of certain soybean cultivars to bean leaf beetle, striped blister beetle, and bollworm. *Journal of Economic Entomology, 65*(6), 1669–1672.

Coelho, M., Godoy, A. F., Baptista, Y. A., Bentivenha, J. P. F., Lourenção, A. L., & Baldin, E. L. L. (2020). Assessing soybean genotypes for resistance to *Helicoverpa armigera* (Lepidoptera: Noctuidae). *Journal of Economic Entomology, 113*, 471–481.

Collard, B. C. Y., & Mackill, D. J. (2008). Marker-assisted selection: An approach for precision plant breeding in the twenty-first century. *Philosophical Transactions of the Royal Society B: Biological Sciences, 363*, 557–572.

Cunningham, J. P., & Zalucki, M. P. (2014). Understanding heliothine (Lepidoptera: Heliothinae) pests: What is a host plant? *Journal of Economic Entomology, 107*, 881–896.

Curtin, S. J., Xiong, Y., Michno, J., Campbell, B. W., Stec, A. O., & Čermák, T. (2018). CRISPR/Cas9 and TALENs generate heritable mutations for genes involved in small RNA processing of *Glycine max* and *Medicago truncatula*. *Plant Biotechnology Journal, 16*, 1125–1137.

Dang, W., & Wei, Z. (2007). An optimized *Agrobacterium*-mediated transformation for soybean for expression of binary insect resistance genes. *Plant Science, 173*, 381–389.

Deveau, H., Garneau, J. E., & Moineau, S. (2010). CRISPR/Cas system and its role in phage-bacteria interactions. *Annual Review of Microbiology, 64*, 475–493.

Devoto, A., & Turner, J. G. (2005). Jasmonate-regulated *Arabidopsis* stress signaling network. *Physiologic Plantarum, 123*, 161–172.

Dodd, A. N., Kudla, J., & Sanders, D. (2010). The language of calcium signaling. *Annual Review of Plant Biology, 61*, 593–620.

Du Plessis, H. (2003). First report of groundnut leafminer, *Aproaerema modicella* (Deventer) (Lepidoptera: Gelechiidae) on groundnut, soybean and lucerne in South Africa. *South African Journal of Plant and Soil, 20*, 48.

Dufourmantel, N., Tissot, G., Goutorbe, F., Garcon, F., Muhr, C., & Jansens, S. (2005). Generation and analysis of soybean plastid transformants expressing *Bacillus thuringiensis* Cry1Ab protoxin. *Plant Molecular Biology, 58*, 659–668.

Erb, M., & Reymond, P. (2019). Molecular interactions between plants and insect herbivores. *Annual Review of Plant Biology, 70*, 527–557.

Feng, H., Wu, X., Wu, B., & Wu, K. (2009). Seasonal migration of *Helicoverpa armigera* (Lepidoptera: Noctuidae) over the Bohai Sea. *Journal of Economic Entomology, 102*, 95–104.

Gain, D., & Kundu, G. G. (1986). Seasonal incidence of the bean stem miner, *Melanagromyza sojae* (Zehntner) in soybean at Delhi (India). *Journal of Entomological Research, 15.*

Gaur, N., & Mogalapu, S. (2018). Pests of soybean. In *Pests and Their Management* (pp. 137–162). Springer.

Ghione, C. E., Lombardo, L. A., Vicentin, I. G., & Heinz, R. A. (2021). Association mapping to identify molecular markers associated with resistance genes to stink bugs in soybean. *Euphytica, 217*, 1–12.

Ghosh, L. K. (2008). *Handbook on Hemipteran Pests in India.* Zoological Survey of India.

Gresshoff, P. M., & Ferguson, B. J. (2017). Molecular signals in nodulation control. *International Journal of Molecular Sciences, 18*(1), 125.

Gulzar, A., Maqsood, A., Munir, A., Tariq, M., Ali, M., & Qureshi, R. (2017). Toxicity, antifeedant, and sub-lethal effects of *Citrullus colocynthis* extracts on cotton bollworm, *Helicoverpa armigera* (Lepidoptera: Noctuidae). *Pakistan Journal of Zoology, 49*(6).

Haile, F., Nowatzki, T., & Storer, N. (2021). Overview of pest status, potential risk, and management considerations of *Helicoverpa armigera* (Lepidoptera: Noctuidae) for U.S. soybean production. *Journal of Integrated Pest Management, 12*, 3.

Hammond, R. B., Higgins, R. A., Mack, T. P., Pedigo, L. P., & Bechinski, E. J. (1991). Soybean pest management. In *CRC Handbook of Pest Management in Agriculture* (Vol. *3*, pp. 341–472).

Harish, G., Patil, R. H., & Patil, P. V. (2009). Identification of resistant sources to major defoliator pests of soybean. *Karnataka Journal of Agricultural Sciences, 22*, 215–217.

Hill, C. B., Kim, K. S., Crull, L., Diers, B. W., & Hartman, G. L. (2009). Inheritance of resistance to the soybean aphid in soybean PI 200538. *Crop Science, 49*, 1193–1200.

Hill, J. H., Alleman, R., Hogg, D. B., & Grau, C. R. (2001). First report of transmission of *Soybean mosaic virus* and *Alfalfa mosaic virus* by *Aphis glycines* in the New World. *Plant Disease, 85*, 561.

Hinchee, M. A. W., Connor-Ward, D. V., Newell, C. A., McDonnell, R. E., Sato, S. J., & Gasser, C. S. (1988). Production of transgenic soybean plants using *Agrobacterium*-mediated DNA transfer. *Biotechnology, 6*, 915–922.

Hodgson, E., & Koch, R. (2018). *Soybean Aphid Field Guide.*

Homrich, M. S., Passaglia, L. M. P., Pereira, J. F., Bertagnolli, P. F., Pasquali, G., & Zaidi, M. A. (2008). Resistance to *Anticarsia gemmatalis* Hübner (Lepidoptera, Noctuidae) in transgenic soybean (*Glycine max* (L.) Merrill Fabales, Fabaceae) cultivar IAS5 expressing a modified Cry1Ac endotoxin. *Genetics and Molecular Biology, 31*, 522–531.

Jayappa, A. H. (2000). *Studies on the Soybean Stem fly M. sojae (Zehnt) with Reference to Crop Loss Assessment, Screening of Varieties, and Management* (Master's thesis). University of Agricultural Sciences, Bangalore.

Joung, J. K., & Sander, J. D. (2013). TALENs: A widely applicable technology for targeted genome editing. *Nature Reviews Molecular Cell Biology, 14*, 49–55.

Jun, T. H., Rouf Mian, M. A., & Michel, A. P. (2012). Genetic mapping revealed two loci for soybean aphid resistance in PI 567301B. *Theoretical and Applied Genetics, 124*, 13–22.

Jun, T. H., Rouf Mian, M. A., & Michel, A. P. (2013). Genetic mapping of three quantitative trait loci for soybean aphid resistance in PI 567324. *Heredity, 111*, 16–22.

Kang, S. T., Mian, M. A. R., & Hammond, R. B. (2008). Soybean aphid resistance in PI 243540 is controlled by a single dominant gene. *Crop Science, 48*, 1744–1748.

Karban, R., Agrawal, A. A., & Mangel, M. (1997). The benefits of induced defenses against herbivores. *Ecology, 78*, 1351–1355.

Karim, S. (2000). Management of *Helicoverpa armigera*: A review and prospectus for Pakistan. *Pakistan Journal of Biological Sciences, 3*, 1213–1222.

Kenis, M., & Cugala, D. (2006). Prospects for the biological control of the groundnut leaf miner, *Aproaerema modicella*, in Africa. *Biocontrol News and Information, 27*, 1–9.

Keshan, R., Singh, I. K., & Singh, A. (2021). Genome-wide investigation of MAPKKKs from *Cicer arietinum* and their involvement in plant defense against *Helicoverpa armigera. Physiological and Molecular Plant Pathology, 115*, 101685.

Kilen, T. C., & Lambert, L. (1986). Evidence for different genes controlling insect resistance in three soybean genotypes. *Crop Science, 26*, 869–871. https://doi.org/10.2135/cropsci1986.0011183x002600050006x.

Kim, K. S., Bellendir, S., Hudson, K. A., Hill, C. B., Hartman, G. L., & Hyten, D. L. (2010). Fine mapping the soybean aphid resistance gene *Rag1* in soybean. *Theoretical and Applied Genetics, 120*, 1063–1071.

King, A. B. S. (1994). *Heliothis/Helicoverpa (Lepidoptera: Noctuidae)*. In *Insect Pests of Cotton.*

Kogan, M., & Ortman, E. F. (1978). Antixenosis—A new term proposed to define Painter's "nonpreference" modality of resistance. *Bulletin of the Entomological Society of America, 24*, 175, 176.

Kogan, M., & Turnipseed, S. G. (1987). Ecology and management of soybean arthropods. *Annual Review of Entomology, 32*, 507–538.

Kulamarva, A. G., Sosle, V. R., & Raghavan, G. S. V. (2009). Nutritional and rheological properties of sorghum. *International Journal of Food Properties, 12*, 55–69.

Kumar, A., Panwar, R., Singh, A., & Singh, I. K. (2020). Role of calcium signaling during plant–herbivore interaction. In *Plant Stress Biology* (pp. 491–510). Springer, Singapore.

Kumar, J., Ramlal, A., Kumar, K., Rani, A., & Mishra, V. (2021). Signaling pathways and downstream effectors of host innate immunity in plants. *International Journal of Molecular Sciences, 22*(16), 9022. https://doi.org/10.3390/ijms22169022.

Kumar, R., Saini, M., Taku, M., Debbarma, P., Mahto, R. K., Ramlal, A., Sharma, D., Rajendran, A., Pandey, R., Gaikwad, K., Lal, S. K., & Talukdar, A. (2023). Identification of quantitative trait loci (QTLs) and candidate genes for seed shape and 100-seed weight in soybean (*Glycine max* (L.) Merr.). *Frontiers in Plant Science, 13*(January), 1–15. https://doi.org/10.3389/fpls.2022.1074245.

LaMantia, J. M., Mian, M. A. R., & Redinbaugh, M. G. (2019). Genetic mapping of soybean aphid biotype 3 and 4 resistance in PI 606390A. *Molecular Breeding, 39*(4).

Lee, J. S., Yoo, M. H., Jung, J. K., Bilyeu, K. D., Lee, J. D., & Kang, S. (2015). Detection of novel QTLs for foxglove aphid resistance in soybean. *Theoretical and Applied Genetics, 128*, 1481–1488.

Li, R., Zhang, J., Li, J., Zhou, G., Wang, Q., & Bian, W. (2015). Prioritizing plant defence over growth through WRKY regulation facilitates infestation by non-target herbivores. *eLife, 4*, e04805.

Li, X., Qin, R., Du, Q., Cai, L., Hu, D., Du, H., et al. (2020). Knockdown of *GmVQ58* encoding a VQ motif-containing protein enhances soybean resistance to the common cutworm (*Spodoptera litura* Fabricius). *Journal of Experimental Botany, 71*, 3198–3210.

Li, Y., Hill, C. B., Carlson, S. R., Diers, B. W., & Hartman, G. L. (2007). Soybean aphid resistance genes in the soybean cultivars Dowling and Jackson map to linkage group M. *Molecular Breeding, 19*, 25–34.

Lin, C., Li, L., Wang, Y., Xun, Z., Zhang, G., & Li, S. (1993). Effects of aphid density on the major economic characters of soybean. *Soybean Science, 12*, 252–254.

Liu, H., Che, Z., Zeng, X., Zhang, G., Wang, H., & Yu, D. (2016). Identification of single nucleotide polymorphisms in soybean associated with resistance to common cutworm (*Spodoptera litura* Fabricius). *Euphytica, 209*, 49–62.

Liu, X., Wu, S., Xu, J., Sui, C., & Wei, J. (2017). Application of CRISPR/Cas9 in plant biology. *Acta Pharmaceutica Sinica B, 7*, 292–302.

Longhini, D. (1996). Surfactants on cleanup detail in paper mills. *Chemical Engineering, 103*.

Major, I. T., & Constabel, C. P. (2006). Molecular analysis of poplar defense against herbivory: Comparison of wound- and insect elicitor-induced gene expression. *New Phytologist, 172*, 617–635.

Maree, J. M., Kallar, S. A., & Khuhro, R. D. (1999). Relative abundance of *Spodoptera litura* F. and *Agrotis ipsilon* Rott. on cabbage. *Pakistan Journal of Zoology, 31*, 31–34.

Marking, S. (1998). The checkoff-supported new-uses payoff begins. *Soybean Digest, 58*, 9–21.

Matsumura, S. (1917). A list of the Aphididae of Japan, with description of new species and genera. *The Journal of the College of Agriculture, Tohoku Imperial University, 7*, 351–414.

McCornack, B. P., Costamagna, A. C., & Ragsdale, D. W. (2008). Within-plant distribution of soybean aphid (Hemiptera: Aphididae) and development of node-based sample units for estimating whole-plant densities in soybean. *Journal of Economic Entomology, 101*, 1488–1500.

McPherson, R. M., Johnson, W. C., Mullinix, B. G., Mills, W. A., & Peebles, F. S. (2003). Influence of herbicide tolerant soybean production systems on insect pest populations and pest-induced crop damage. *Journal of Economic Entomology, 96*, 690–698.

Meng, F. L., Ran, R. X., Li, Y., Li, N., Li, H. Z., & Wang, Z. K. (2017). RNA interference-mediated serine protease gene (*Spbtry1*) knockdown affects growth and mortality in the soybean pod borer (Lepidoptera: Olethreutidae). *Florida Entomologist, 100*, 607–615.

Meng, F., Li, Y., Zang, Z., Li, N., Ran, R., Cao, Y., et al. (2017). Expression of the double-stranded RNA of the soybean pod borer *Leguminivora glycinivorella* (Lepidoptera: Tortricidae) ribosomal protein P0 gene enhances the resistance of transgenic soybean plants. *Pest Management Science, 73*, 2447–2455.

Miklos, J. A., Alibhai, M. F., Bledig, S. A., Connor-Ward, D. C., Gao, A. G., & Holmes, B. A. (2007). Characterization of soybean exhibiting high expression of a synthetic *Bacillus thuringiensis cry1A* transgene that confers a high degree of resistance to lepidopteran pests. *Crop Science, 47*, 148–157.

Miller, G., Schlauch, K., Tam, R., Cortes, D., Torres, M. A., & Shulaev, V. (2009). The plant NADPH oxidase RBOHD mediates rapid systemic signaling in response to diverse stimuli. *Science Signaling, 45*.

Mohammad, A. (1981). The groundnut leafminer, *Aproaerema modicella* Deventer (= *Stomopteryx subsecivella* Zeller) (Lepidoptera: Gelechiidae): A review of world literature.

Mohanta, T. K., Bashir, T., Hashem, A., Abd_Allah, E. F., & Bae, H. (2017). Genome editing tools in plants. *Genes, 8*, 399.

Naresh, R. K., Dhaliwal, S. S., Chaudhary, M., Chandra, M. S., & Kumar, A. (2019). Chapter-5 Legumes: An option to enhance productivity and soil health sustainability. *Agronomy, 93*.

Naseri, B., Fathipour, Y., Moharramipour, S., & Hosseininaveh, V. (2011). Comparative reproductive performance of *Helicoverpa armigera* (Hübner) (Lepidoptera: Noctuidae) reared on thirteen soybean varieties. *Journal of Agricultural Science and Technology, 13*, 17–26.

Ngaki, M. N., Sahoo, D. K., Wang, B., & Bhattacharyya, M. K. (2021). Overexpression of a plasma membrane protein generated broad-spectrum immunity in soybean. *Plant Biotechnology Journal, 19*, 502–516.

Oerke, E. C. (2006). Crop losses to pests. *Journal of Agricultural Science, 144*, 31–43.

Onkokesung, N., Reichelt, M., van Doorn, A., Schuurink, R. C., van Loon, J. J. A., & Dicke, M. (2014). Modulation of flavonoid metabolites in *Arabidopsis thaliana* through overexpression of the MYB75 transcription factor. *Journal of Experimental Botany, 65*, 2203–2217.

Parrott, W. A., All, J. N., Adang, M. J., Bailey, M. A., Boerma, H. R., & Stewart, C. N. (1994). Recovery and evaluation of soybean plants transgenic for a *Bacillus thuringiensis* var. *Kurstaki* insecticidal gene. *In Vitro Cellular & Developmental Biology–Plant, 30*, 144–149.

Patankar, A. G., Giri, A. P., Harsulkar, A. M., Sainani, M. N., Deshpande, V. V., & Ranjekar, P. K. (2001). Complexity in specificities and expression of *Helicoverpa armigera* gut proteinases explains polyphagous nature of the insect pest. *Insect Biochemistry and Molecular Biology, 31*, 453–464.

Patil, G., Vuong, T. D., Kale, S., Valliyodan, B., Deshmukh, R., & Zhu, C. (2018). Dissecting genomic hotspots underlying seed protein, oil, and sucrose content in an interspecific mapping population of soybean using high-density linkage mapping. *Plant Biotechnology Journal, 16*, 1939–1953.

Pieterse, C. M. J., van der Does, A., Zamioudis, C., Leon-Reyes, A., & Van Wees, S. C. M. (2012). Hormonal modulation of plant immunity. *Annual Review of Cell and Developmental Biology, 28*, 489–521.

Pogue, M. G. (2006). *World Spodoptera Database (Lepidoptera: Noctuidae)*. Retrieved from: http://www.sel.barc.usda.gov.

Punithavalli, M., Sharma, A. N., & Rajkumar, M. B. (2014). Seasonality of the common cutworm *Spodoptera litura* in a soybean ecosystem. *Phytoparasitica, 42*, 213–222.

Qin, D., Liu, X. Y., Miceli, C., Zhang, Q., & Wng, P. W. (2019). Soybean plants expressing the *Bacillus thuringiensis cry8*-like gene show resistance to *Holotrichia parallela. BMC Biotechnology, 19*, 1–12.

Qin, H. G., & Ye, Z. H. (2007). Occurrence and management of the common cutworm, *Spodoptera litura*, in China. *Journal of Agricultural Science and Technology Press*.

Ragsdale, D. W., McCornack, B. P., Venette, R. C., Potter, B. D., MacRae, I. V., & Hodgson, E. W. (2007). Economic threshold for soybean aphid (*Hemiptera: Aphididae*). *Journal of Economic Entomology, 100*, 1258–1267.

Ragsdale, D. W., Voegtlin, D. J., & O'Neil, R. J. (2004). Soybean aphid biology in North America. *Annals of the Entomological Society of America, 97*, 204–208.

Rajendran, A., Lal, S. K., Raju, D., & Ramlal, A. (2022). Associations of direct and indirect selection for pregermination anaerobic stress tolerance in soybean (*Glycine max*). *Plant Breeding, 141*(5), 634–643. https://doi.org/10.1111/pbr.13048.

Rajendran, A., Lal, S. K., Raju, D., Mallikarjun, B. P., Ramlal, A., & Sharma, D. (2023). Waterlogging tolerance evaluation methods for soybean (*Glycine max* (L.) Merr.) at the pregermination stage. *Genetic Resources and Crop Evolution, 70,* 2451–2461. https://doi.org/10.1007/s10722-023-01573-0.

Ramlal, A., Nautiyal, A., Baweja, P., Kumar, V., Mehta, S., Mahto, R. K., Tripathi, S., Shanmugam, A., Pujari Mallikarjuna, B., Raman, P., Lal, S. K., Raju, D., & Rajendran, A. (2022). Angiotensin-converting enzyme inhibitory peptides and isoflavonoids from soybean (*Glycine max* (L.) Merr.). *Frontiers in Nutrition, 9,* 1068388. https://doi.org/10.3389/fnut.2022.1068388.

Ramlal, A., Nautiyal, A., Baweja, P., Mahto, R. K., Mehta, S., Mallikarjun, B. P., Vijayan, R., Saluja, S., Kumar, V., & Dhiman, S. K. (2022). Harnessing heterosis and male sterility in soybean (*Glycine max* (L.) Merrill): A critical revisit. *Frontiers in Plant Science, 13,* 981768.

Rector, B. G., All, J. N., Parrott, W. A., & Boerma, H. R. (1998). Identification of molecular markers linked to quantitative trait loci for soybean resistance to corn earworm. *Theoretical and Applied Genetics, 96*(6, 7), 786–790. https://doi.org/10.1007/s001220050803.

Rector, B. G., All, J. N., Parrott, W. A., & Boerma, H. R. (1999). Quantitative trait loci for antixenosis resistance to corn earworm in soybean. *Crop Science, 39*(2), 531–538. https://doi.org/10.2135/cropsci1999.0011183X003900020038x.

Rector, B. G., All, J. N., Parrott, W. A., & Boerma, H. R. (2000). Quantitative trait loci for antibiosis resistance to corn earworm in soybean. *Crop Science, 40*(1), 233–238.

Reigada, C., Guimarães, K. F., & Parra, J. R. P. (2016). Relative fitness of *Helicoverpa armigera* (*Lepidoptera: Noctuidae*) on seven host plants: A perspective for IPM in Brazil. *Journal of Insect Science, 16.*

Savajji, K. (2006). *Biology and Management of Soybean Stem Fly (Melanagromyza sojae) (Zehntner) (Diptera: Agromyzidae)* (Master's thesis, University of Agricultural Sciences).

Schuman, M. C., & Baldwin, I. T. (2016). The layers of plant responses to insect herbivores. *Annual Review of Entomology, 61*, 373–394.

Schweizer, F., Bodenhausen, N., Lassueur, S., Masclaux, F. G., & Reymond, P. (2013). Differential contribution of transcription factors to *Arabidopsis thaliana* defense against *Spodoptera littoralis*. *Frontiers in Plant Science, 4*, 13.

Seagraves, M. P., McPherson, R. M., & Ruberson, J. R. (2004). Impact of *Solenopsis invicta* Buren suppression on arthropod ground predators and pest species in soybean. *Journal of Entomological Science, 39*, 433–443.

Shanower, T. G., Wightman, J. A., & Gutierrez, A. P. (1993). Biology and control of the groundnut leafminer, *Aproaerema modicella* (*Deventer*) (*Lepidoptera: Gelechiidae*). *Crop Protection, 12*, 3–10.

Singh, A., & Singh, I. K. (2018). *Molecular Aspects of Plant-Pathogen Interaction* (Vol. 10, pp. 978–981). Springer, Singapore.

Singh, A., Jain, D., Tyagi, C., Singh, S., Kumar, S., & Singh, I. K. (2018). *In silico* prediction of active site and *in vitro* DNase and RNase activities of *Helicoverpa*-inducible pathogenesis-related-4 protein from *Cicer arietinum*. *International Journal of Biological Macromolecules, 113*, 869–880.

Singh, A., Singh, I. K., & Verma, P. K. (2008). Differential transcript accumulation in *Cicer arietinum* L. in response to a chewing insect *Helicoverpa armigera* and defence regulators correlate with reduced insect performance. *Journal of Experimental Botany, 59*, 2379–2392.

Singh, A., Singh, S., & Singh, I. K. (2016). Recent insights into the molecular mechanism of jasmonate signaling during insect-plant interaction. *Australasian Plant Pathology, 45*, 123–133.

Singh, A., Singh, S., Singh, R., Kumar, S., Singh, S. K., & Singh, I. K. (2021). Dynamics of *Zea mays* transcriptome in response to a polyphagous herbivore, *Spodoptera litura*. *Functional & Integrative Genomics, 21*, 571–592.

Singh, S. (1982). *Ecology of the Agromyzidae (Diptera) Associated with Leguminous Crops in India* [Doctoral dissertation, St. Johns College].

Spencer, K. A. (1973). *Agromyzidae (Diptera) of Economic Importance*. Springer Science and Business Media.

Stewart, C. N., Adang, M. J., All, J. N., Boerma, H. R., Cardineau, G., & Tucker, D. (1996). Genetic transformation, recovery, and characterization of fertile soybean transgenic for a synthetic *Bacillus thuringiensis cryIAc* gene. *Plant Physiology, 112*, 121–129.

Stintzi, A., Weber, H., Reymond, P., Browse, J., & Farmer, E. E. (2001). Plant defense in the absence of jasmonic acid: The role of cyclopentenones. *Proceedings of the National Academy of Sciences, 98*, 12837–12842.

Taki, N., Sasaki-Sekimoto, Y., Obayashi, T., Kikuta, A., Kobayashi, K., & Ainai, T. (2005). 12-oxo-phytodienoic acid triggers expression of a distinct set of genes and plays a role in wound-induced gene expression in *Arabidopsis*. *Plant Physiology, 139*, 1268–1283.

Talekar, N. S. (1990). *Agromyzid Flies of Food Legumes in the Tropics*.

Tanksley, S. D. (1993). Mapping polygenes. *Annual Review of Genetics, 27*, 205–233.

Tay, W. T., Walsh, T. K., Downes, S., Anderson, C., Jermiin, L. S., & Wong, T. K. F. (2017). Mitochondrial DNA and trade data support multiple origins of *Helicoverpa armigera* (Lepidoptera, Noctuidae) in Brazil. *Scientific Reports, 7*, 1–10.

Thapa, R. B. (2012). Redescription of *Melanagromyza sojae* (Zehntner) from India and Nepal. *Nepalese Journal of Biosciences, 2*, 64–70.

Thines, B., Katsir, L., Melotto, M., Niu, Y., Mandaokar, A., & Liu, G. (2007). JAZ repressor proteins are targets of the SCFCOI1 complex during jasmonate signaling. *Nature, 448*, 661–665.

Tian, W., Wang, C., Gao, Q., Li, L., & Luan, S. (2020). Calcium spikes, waves, and oscillations in plant development and biotic interactions. *Nature Plants, 6*, 750–759.

Tuberosa, R., Salvi, S., Sanguineti, M. C., Landi, P., Maccaferri, M., & Conti, S. (2002). Mapping QTLs regulating morpho-physiological traits and yield: Case studies, shortcomings, and perspectives in drought-stressed maize. *Annals of Botany, 89*, 941–963.

Turnipseed, S. G., & Kogan, M. (1976). Soybean entomology. *Annual Review of Entomology, 21*, 247–282.

Van Duyn, J. W., Turnipseed, S. G., & Maxwell, J. D. (1971). Resistance in soybeans to the Mexican bean beetle: I. Sources of resistance. *Crop Science, 11*(4), 572–573.

Van Duyn, J. W., Turnipseed, S. G., & Maxwell, J. D. (1972). Resistance in soybeans to the Mexican bean beetle: II. Reactions of the beetle to resistant plants. *Crop Science, 12*(5), 561–562.

Voegtlin, D. J., Halbert, S. E., & Qiao, G. (2004). A guide to separating *Aphis glycines* Matsumura and morphologically similar species that share its hosts. *Annals of the Entomological Society of America, 97*, 227–232.

Walker, D. R., Narvel, J. M., Boerma, H. R., All, J. N., & Parrott, W. A. (2004). A QTL that enhances and broadens Bt insect resistance in soybean. *Theoretical and Applied Genetics, 109*, 1051–1057.

Walker, D., Boerma, H. R., All, J., & Parrott, W. (2002). Combining *cry1Ac* with QTL alleles from PI 229358 to improve soybean resistance to lepidopteran pests. *Molecular Breeding, 9*, 43–51.

Walt, A., van der Berg, J., & du Plessis, H. (2008). Using morphological characteristics to distinguish between male and female larvae and pupae of the groundnut leafminer, *Aproaerema modicella* (Deventer) (Lepidoptera: Gelechiidae). *South African Journal of Plant and Soil, 25*, 182–184.

Wang, H., Yan, H., Du, H., Chao, M., Gao, Z., & Yu, D. (2015). Mapping quantitative trait loci associated with soybean resistance to common cutworm and soybean compensatory growth after defoliation using SNP marker-based genome-wide association analysis. *Molecular Breeding, 35*, 1–15.

Wang, X., Fang, Y., Lin, Z., Zhang, L., & Wang, H. (1994). A study on the damage and economic threshold of the soybean aphid at the seedling stage. *Plant Protection (Institute of Plant Protection, CAAS, China), 20*, 12, 13.

Wilson, R. F. (2008). Soybean: Market-driven research needs. In *Genetics and Genomics of Soybean* (pp. 3–15). Springer.

Wu, Q., Wang, H., Wu, J., Wang, D., Wang, Y., & Zhang, L. (2015). Soybean *GmAOC3* promotes plant resistance to the common cutworm by increasing the expression of genes involved in resistance and volatile substance emission in transgenic tobaccos. *Journal of Plant Biology, 58*, 242–251.

Xiao, L., Zhong, Y. P., Wang, B., & Wu, T. L. (2014). Mapping an aphid resistance gene in soybean (*Glycine max* (L.) Merr.) P746. *Genetics and Molecular Research, 13*, 9152–9160.

Yadav, M., Pandey, J., Chakraborty, A., et al. (2022). A comprehensive analysis of calmodulin-like proteins of *Glycine max* indicates their role in calcium signaling and plant defense against insect attack. *Frontiers in Plant Science, 13,* 817950.

Yan, S., Qian, J., Cai, C., Ma, Z., Li, J., & Yin, M. (2020). Spray method application of transdermal dsRNA delivery system for efficient gene silencing and pest control on soybean aphid *Aphis glycines*. *Journal of Pest Science, 93*, 449–459.

Yücel, S., & Hanife, G. (2018). Laboratory rearing of cotton bollworm, *Helicoverpa armigera* (Hübner) (Lepidoptera: Noctuidae) wild colony on different artificial diets. *Journal of Applied Biological Sciences, 12*, 26–32.

Zaidi, S. S. A., Vanderschuren, H., Qaim, M., Mahfouz, M. M., Kohli, A., & Mansoor, S. (2019). New plant breeding technologies for food security. *Science, 363*, 1390, 1391.

Zhang, G., Gu, C., & Wang, D. (2009). Molecular mapping of soybean aphid resistance genes in PI 567541B. *Theoretical and Applied Genetics, 118*, 473–482.

Zhang, G., Gu, C., & Wang, D. (2010). A novel locus for soybean aphid resistance. *Theoretical and Applied Genetics, 120*, 1183–1191.

Zhang, S., Zhang, Z., Bales, C., Gu, C., DiFonzo, C., & Li, M. (2017). Mapping novel aphid resistance QTL from wild soybean, *Glycine soja* 85–32. *Theoretical and Applied Genetics, 130*, 1941–1952.

Zhao, G., Jiang, Z., Li, D., Han, Y., Hu, H., & Wu, L. (2015). Molecular loci associated with seed isoflavone content may underlie resistance to soybean pod borer (*Leguminivora glycinivorella*). *Plant Breeding, 134*, 78–84.

CHAPTER 12

Abiotic Factors Influencing the Yield and Productivity of Soybean

PONNUCHAMY MAGUDEESWARI,[1] M. VANI PRAVEENA,[2] ARAVINDAN SHANMUGAM,[3] BALAMURUGAN MOHANAPRIYA,[3] and RAMAN PUSHPA[4,*]

[1]*School of Crop Improvement, College of Post Graduate Studies in Agricultural Sciences, Central Agricultural University, Imphal, Manipur, India*

[2]*Department of Genetics and Plant Breeding, Acharya N.G. Ranga Agricultural University, Andhra Pradesh, India*

[3]*Department of Genetics and Plant Breeding, Tamil Nadu Agricultural University, Coimbatore, Tamil Nadu, India*

[4]*Tamil Nadu Rice Research Institute, Tamil Nadu Agricultural University, Aduthurai, Tamil Nadu, India*

[*]Corresponding author

ABSTRACT

Soybean is a seed legume and is often called a "miracle crop" because of its versatile nutritional profile, vegetable protein, and oil content. Among oilseed crops, soybean is cultivated on approximately 6% of the global area. Abiotic stress—such as drought, flooding, high temperature, cold, soil salinity, and soil acidity—affects seed germination, crop growth, and development by disrupting cellular and metabolic activities, resulting in yield loss. Genetic improvement, specifically breeding for tolerant varieties, offers a cost-effective approach to mitigate these challenges. Conventional breeding has made significant contributions, and with

Soybeans, Volume 2: Genome, Biotechnology, Stress, and Disease.
R. Ambika Rajendran, Ayyagari Ramlal, Sanjay Kumar Lal, and Godfree Chigeza (Eds.)

DOI: 10.1201/9781998511730-15

the advent of next-generation sequencing (NGS) technologies, several advanced molecular breeding approaches have been adopted to develop stress-tolerant varieties. Numerous QTLs and genes associated with tolerance to abiotic stresses have been identified. However, characterizing and validating a large number of lines for each trait remains labor-intensive, especially in evaluating physiological and biochemical changes under stress conditions. High-throughput phenotyping (HTP) tools help reduce this workload by detecting subtle physiological changes and delivering precise results. The integration of genomic tools with HTP will support the development of soybean varieties with improved abiotic stress tolerance. In this context, the chapter reviews the impact of abiotic stresses on soybean yield and explores tools to mitigate stress for sustainable production.

12.1 INTRODUCTION

Soybean is a member of the legume family Leguminosae in the genus *Glycine*. It is divided into two subgenera: *Glycine* (Perennial) and *Soja* (Annual). The subgenus *Glycine* has 22 species, including *G. tabacina* and *G. tomentella*. The subgenus *Soja* contains *G. max* (cultivated soybean) and *G. soja* (wild annual soybean). The cultivated soybean is diploid, with a chromosome number of 2n = 20 and a genome size of 1013.2 Mb (Xie et al., 2019). Soybean is a dual-purpose crop and the largest source of vegetable oil (20%), with the highest seed protein content of 42%, carbohydrate content of 35%, and 5% ash (Liu, 1997; Ramlal et al., 2022a, b). Among the oilseed crops, soybean stands second as the highest oil-containing legume (20–22%) after groundnut (Sugiyama et al., 2015). With a registered yield of 9278 kg/ha and production of 11.2 million metric tons, India ranks fifth in soybean production. The US, Brazil, and Argentina are the leading soybean producers, with 119.8, 114, and 52 million metric tons, respectively, in 2021–2022 (SOPA). Although soybean is native to China, the production is around 19 million metric tons. The percent contribution of major countries to the overall production is given in Figure 12.1.

Since soybean is an important legume crop, it has been cultivated in a wide range of climatic conditions. However, crop growth, development, and yield are highly affected by various abiotic stress conditions like drought, flooding, heat stress, salinity, cold, and soil acidity. Genetic variability is a key resource for varietal improvement against various abiotic stresses. The exploitation and utilization of wild germplasm will help in

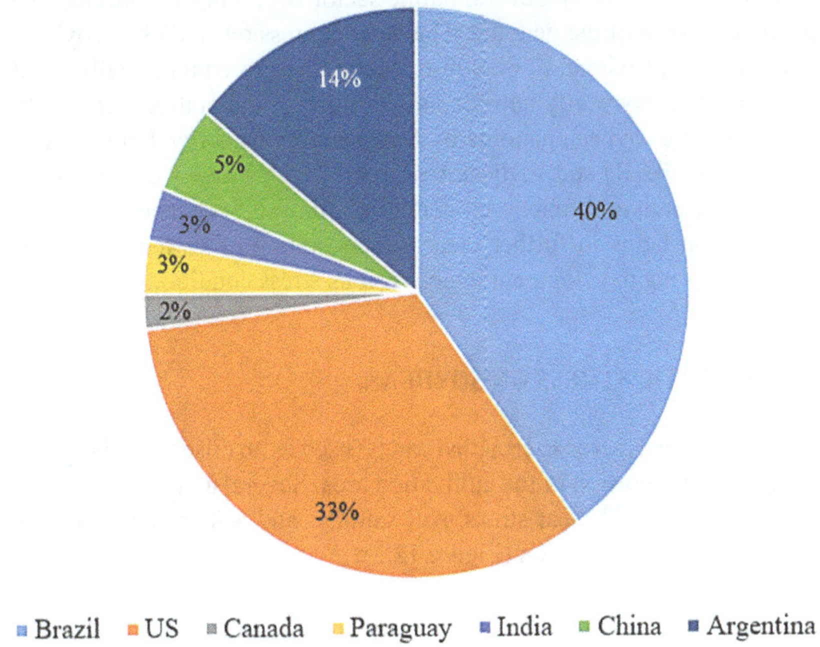

FIGURE 12.1 Percent contribution of major countries to the overall soybean production.

the improvement against various stresses. As of now, more than 170,000 *G. max* soybean accessions are maintained in 160 institutions. In addition to that, around 10,000 accessions of *G. soja* and 3500 accessions of *Glycine* species are maintained (www.ipgri.org). In the world, China has the largest soybean germplasm collection, with about 26,000 accessions of *G. max*, followed by the US with about 17,000 accessions of *G. max* in the USDA soybean germplasm collection (Carter et al., 2004; Chang et al., 1999).

12.2 GENETIC RESOURCES AVAILABLE FOR SOYBEAN

The soybean genome consists of 46–60% repetitive sequences and heterochromatic regions (Goldberg, 1978; Singh and Hymowitz, 1988). The classical map represents only 67 loci on 19 linkage groups (LGs), whereas the molecular map represents all 20 LGs based on SSR and RFLP markers mapped on five different populations (Hedges and Palmer,

1993; Song et al., 2004). Several public sector BAC libraries provide over 35-fold coverage of the genome. The gene expression studies of soybean, comprising approximately 60,000 transcripts, were commercially available on http://www.affymetrix.com. Further collaborative efforts for sequencing the soybean genome by Department of Energy-Joint Genome Institute (DOE-JGI) and soybean research provided a new dimension for soybean research and functional genomics studies. Sequence-based gene identification helps in further clear understanding of various biological processes during development under various stress conditions.

12.3 ABIOTIC STRESS ON SOYBEAN

Soybean is a dual-purpose crop that can be used as an oilseed, and legumes occupy a major portion of the cultivation area. Several abiotic factors, like drought, flood, cold, heat stress, soil salinity, and acidity, tend to affect crop growth and production (Figure 12.2).

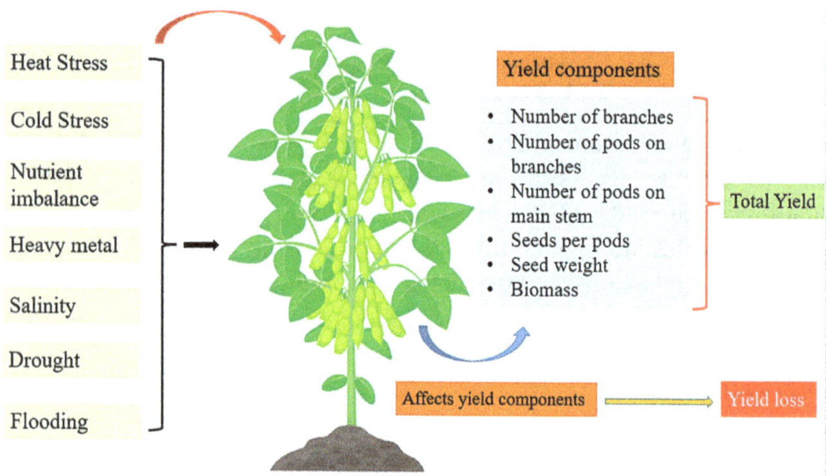

FIGURE 12.2 Different abiotic factors that affect crop growth and yield.

12.3.1 DROUGHT

Drought stress is one of the most devastating abiotic stresses affecting productivity by up to 40% (Le et al., 2012) in soybean. High temperature, water deficiency, and low humidity in the atmosphere are the main reasons for drought tolerance, which affects germination, early seedling vigor, plant growth, carbon assimilation, and the phenology of the plant (Hirt and Shinozaki, 2004; Kaya et al., 2006; Szilagyi, 2003). During drought conditions, the turgor pressure of the cells declines, which reduces the cell elongation rate (Tripathy et al., 2000). Under stress conditions, plants adopt several mitigation strategies like escape, tolerance, and avoidance (Turner et al., 2001). In the first mechanism, the plants complete their life cycle earlier than normal with very few seeds (Boerma and Specht, 2004). To overcome stress, plants also adopt avoidance mechanisms, including the reduction of evapotranspiration and efficient absorption of water through the root system. The tolerance mechanism mainly depends on the production of osmoprotectants and osmolytes (Kavar et al., 2008).

Although plants adopt different mitigation strategies to overcome drought, the development of drought-tolerant cultivars is one of the best alternatives to overcome stress. Several studies were conducted on root and shoot traits to improve drought tolerance in soybean. The presence of long root length, taproot, and dense root system during the vegetative stage helps to withstand drought stress conditions (Hirasawa, 1994; Hoogenboom et al., 1987; Huck et al., 1983). These root traits are key indicators of drought stress, and a large number of germplasm resources are available for improving the root traits. However, it is difficult and laborious to study the root morphology and physiological changes during drought stress. Molecular and biotechnological approaches facilitate the improvement of soybean for drought stress. Under drought stress, the major genes RAB-18, GOLS, bZIP, and DERB were expressed differentially (Marcolina-Gomes et al., 2014). Regarding shoot traits, stomatal conductance (Earl, 2003; Liu et al., 2005), epidermal conductance (Hufstetler et al., 2007; James et al., 2009), water use efficiency (Liu et al., 2005), and osmotic adjustment were studied, and genetic variation for these traits was observed in soybean germplasm.

12.3.1.1 MOLECULAR STUDIES FOR DROUGHT TOLERANCE IN SOYBEAN

Drought stress involves complex mechanisms with the interaction of several genes that are upregulated and downregulated under stress conditions (Kanvar et al., 2008). Structural and functional genomic approaches were used to study different mechanisms in soybean. With the advent of sequencing technologies and the availability of WGS information, thousands of SSRs and millions of SNPs were identified, and they were associated with seed quality traits, yield, and biotic and abiotic stress tolerance (Bisht et al., 2022; Patil et al., 2016; Vidal et al., 2012).

Next-generation sequencing approaches and advanced microarray techniques provide cost-effective and simple means of obtaining genomic information. In soybean, several QTLs have been reported for yield and tolerance to biotic and abiotic stress (www.soybase.org), especially for drought, from different studies independently (Dhungana et al., 2021; Liu et al., 2005; Manavalan et al., 2009; Ouyang et al., 2022; Ren et al., 2020). Using a RIL population derived from Kefeng1 × Nannong 1138-2, 10 QTLs were identified and associated with drought tolerance (Du et al., 2009). Carpentieri-Pipolo et al. (2012) identified four QTLs related to transpiration rate using the RIL population derived from the PI416937 × Benning cross and explained around 17.7% to 24.7% of phenotypic variation. Using a GWAS approach, several drought tolerance QTLs have been identified on chromosomes 5, 16, and 17, which explain around 88.55% to 95.22% of phenotypic variation (Khan et al., 2020; Liu et al., 2020; Zhang et al., 2022). Around 21 genes are involved in the drought response pathway given in the soybean database, and they belong to the drought response protein family.

Since the availability of the soybean reference genome Williams 82, released in 2010, several novel genes, SNPs, and InDels have been discovered by comparing different soybean cultivars. Several genes, like *gmderb1, gmGoLs, gmpip1b, Gmereb,* and *Gmdefensin* (Stolf-Moreira et al., 2010), *LEA-D11* (Savitri et al., 2013), and *GmNAC* (Thao et al., 2013), were identified and can be utilized in breeding programs. Besides, transcription factors (TFs) like GmDREB, GmDREB2, GmERF3b, GmZIP1, GmGT-2A, GsZFP1, GmHdz4, GmWRKY54, GmWRKY20, and GmMYBJ1 were identified in different studies and found to be upregulated under drought stress conditions (Shaheen et al., 2016). Breeding for drought resistance includes high-throughput phenotyping (HTP), selection, marker-assisted breeding, and field evaluation (Fenta et al., 2014; Lightfoot, 2015; Thao and Tran, 2012). Marker-assisted breeding is an important approach to developing drought-tolerant cultivars and also helps in the precise selection of the genotypes/lines for traits of interest. The presence of linked/gene-specific markers increases the accuracy of selection.

Several genetic engineering approaches (Shinwari et al., 2020) and CRISPR/Cas studies (Zhong et al., 2022) have been utilized recently to develop drought tolerance in soybean.

12.3.2 SUBMERGENCE

Floods cover the developing plants partially or entirely; hence, flooding stress is characterized as waterlogging, partial submergence, or full submergence based on the water level relative to the flooded plant tissues (Voesenek and Bailey-Serres, 2015). Soybean is highly sensitive to flooding at germination, early vegetative, and reproductive stages. The yield of crops is significantly affected by submergence/flooding stress (Heng et al., 2018; Hou and Thseng, 1991). Molecular and genetic mechanisms of flooding tolerance have been studied extensively in rice and thale grass, with a detailed report on the function of group 7 ethylene response factor (*ERF*) genes, including SUB1A. The available genetic variation for *ERF* genes determines the escape strategies, such as stem elongation, growing out of flood water to get oxygen for aeration, and quiescence responses like leaf elongation and growth suppression. However, the physiological mechanisms underlying flood tolerance in soybean need to be further studied. Molecular linkage analysis could assist researchers in figuring out the physiological processes involved in flooding tolerance.

Negative effects of flooding include a relative reduction of ureide-N content, the number and dry weight of nodules, leaf area and color, canopy height, canopy mass, dry weight of root and shoot, and finally terminal yield loss (Youn et al., 2008). Yield reduction in soybean crops varies based on the growth stage of the crop affected by flooding; it ranges from 15% to 45% at the early vegetative growth stage and 50% to 56% at the reproductive stage (Linkemer et al., 1998). However, flooding for a period of 2 continuous days will reduce the yield by 27%. Soybean has evolved a tolerance mechanism to flooding that includes increasing the availability of soluble sugars (Kim et al., 2015), aerenchyma formation, and adventitious root formation (Valliyodan et al., 2014), increased rate of glycolysis (Valliyodan et al., 2017), and antioxidant defense against oxidative stress (Ahmed et al., 2013).

Several donor lines for flooding tolerance have been identified, that is, Archer, Misuzudaiz, PI 408105A, PI 561271, PI 567651, PI 407184, PI

603910C, and PI 567394B belonging to *Glycine max* type, and lines PI 467162, PI 479751, PI 407229, PI 597459C, PI 424082, PI 378699A, PI 424107A, and PI 366124 belonging to *Glycine soja* type were identified. The lines S12-1362, S13-15764, and S13-15999 are high-yield advanced breeding lines developed in the soybean breeding programs at the University of Missouri, USA. The genotypes EC471920, EC471972, and EC472119 were identified as pre-germination anaerobic stress-tolerant genotypes after four levels of screening (Rajendran et al., 2022, 2023). Till now, around 27 QTLs have been reported for submergence tolerance in soybean at the early vegetative stage, vegetative stage, and early reproductive stage (Ahmed et al., 2012; Cornelious et al., 2005; Githiri et al., 2006; Van Toai et al., 2001; Zhang et al., 2022) and were documented in the Soybase database.

Besides, several candidate genes involved in submergence tolerance were identified (Ali et al., 2020; Sharmin et al., 2020; Valliodan et al., 2014; Ye et al., 2018; Yu et al., 2019). Based on the available QTL information, 33 candidate genes were identified that confer submergence tolerance either directly or indirectly. The candidate gene *Glyma10g34490* in the QTL *RSL-a-10-2* was involved directly in submergence tolerance (Khan et al., 2015), along with *Glyma01g32090.1* (Kamal and Komatsu, 2016) and *Glyma0411211* (Wang et al., 2014).

12.3.3 SALINITY TOLERANCE

Although soybean is considered moderately tolerant to salt stress, more than 8 dsm/m of soil salinity affects germination, crop growth, and pod development. In addition, it also affects seed oil and protein content (El-Sabagh et al., 2015). Four major mechanisms play a vital role in salt tolerance, *viz.*, (i) maintaining ion homeostasis; (ii) osmotic adjustment; (iii) restoration of oxidative balance; and (iv) metabolic and structural adaptations.

1. **Maintaining Ion Homeostasis:** Retaining Na^+ or Cl^- in the roots of *Glycine max* and *Glycine soja*, respectively, the most studied salt tolerance mechanism is Luo et al. (2005). Inter- and intra-cellular compartmentalization of Na^+ also contributes to salt tolerance in soybean by maintaining ion homeostasis. Transporters like Na^+/H^+ and K^+/H^+ antiporters actively pump out Na^+ in exchange for K^+.

Tonoplast-localized Na^+–H^+ exchangers (NHXs) are involved in the sequestration of cytosolic Na^+ into vacuoles, thus overcoming the buildup of toxic Na^+ levels in the cytosol. Expression of *GmNHX1* in the leaves and roots of tolerant soybean cultivars was observed under salt stress (Sun et al., 2006). Recently, two NHX homologs, *GmNHX1* and *GmNHX2*, were cloned from soybean (Li et al., 2006; Sun et al., 2006; Zhou et al., 2009) to identify their role in the sequestration of Na^+ into the vacuole. *GmCAX1*, a putative cation/H^+ exchanger gene, was identified and functionally described in soybean for the first time by Lee et al. (2005), which is known to impart salt tolerance by increasing the K^+/Na^+ ratio while regulating the accumulation and transport of Cl^- in soybean (Qi et al., 2014).

2. **Osmo-Protectants:** Glycine betaine, trigonelline (TRG), late embryogenesis abundant (LEA) proteins, proline, and pinitol are major components produced in all plant parts of soybean in response to stress.
3. **Restoration of Oxidative Balance:** Reactive oxygen species (ROS), which are produced from chloroplasts and mitochondria in response to metabolic activities due to different biotic and abiotic stresses, have unfavorable toxic effects, namely inhibiting enzymatic activity, degradation of chlorophyll, and lipid peroxidation. Plants defend themselves against these ROS molecules by producing antioxidant enzymes like superoxide dismutase (SOD), catalase (CAT), peroxidase, and antioxidant products such as carotenoids, ascorbic acid (AA), and glutathione (Phang et al., 2008; Sajedi et al., 2011).
4. **Structural Modifications:** Structural modifications include salt gland-like structures and glandular hairs, cell wall modifications, and cell membrane modifications involved in salt stress tolerance.

12.3.3.1 MOLECULAR APPROACHES TO SALINITY TOLERANCE

The first major QTL for salt tolerance in soybean, near the SSR marker Sat_091 on linkage group N, with a genotypic variation of 41%, was discovered using an $F_{2:5}$ mapping population of S-100/Tokyo (Lee et al.,

2004). A total of 184 RILs derived from Kefeng No. 1/ Nannong1138-2 were subjected to QTL mapping using 221 SSR markers, which resulted in the identification of *qppsN.1* between SSR markers Sat_164 and Sat_358 on linkage group G (Chen et al., 2008). The marker *Satt* 339 was identified for salt tolerance rating and plant leaf SPAD value using the F_2 mapping population from the cross PI548657 × JWS156-1 (Hamwieh and Xu, 2008). Hamwieh et al. (2011) developed RILs from two different crosses, that is, FT-Abyara 9/C01 and Jin Dou No. 6/90197, to identify the salt tolerance QTL through linkage mapping. They identified one major QTL for salt tolerance rating and plant leaf SPAD value on linkage group N, with 44 to 47% phenotypic variation for the traits. Ha et al. (2013) developed the RIL population from the cross PI 483463 × Hutcheson and identified the marker Satt255-BARC-038333-10036 for leaf scorch. A RIL population of Jidou12/Ji NF 58 was subjected to QTL mapping by Shi et al. (2018) and identified QTL for salt tolerance rating (R^2 = 44.3–44.7%) and SPAD (R^2 = 27–27.6%) on chromosome 3, flanked by the markers GMABAB and Barcsoyssr_03_1421. QTL mapping on 269 $F_{2:3}$ families of Ozark/ Jake using the SoySNP 6K chip revealed one major QTL for salt tolerance on chromosome 3 and identified a new minor QTL on chromosome 19 (Lopez et al., 2018). Further fine mapping of chromosome 3 was done, and the *Glyma03g32900.1* candidate gene in the QTL *GmSALT3* (Guan et al., 2014) was identified. Similarly, QTL identified from the cross FT-Abyara and C011 were subjected to fine mapping and identified the candidate gene *Glyma03g32900* within the 16.6 kb region of QTL Ncl, which was located between SSR25.8 and CAPS42.4. Moreover, a 3.8-kb insertion of the Ty1/Copia type retrotransposon was found to be responsible for the loss of function of the salt-tolerant gene *Glyma03g23900* (Do et al., 2016). Qi et al. (2014) identified a novel ion transporter gene, *GmCHX1*, using a combination of de novo sequencing data and their previous germplasm resequencing data. The ion transporter gene *GmCHX1* was confirmed as the causal locus of the gene *Glyma03g32900*.

12.3.4 SOIL ACIDITY

Most of the world's agricultural soils are severely affected by mineral deficiencies and toxicities due to their low mobility and fixation in soils. Among the minerals, phosphorus (P) is one such mineral required for the synthesis of nucleic acids, cellular and energy storage molecules (ATP), and phosphorylation events (Johnston et al., 2000; Khan et al., 2009). Soil acidification is one of the main reasons that affect the availability of P to plants. In acid soils, P gets easily fixed when it interacts with iron and aluminum and forms organic complexes. Compared to other non-leguminous crops, soybean requires a relatively large amount of P for initial growth and pod formation (Li et al., 2011). A deficiency of P leads to reduced nodule formation, affects microbial interaction and colonization, decreases nitrogen fixation, causes flower bud abscission, impairs overall growth and development, and ultimately results in a reduction in yield, as reported (Cassman et al., 1980; Olivera et al., 2004; Zhang et al., 2010, 2014, 2016). In the past, several QTLs were identified for low P tolerance in soybean (Li et al., 2005; Liang et al., 2010; Zhang et al., 2009, 2010). However, very few candidate genes were identified, and few of them were utilized in the breeding programs. The gene *GmACP1* is one such gene cloned from QTL (qPE8) that encodes acid phosphatase and enhances P efficiency in soybean (Zhang et al., 2014). Wang et al. (2022) reported 14 phosphate starvation response genes, among which *GmPHR14* and *GmPHR32* were not reported earlier and positively regulate low P response in soybean. Besides, the purple acid phosphatase gene (*GmPAP17*) significantly enhanced P tolerance by enhancing root growth and P use efficiency (Xu et al., 2022). Liu et al. (2022) reported that the TF GmWRKY46 negatively regulates P tolerance through modifications in root systems.

12.3.5 HEAT TOLERANCE

Heat stress mainly occurs due to an increase in global average temperature. According to a report by the IPCC in 2022, the global temperature will increase from 1.5°C to 4°C in the next two decades (IPCC-2022). Excessive temperature affects growth, development, metabolism, and yield loss, and sometimes causes the extinction of plant species. An increase of 1°C in temperature causes a yield loss of up to 17% (Ozturk et al., 2015). Soybean is a cool climate crop and is highly sensitive to heat stress. Heat stress directly affects the photosynthesis rate, transpiration, and ultimately leads to yield loss (Puteh et al., 2013). Earlier studies showed that the temperature range between 15°C and 25°C is suitable for soybean germination, growth, and establishment, which affects plant growth (Tyagi and Tripathi, 1983). Heat stress interferes with germination, seedling vigor (Anto and Jayaram, 2010), reduction in plant height (PH), leaf area, total biomass (Kumar et al., 2008), flowering (Sapra and Anaek, 1991), nitrogen-fixing capacity (Djanaguiraman et al., 2011), and photosynthesis (Djanaguiraman and Prasad, 2010).

During stress conditions, several osmoprotectants, polyamines, antioxidants, and heat shock proteins were induced. Heat shock proteins were identified in different plant species and proven to be effective against heat stress in plants. The HSP can be synthesized at any growth stage when the plant is exposed to stress. The HSP 101 was absent in normal soybean seedlings and was upregulated when the plants were exposed to heat stress, providing thermotolerance (Adam et al., 2001; Agarwal et al., 2001; Lee et al., 1994; Keeler et al., 2000). Similarly, several heat shock proteins like GmHSP17.9 and GmDNJ1 were reported for thermotolerance (Li et al., 2020; Yang et al., 2022). Heat shock protein 90A2 (*GmHSP90A2*) was identified as a positive regulator for thermotolerance in soybean (Huang et al., 2019). Several studies were conducted with different soybean genotypes for heat stress tolerance and identified the most tolerant genotypes: PI 471938 and D 88-5320 (Nahar et al., 2016).

12.3.6 COLD STRESS

Soybean is highly sensitive to high and low temperatures from germination to maturity, but the major critical period observed for low temperatures is the flowering period. The minimum temperature that a plant can withstand is 17–20°C (Gass et al., 1996). The average air temperature drops below 15°C, causing plant growth retardation and inhibition of new leaf and shoot development. Sometimes, when the temperature drops below 10°C, the plant may fail to enter the reproductive stage (Camara et al., 1997). Cold stress distorts the plants' bioenergetic processes, metabolism, and affects cellular structures, causing the stunted plants to lead to poor plant growth, yield, and quality (Hou et al., 2006). The chemical composition of soybean seeds will alter based on genetic factors and environmental factors like temperature ranges. Mourtzinis et al. (2017) reported an increase in crude fat content in soybean seeds under higher temperature conditions, while Kumar et al. (2006) proved a positive relationship between temperature and protein content, and a negative relationship between temperature and fat concentration. Several cold regulator genes (COR) have been identified as positive regulators for cold stress in soybean (Liao et al., 2008; Yu et al., 2014; Zhou et al., 2008). Dong et al. (2021) identified a gene *GmTCF1a* as a homologue to *AtTCF1* in *Arabidopsis thaliana* that mediates the plant's response to cold stress.

12.4 HIGH THROUGHPUT GENOTYPING

In any crop breeding program, the development of varieties with resistance/tolerance to abiotic stress along with yield is essential. In soybean, the yield per unit area has not improved significantly in the past (Liu et al., 2020) due to limitations in precise phenotyping. Innovative genotyping platforms and sequencing have facilitated the selection of desirable lines, as well as the identification, evaluation, and validation of different genes and markers (Bailey-Serres et al., 2019; Viquez-Zamora et al., 2013; Yu et al., 2014). Publication of the soybean reference genome has made the availability of single nucleotide polymorphisms (SNPs) effectively utilized in breeding programs. The low-density SNP arrays were mainly used in mapping studies, association analysis, and diversity analysis. Recently, 50K high-density SNP chips were used to genotype the germplasms, elite lines, and wild accessions to study evolution and domestication (Song et al., 2013). This genotyping information was utilized in trait association analysis, such as GWAS studies for flooding (Patil et al., 2016; Sharmin et al., 2021; Wu et al., 2019). High-density marker genotyping arrays were further developed, including the 180K Axiom (Lee et al., 2015) and NJAU 355K SoySNP array (Wang et al., 2016), which provide insight for more in-depth inquiry into genomic diversity in soybean (Saleem et al., 2021).

12.5 HIGH THROUGHPUT PHENOTYPING

In crop improvement programs, breeders and geneticists continuously evaluate the breeding materials for different qualitative and quantitative traits. In order to characterize the breeding materials or parents for different biotic and abiotic stresses, multiple trial evaluations are required with experienced personnel over countless days. This has led to the requirement and development of HTP facilities. The HTP includes various sensors, unmanned aerial vehicles, remote sensors, and robotics. The advancement in machine learning approaches has made utilization and data processing easy. HTP of soybean seed shape and color using image-based analysis has enabled precise phenotyping of a large set of seeds in a short period (Baek et al., 2020). A multisensory system is used for phenotyping the canopy height, temperature range, and reflectance (Bai et al., 2016). Santana et al. (2022) phenotyped the soybean materials for earliness and made an effective selection for earliness with high yield.

12.6 CONCLUSION AND FUTURE PROSPECTS

Soybean is a widely cultivated and highly consumed dual-purpose crop and is prone to several biotic and abiotic stress factors. Several molecular breeding approaches have been used for the detection of QTLs/genes responsible for drought, flood, salinity, cold, and acidity stress. Identification of candidate genes and linked markers for these traits can help to incorporate QTLs/genes into elite breeding lines through molecular breeding approaches like marker-assisted selection and backcrossing. Furthermore, the availability of sequencing and SNP information can be effectively utilized for detecting SMPs in breeding programs.

KEYWORDS

- **abiotic stress tolerance**
- **high-throughput phenotyping**
- **marker-assisted selection**
- **molecular breeding**
- **next-generation sequencing**
- **quantitative trait loci**
- **soybean yield stability**

REFERENCES

Adam, Z., Adamska, I., Nakabayashi, K., Ostersetzer, O., Haussuhl, K., Manuell, A., Zheng, B., Vallon, O., Rodermel, S. R., Shinozaki, K., & Clarke, A. K. (2001). Chloroplast and mitochondrial proteases in *Arabidopsis*: A proposed nomenclature. *Plant Physiology, 125*, 1912–1918.

Agarwal, M., Katiyar-Agarwal, S., Sahi, C., Gallie, D. R., & Grover, A. (2001). *Arabidopsis thaliana* Hsp100 proteins: Kith and kin. *Cell Stress & Chaperones, 6*, 219–224.

Ahmed, F., Rafii, M. Y., Ismail, M. R., Juraimi, A. S., Rahim, H. A., Asfaliza, R., & Latif, M. A. (2013). Waterlogging tolerance of crops: Breeding, mechanism of tolerance, molecular approaches, and future prospects. *BioMed Research International, 2013,* 963525.

Ali, M. J., Xing, G., He, J., Zhao, T., & Gai, J. (2020). Detecting the QTL-allele system controlling seed-flooding tolerance in a nested association mapping population of soybean. *The Crop Journal, 8*(5), 781–792.

Anto, K. B., & Jayaram, K. M. (2010). Effect of temperature treatment on seed water content and viability of green pea (*Pisum sativum* L.) and soybean (*Glycine max* L. Merr.) seeds. *International Journal of Botany, 6*, 122–126.

Baek, J., Lee, E., Kim, N., Kim, S. L., Choi, I., Ji, H., Chung, Y. S., Choi, M. S., Moon, J. K., & Kim, K. H. (2020). High throughput phenotyping for various traits on soybean seeds using image analysis. *Sensors, 20*(1), 248.

Bai, G., Ge, Y., Hussain, W., Baenziger, P. S., & Graef, G. (2016). A multi-sensor system for high throughput field phenotyping in soybean and wheat breeding. *Computers and Electronics in Agriculture, 128*, 181–192.

Bailey-Serres, J., Parker, J. E., Ainsworth, E. A., Oldroyd, G. E. D., & Schroeder, J. I. (2019). Genetic strategies for improving crop yields. *Nature, 575*, 109–118.

Bisht, A., Saini, D. K., Kaur, B., Batra, R., Kaur, S., Kaur, I., Jindal, S., Malik, P., Sandhu, P. K., Kaur, A., & Gill, B. S. (2022). Multi-omics assisted breeding for biotic stress resistance in soybean: Challenges and opportunities. *Molecular Biology Reports, 50*(4), 3787–3814.

Boerma, H. R., & Specht, J. E. (2004). *Soybeans: Improvement, Production and Uses.* American Society of Agronomy and Crop Science, Society of America Soil Science.

Çamara, G. M. S., Sediyama, T., Dourado-Neto, D., & Bernardes, M. S. (1997). Influence of photoperiod and air temperature on the growth, flowering and maturation of soybean (*Glycine max* (L.) Merrill). *Scientia Agricola, 54*, 149–154.

Carpentieri Pipolo, V., Pipolo, A. E., Abdel-Haleem, H., Boerma, H. R., & Sinclair, T. R. (2012). Identification of QTLs associated with limited leaf hydraulic conductance in soybean. *Euphytica, 186*(3), 679–686.

Cassman, K. G., Whitney, A. S., & Stockinger, K. (1980). Root growth and dry matter distribution of soybean as affected by phosphorus stress, nodulation, and nitrogen source. *Crop Science, 20*, 239–244.

Chang, R. Z. (1999). Effect of salt on agricultural characters and chemical quality of seed in soybean. *Soybean Science, 13*, 101–105.

Chen, H., Cui, S., Fu, S., Gai, J., & Yu, D. (2008). Identification of quantitative trait loci associated with salt tolerance during seedling growth in soybean (*Glycine max* L.). *Australian Journal of Agricultural Research, 59*(12), 1086–1091.

Cornelious, B., Chen, P., Chen, Y., De Leon, N., Shannon, J. G., & Wang, D. (2005). Identification of QTLs underlying water-logging tolerance in soybean. *Molecular Breeding, 16*(2), 103–112.

Dhungana, S. K., Park, J. H., Oh, J. H., Kang, B. K., Seo, J. H., Sung, J. S., Kim, H. S., Shin, S. O., Baek, I. Y., & Jung, C. S. (2021). Quantitative trait locus mapping for drought tolerance in soybean recombinant inbred line population. *Plants, 10*(9), 1816.

Djanaguiraman, M., & Prasad, P. V. V. (2010). Ethylene production under high temperature stress causes premature leaf senescence in soybean. *Functional Plant Biology, 37*, 1071–1084.

Djanaguiraman, M., Prasad, P. V. V., & Al-Khatib, K. (2011). Ethylene perception inhibitor 1-MCP decreases oxidative damage of leaves through enhanced antioxidant defense mechanisms in soybean plants grown under high temperature stress. *Environmental and Experimental Botany, 71*, 215–223.

Do, T. D., Chen, H., Vu, H. T. T., Hamwieh, A., Yamada, T., Sato, T., Yan, Y. L., Cong, H., Shono, M., Suenaga, K., & Xu, D. (2016). NaCl synchronously regulates Na+, K+, and Cl− in soybean and greatly increases the grain yield in saline field conditions. *Scientific Reports, 6*, 19147.

Dong, Z., Wang, H., Li, X., & Ji, H. (2021). Enhancement of plant cold tolerance by soybean RCC1 family gene GmTCF1a. *BMC Plant Biology, 21*(1), 1–16.

Du, W., Wang, M., Fu, S., & Yu, D. (2009). Mapping QTLs for seed yield and drought susceptibility index in soybean (*Glycine max* L.) across different environments. *Journal of Genetics and Genomics, 36*, 721–731.

Earl, H. J. (2003). A precise gravimetric method for simulating drought stress in pot experiments. *Crop Science, 43*, 1868–1873.

El-Sabagh, A., Sorour, S., Ueda, A., Saneoka, H., & Barutcular, C. (2015). Evaluation of salinity stress effects on seed yield and quality of three soybean cultivars. *Azarian Journal of Agriculture, 2*(5), 138–141.

Fenta, B. A., Beebe, S. E., Kunert, K. J., Burridge, J. D., Barlow, K. M., Lynch, J. P., & Foyer, C. H. (2014). Field phenotyping of soybean roots for drought stress tolerance. *Agronomy, 4*(3), 418–435.

Gass, T., Schori, A., Fossati, A., Soldati, A., & Stamp, P. (1996). Cold tolerance of soybean (*Glycine max* (L.) Merr.) during the reproductive phase. *European Journal of Agronomy, 5*, 71–88.

Githiri, S. M., Watanabe, S., Harada, K., & Takahashi, R. (2006). QTL analysis of flooding tolerance in soybean at an early vegetative growth stage. *Plant Breeding, 125*(6), 613–618.

Goldberg, R. B. (1978). DNA sequence organization in the soybean plant. *Biochemical Genetics, 16*, 45–68.

Guan, R., Qu, Y., Guo, Y., Yu, L., Liu, Y., Jiang, J., et al. (2014). Salinity tolerance in soybean is modulated by natural variation in GmSALT3. *The Plant Journal, 80*(6), 937–950.

Ha, B. K., Vuong, T. D., Velusamy, V., Nguyen, H. T., Grover Shannon, J., & Lee, J. D. (2013). Genetic mapping of quantitative trait loci conditioning salt tolerance in wild soybean (*Glycine soja*) PI 483463. *Euphytica, 193*(1), 79–88.

Hamwieh, A., & Xu, D. (2008). Conserved salt tolerance quantitative trait locus (QTL) in wild and cultivated soybeans. *Breeding Science, 58*(4), 355–359.

Hamwieh, A., Tuyen, D. D., Cong, H., Benitez, E. R., Takahashi, R., & Xu, D. H. (2011). Identification and validation of a major QTL for salt tolerance in soybean. *Euphytica, 179*(3), 451–459.

Hedges, B. R., & Palmer, R. G. (1993). Mapping the *w4* locus in soybean. *Soybean Genetics Newsletter, 20*, 20–26.

Hirasawa, T. (1994). Effects of pre-flowering soil moisture deficits on dry matter production and eco-physiological characteristics in soybean plants under drought conditions during grain filling. *Japanese Journal of Crop Science, 63*, 721–730.

Hirt, H., & Shinozaki, K. (2004). *Plant Responses to Abiotic Stress.* Springer.

Hoogenboom, G., Huck, M., & Peterson, C. M. (1987). Root growth rate of soybean as affected by drought stress. *Agronomy Journal, 79*, 607–614.

Hou, F. F., & Thseng, F. S. (1991). Studies on the flooding tolerance of soybean seed: Varietal differences. *Euphytica, 57*(2), 169–173.

Hou, G., Ablett, G. R., Pauls, K. P., & Rajcan, I. (2006). Environmental effects on fatty acid levels in soybean seed oil. *Journal of the American Oil Chemists' Society, 83*, 759–763.

Huang, Y., Xuan, H., Yang, C., Guo, N., Wang, H., Zhao, J., & Xing, H. (2019). *GmHsp90A2* is involved in soybean heat stress as a positive regulator. *Plant Science, 285*, 26–33.

Huck, M. G., Ishihara, K., Peterson, C. M., & Ushijima, T. (1983). Soybean adaptation to water stress at selected stages of growth. *Plant Physiology, 73*, 422–427.

Hufstetler, E. V., Boerma, H. R., Carter, T. E., & Earl, H. J. (2007). Genotypic variation for three physiological traits affecting drought tolerance in soybean. *Crop Science, 47*, 25–35.

James, A., Lawn, R., & Cooper, M. (2008). Genotypic variation for drought stress response traits in soybean. I. Variation in soybean and wild *Glycine* spp. for epidermal conductance, osmotic potential, and relative water content. *Australian Journal of Agricultural Research, 59*, 656–669.

Johnston, A., & Steen, I. (2000). *Understanding Phosphorus and its Use in Agriculture.* European Fertilizer Manufacturers Association.

Kamal, A. H. M., & Komatsu, S. (2016). Proteins involved in biophoton emission and flooding-stress responses in soybean under light and dark conditions. *Molecular Biology Reports, 43*(2), 73–89.

Kavar, T., Maras, M., Kidric, M., Sustar-Vozlic, J., & Meglic, V. (2008). Identification of genes involved in the response of leaves of *Phaseolus vulgaris* to drought stress. *Molecular Breeding, 21*, 159–172.

Kaya, M. D., Okçu, G., Atak, M., Çikili, Y., & Kolsarici, Ö. (2006). Seed treatments to overcome salt and drought stress during germination in sunflower (*Helianthus annuus* L.). *European Journal of Agronomy, 24*, 291–295.

Keeler, S. J., Boettger, C. M., Haynes, J. G., Kuches, K. A., Johnson, M. M., Thureen, D. L., Keeler, Jr, C. L., & Kitto, S. L. (2000). Acquired thermotolerance and expression of the *HSP100/ClpB* genes of lima bean. *Plant Physiology, 123*, 1121–1132.

Khan, A. A., Jilani, G., Akhtar, M. S., Naqvi, S. S., & Rasheed, M. (2009). Phosphorus-solubilizing bacteria: Occurrence, mechanisms, and their role in crop production. *Journal of Agricultural and Biological Science, 1*, 48–58.

Khan, M. A., Tong, F., Wang, W., He, J., Zhao, T., & Gai, J. (2020). Molecular characterization of QTL-allele system for drought tolerance at seedling stage and optimal genotype design using multi-locus multi-allele genome-wide association analysis in a half-sib population of soybean (*Glycine max* (L.) Merr.). *Plant Genetic Resources, 18*(5), 295–306.

Khan, M. N., Sakata, K., & Komatsu, S. (2015). Proteomic analysis of soybean hypocotyl during recovery after flooding stress. *Journal of Proteomics, 121*, 15–27.

Kim, Y. H., Hwang, S. J., Waqas, M., Khan, A. L., Lee, J. H., Lee, J. D., Nguyen, H. T., & Lee, I. J. (2015). Comparative analysis of endogenous hormone levels in two soybean (*Glycine max* L.) lines differing in waterlogging tolerance. *Frontiers in Plant Science, 6*, 714.

Kumar, A., Pandey, V., Shekh, A. M., & Kumar, M. (2008). Growth and yield response of soybean (*Glycine max* L.) in relation to temperature, photoperiod, and sunshine duration at Anand, Gujarat, India. *American-Eurasian Journal of Agronomy, 1*, 45–50.

Kumar, V., Rani, A., Solanki, S., & Hussain, S. M. (2006). Influence of growing environment on the biochemical composition and physical characteristics of soybean seed. *Journal of Food Composition and Analysis, 19*, 188–195.

Le, D. T., Nishiyama, R., Watanabe, Y., Tanaka, M., Seki, M., Yamaguchi-Shinozaki, K., Shinozaki, K., & Tran, L. S. P. (2012). Differential gene expression in soybean leaf tissues at late developmental stages under drought stress revealed by genome-wide transcriptome analysis. *PLoS ONE, 7*, e49522.

Lee, G. J., Boerma, H. R., Villagarcia, M. R., Zhou, X., Carter, T. E., Li, Z., & Gibbs, M. O. (2005). A major QTL conditioning salt tolerance in S-100 soybean and descendent cultivars. *Theoretical and Applied Genetics, 109*(8), 1610–1619.

Lee, Y. G., Jeong, N., Kim, J. H., Lee, K., Kim, K. H., Pirani, A., Ha, B. K., Kang, S. T., Park, B. S., & Moon, J. K. (2015). Development, validation, and genetic analysis of a large soybean SNP genotyping array. *Plant Journal, 81*, 625–636.

Lee, Y. R. J., Nagao, R. T., & Key, J. L. (1994). A soybean 101-kD heat shock protein complements a yeast *HSP104* deletion mutant in acquiring thermotolerance. *Plant Cell, 6*, 1889–1897.

Li, K. P., Wong, C. H., Cheng, C. C., Cheng, S. S., Li, M. W., Mansveld, S., Bergsma, A., Huang, T., van Eijk, M. J., & Lam, H. M. (2021). *GmDNJ1*, a type-I heat shock protein 40 (*HSP40*), is responsible for both growth and heat tolerance in soybean. *Plant Direct, 5*(1), e00298.

Li, W. Y. F., Wong, F. L., Tsai, S. N., Phang, T. H., Shao, G., & Lam, H. M. (2006). Tonoplast-located *GmCLC1* and *GmNHX1* from soybean enhance NaCl tolerance in transgenic bright yellow (BY)-2 cells. *Plant, Cell & Environment, 29*(6), 1122–1137.

Li, X., Chang, W., & Zhang, C. (2011). Advances of soybean (*Glycine max* L.) phosphorus nutrition and high P-efficient germplasms screening in China. *Soybean Science, 30*, 322–327.

Li, X., Liu, X., Gu, M., Pi, B., Sun, S., Zhao, T., & Yu, B. (2022). Overexpression of *GsCNGC20-f* from *Glycine soja* confers submergence tolerance to hairy-root composite soybean plants and *Arabidopsis* seedlings by enhancing anaerobic respiration. *Environmental and Experimental Botany, 199*, 104901.

Li, Y., Wang, Y., Tong, Y., Gao, J., Zhang, J., & Chen, S. (2005). QTL mapping of phosphorus deficiency tolerance in soybean (*Glycine max* L. Merr.). *Euphytica, 142*, 137–142.

Liang, Q., Cheng, X., Mei, M., Yan, X., & Liao, H. (2010). QTL analysis of root traits as related to phosphorus efficiency in soybean. *Annals of Botany, 106*, 223–234.

Liao, Y., Zou, H. F., Wang, H. W., Zhang, W. K., Ma, B., Zhang, J. S., et al. (2008). Soybean *GmMYB76, GmMYB92,* and *GmMYB177* genes confer stress tolerance in transgenic *Arabidopsis* plants. *Cell Research, 18*(10), 1047–1060.

Lightfoot, D. A. (2015). Two decades of molecular marker-assisted breeding for resistance to soybean sudden death syndrome. *Crop Science, 55*(4), 1460–1484.

Linkemer, G., Board, J. E., & Musgrave, M. E. (1998). Waterlogging effects on growth and yield components in late-planted soybean. *Crop Science, 38*(6), 1576–1584.

Liu, K. (1997). *Soybeans: Chemistry, Technology and Utilization.* Aspen Publishers.

Liu, S. L., Zhang, M., Feng, F., & Tian, Z. X. (2020). Toward a "Green Revolution" for soybean. *Molecular Plant, 13*, 688–697.

Liu, X., Yang, Y., Wang, R., Cui, R., Xu, H., Sun, C., Wang, J., Zhang, H., Chen, H., & Zhang, D. (2022). *GmWRKY46*, a WRKY transcription factor, negatively regulates phosphorus tolerance primarily through modifying root morphology in soybean. *Plant Science, 315*, 111148.

Liu, Y., Gai, J. Y., Lu, H., Wang, Y. J., & Chen, S. Y. (2005). Identification of drought tolerant germplasm and inheritance and QTL mapping of related root traits in soybean (*Glycine max* L. Merr.). *Yi Chuan Xue Bao, 32*, 855–863.

Liu, Z., Li, H., Gou, Z., Zhang, Y., Wang, X., Ren, H., Wen, Z., Kang, B. K., Li, Y., Yu, L., & Gao, H. (2020). Genome-wide association study of soybean seed germination under drought stress. *Molecular Genetics and Genomics, 295*(3), 661–673.

Luo, Q., Yu, B., & Liu, Y. (2005). Differential sensitivity to chloride and sodium ions in seedlings of *Glycine max* and *G. soja* under NaCl stress. *Journal of Plant Physiology, 162*(9), 1003–1012.

Manavalan, L. P., Guttikonda, S. K., Phan Tran, L. S., & Nguyen, H. T. (2009). Physiological and molecular approaches to improve drought resistance in soybean. *Plant & Cell Physiology, 50*(7), 1260–1276.

Marcolino-Gomes, J., Rodrigues, F. A., Fuganti-Pagliarini, R., Bendix, C., Nakayama, T. J., Celaya, B., Molinari, H. B. C., De Oliveira, M. C. N., Harmon, F. G., & Nepomuceno, A. (2014). Diurnal oscillations of soybean circadian clock and drought responsive genes. *PLoS ONE, 9*, e86402.

Mourtzinis, S., Gaspar, A. P., Naeve, S. L., & Conley, S. P. (2017). Planting date, maturity, and temperature effects on soybean seed yield and composition. *Journal of Agronomy, 109*, 2040–2049.

Nahar, K., Hasanuzzaman, M., & Fujita, M. (2016). Heat stress responses and thermotolerance in soybean. In *Abiotic and Biotic Stresses in Soybean Production* (pp. 261–284). Academic Press.

Olivera, M., Tejera, N., Iribarne, C., Ocana, A., & Lluch, C. (2004). Growth, nitrogen fixation, and ammonium assimilation in common bean (*Phaseolus vulgaris*): Effect of phosphorus. *Physiologic Plantarum, 121*, 498–505.

Ouyang, W., Chen, L., Ma, J., Liu, X., Chen, H., Yang, H., Guo, W., Shan, Z., Yang, Z., Chen, S., & Zhan, Y. (2022). Identification of quantitative trait locus and candidate genes for drought tolerance in a soybean recombinant inbred line population. *International Journal of Molecular Sciences, 23*(18), 10828.

Ozturk, M., Hakeem, K. R., Faridah-Hanum, I., & Efe, R. (2015). *Climate Change Impacts on High-Altitude Ecosystems* (Vol. 736). Springer.

Patil, G., Do, T., Vuong, T. D., Valliyodan, B., Lee, J. D., Chaudhary, J., Shannon, J. G., & Nguyen, H. T. (2016). Genomic-assisted haplotype analysis and the development of high-throughput SNP markers for salinity tolerance in soybean. *Scientific Reports, 6*(1), 1–13.

Phang, T., Shao, G., & Lam, H. (2008). Salt tolerance in soybean. *Journal of Integrative Plant Biology, 50*, 1196–1212.

Puteh, A. B., ThuZar, M., Mondal, M. M. A., Abdullah, N. A. P. B., & Halim, M. R. A. (2013). Soybean (*Glycine max* L. Merrill) seed yield response to high temperature stress during reproductive growth stages. *Australian Journal of Crop Science, 7*, 1472–1479.

Qi, X., Li, M. W., Xie, M., Liu, X., Ni, M., Shao, G., Song, C., Yim, A. K. Y., Tao, Y., Wong, F. L., & Isobe, S. (2014). Identification of a novel salt tolerance gene in wild soybean by whole-genome sequencing. *Nature Communications, 5*, 4340.

Rajendran, A., Lal, S. K., Raju, D., & Ramlal, A. (2022). Associations of direct and indirect selection for pregermination anaerobic stress tolerance in soybean (*Glycine max*). *Plant Breeding, 141*(5), 634–643.

Rajendran, A., Lal, S. K., Raju, D., Mallikarjun, B. P., Ramlal, A., & Sharma, D. (2023). Waterlogging tolerance evaluation methods for soybean (*Glycine max* L. Merr.) at the pregermination stage. *Genetic Resources and Crop Evolution, 70,* 2451–2461.

Ramlal, A., Nautiyal, A., Baweja, P., Kumar Mahto, R., Mehta, S., Pujari Mallikarunja, B., Vijayan, R., Saluja, S., Kumar, V., Kumar Dhiman, S., & Lal, S. K. (2022a). Harnessing heterosis and male sterility in soybean [*Glycine max* (L.) Merrill]: A critical revisit. *Frontiers in Plant Science, 13*, 981768.

Ramlal, A., Nautiyal, A., Baweja, P., Kumar, V., Mehta, S., Mahto, R. K., Tripathi, S., Shanmugam, A., Pujari Mallikarjuna, B., Raman, P., & Lal, S. K. (2022b). Angiotensin-converting enzyme inhibitory peptides and isoflavonoids from soybean [*Glycine max* (L.) Merr.]. *Frontiers in Nutrition, 9*, 1068388.

Ren, H., Han, J., Wang, X., Zhang, B., Yu, L., Gao, H., Hong, H., Sun, R., Tian, Y., Qi, X., & Liu, Z. (2020). QTL mapping of drought tolerance traits in soybean with SLAF sequencing. *Crop Journal, 8*(6), 977–989.

Sajedi, N. A., Ardakani, M. R., Madani, H., Naderi, A., & Miransari, M. (2011). The effects of selenium and other micronutrients on the antioxidant activities and yield of corn (*Zea mays* L.) under drought stress. *Physiology and Molecular Biology of Plants, 17*(3), 215–222.

Saleem, A., Muylle, H., Aper, J., Ruttink, T., Wang, J., Yu, D., & Roldán-Ruiz, I. (2021). A genome-wide genetic diversity scan reveals multiple signatures of selection in a European soybean collection compared to Chinese collections of wild and cultivated soybean accessions. *Frontiers in Plant Science, 12,* 631767.

Santana, D. C., de Oliveira Cunha, M. P., Dos Santos, R. G., Cotrim, M. F., Teodoro, L. P. R., da Silva Junior, C. A., Baio, F. H. R., & Teodoro, P. E. (2022). High-throughput phenotyping allows the selection of soybean genotypes for earliness and high grain yield. *Plant Methods, 18*(1), 1–11.

Sapra, V. T., & Anaek, A. O. (1991). Screening soya bean genotypes for drought and heat tolerance. *Journal of Agronomy and Crop Science, 167*, 96–102.

Savitri, E. S., Basuki, N., Aini, N., & Arumingtyas, E. L. (2013). Identification and characterization of drought tolerance of gene LEA-D11 soybean (*Glycine max* L. Merr.) based on PCR sequencing. *American Journal of Molecular Biology, 3*, 32.

Shaheen, T., Riaz, M. S., & Zafar, Y. (2016). Soybean production and drought stress. In *Abiotic and Biotic Stresses in Soybean Production* (pp. 177–196). Academic Press.

Sharmin, R. A., Bhuiyan, M. R., Lv, W., Yu, Z., Chang, F., Kong, J., Bhat, J. A., & Zhao, T. (2020). RNA-Seq based transcriptomic analysis revealed genes associated with seed-flooding tolerance in wild soybean (*Glycine soja* Sieb. & Zucc.). *Environmental and Experimental Botany, 171*, 103906.

Sharmin, R. A., Karikari, B., Chang, F., Al Amin, G. M., Bhuiyan, M. R., Hina, A., Lv, W., Chunting, Z., Begum, N., & Zhao, T. (2021). Genome-wide association study uncovers major genetic loci associated with seed flooding tolerance in soybean. *BMC Plant Biology, 21*, 497.

Shi, X., Yan, L., Yang, C., Yan, W., Moseley, D. O., Wang, T., Liu, B., Di, R., Chen, P., & Zhang, M. (2018). Identification of a major quantitative trait locus underlying salt tolerance in 'Jidou 12' soybean cultivar. *BMC Research Notes, 11*(1), 1–6.

Shinwari, Z. K., Jan, S. A., Nakashima, K., & Yamaguchi-Shinozaki, K. (2020). Genetic engineering approaches to understanding drought tolerance in plants. *Plant Biotechnology Reports, 14*(2), 151–162.

Singh, R. J., & Hymowitz, T. (1988). The genomic relationship between *Glycine max* (L.) Merr. and *G. soja* Sieb. and Zucc. as revealed by pachytene chromosomal analysis. *Theoretical and Applied Genetics, 76*, 705–711.

Song, Q. J., Marek, L. F., Shoemaker, R. C., Lark, K. G., Concibido, V. C., Delannay, X., Specht, J. E., & Cregan, P. B. (2004). A new integrated genetic linkage map of the soybean. *Theoretical and Applied Genetics, 109*, 122–128.

Song, Q., Hyten, D. L., Jia, G., Quigley, C. V., Fickus, E. W., Nelson, R. L., & Cregan, P. B. (2013). Development and evaluation of SoySNP50K, a high-density genotyping array for soybean. *PLoS One, 8*, e54985.

SOPA (Soybean Processors Association of India). (2022). *World Soybean Production.*

Stolf-Moreira, R., Medri, M., Neumaier, N., Lemos, N., Brogin, R., Marcelino, F., De Oliveira, M., Farias, J., Abdelnoor, R., & Nepomuceno, A. (2010). Cloning and quantitative expression analysis of drought-induced genes in soybean. *Genetics and Molecular Research, 9*, 858–867.

Sugiyama, A., Ueda, Y., Takase, H., & Yazaki, K. (2015). Do soybeans select specific species of *Bradyrhizobium* during growth? *Communicative & Integrative Biology, 8*(1), e992734.

Sun, Y., Wang, D., Bai, Y., & Wang, N. (2006). Studies on the overexpression of the soybean *GmNHX1* in *Lotus corniculatus*: The reduced Na+ level is the basis of the increased salt tolerance. *Chinese Science Bulletin, 51*(11), 1306–1315.

Szilagyi, L. (2003). Influence of drought on seed yield components in common bean. *Bulgarian Journal of Plant Physiology, Special Issue*, 320–330.

Thao, N. P., & Tran, L. S. P. (2012). Potentials toward genetic engineering of drought-tolerant soybean. *Critical Reviews in Biotechnology, 32*(4), 349–362.

Thao, N. P., Thu, N. B. A., Hoang, X. L. T., Ha, C. V., & Tran, L. S. P. (2013). Differential expression analysis of a subset of drought-responsive *GmNAC* genes in two soybean cultivars differing in drought tolerance. *International Journal of Molecular Sciences, 14*, 23828–23841.

Tripathy, J., Zhang, J., Robin, S., Nguyen, T. T., & Nguyen, H. (2000). QTLs for cell-membrane stability mapped in rice (*Oryza sativa* L.) under drought stress. *Theoretical and Applied Genetics, 100*, 1197–1202.

Turner, N. C., Wright, G. C., & Siddique, K. (2001). Adaptation of grain legumes (pulses) to water-limited environments. *Advances in Agronomy, 71*, 194–233.

Tyagi, S. K., & Tripathi, R. P. (1983). Effect of temperature on soybean germination. *Plant and Soil, 74*, 273–280.

Valliyodan, B., Van Toai, T. T., Alves, J. D., de Fátima P. Goulart, P., Lee, J. D., Fritschi, F. B., Rahman, M. A., Islam, R., Shannon, J. G., & Nguyen, H. T. (2014). Expression of root-related transcription factors associated with flooding tolerance of soybean (*Glycine max*). *International Journal of Molecular Sciences, 15*(10), 17622–17643.

Valliyodan, B., Ye, H., Song, L., Murphy, M., Shannon, J. G., & Nguyen, H. T. (2017). Genetic diversity and genomic strategies for improving drought and waterlogging tolerance in soybeans. *Journal of Experimental Botany, 68*(8), 1835–1849.

VanToai, T. T., St. Martin, S. K., Chase, K., Boru, G., Schnipke, V., Schmitthenner, A. F., & Lark, K. G. (2001). Identification of a QTL associated with tolerance of soybean to soil waterlogging. *Crop Science, 41*(4), 1247–1252.

Vidal, R. O., Nascimento, L. C. D., Costa Mondego, J. M., Pereira, G. A. G., & Carazzolle, M. F. (2012). Identification of SNPs in RNA-seq data of two cultivars of *Glycine max* (soybean) differing in drought resistance. *Genetics and Molecular Biology, 35*, 331–334.

Viquez-Zamora, M., Vosman, B., van de Geest, H., Bovy, A., Visser, R. G. F., Finkers, R., & van Heusden, A. W. (2013). Tomato breeding in the genomics era: Insights from a SNP array. *BMC Genomics, 14*, 354.

Wang, J., Chu, S., Zhang, H., Zhu, Y., Cheng, H., & Yu, D. (2016). Development and application of a novel genome-wide SNP array reveals domestication history in soybean. *Scientific Reports, 6*, 20728.

Wang, Q., Du, W., Yu, W., Zhang, W., Huang, F., Cheng, H., & Yu, D. (2022). Genome-wide association analysis discovered new loci and candidate genes associated with low-phosphorus tolerance based on shoot mineral elements concentrations in soybean. *Molecular Genetics and Genomics, 297*, 1–16.

Wang, Y., Lu, J., Chen, S., Shu, L., Palmer, R. G., Xing, G., Li, Y., Yang, S., Yu, D., Zhao, T., & Gai, J. (2014). Exploration of presence/absence variation and corresponding polymorphic markers in soybean genome. *Journal of Integrative Plant Biology, 56*(10), 1009–1019.

Wu, C., Mozzoni, L. A., Moseley, D., Hummer, W., Ye, H., Chen, P., Shannon, G., & Nguyen, H. (2019). Genome-wide association mapping of flooding tolerance in soybean. *Molecular Breeding, 40*, 4.

Xie, M., Chung, C. Y., Li, M. W., Wong, F. L., Wang, X., Liu, A., Wang, Z., Leung, A. K., Wong, T. H., Tong, S. W., Xiao, Z., Fan, K., Ng, M. S., Qi, X., Yang, L., Deng, T., He, L., Chen, L., Fu, A., & Lam, H. M. (2019). A reference-grade wild soybean genome. *Nature Communications, 10*(1), 1216.

Xu, H., Zhang, H., Fan, Y., Wang, R., Cui, R., Liu, X., Chu, S., Jiao, Y., Zhang, X., & Zhang, D. (2022). The purple acid phosphatase *GmPAP17* predominantly enhances phosphorus use efficiency in soybean. *Plant Science, 320*, 111283.

Yang, Z., Du, H., Xing, X., Li, W., Kong, Y., Li, X., & Zhang, C. (2022). A small heat shock protein, *GmHSP17.9*, from nodule confers symbiotic nitrogen fixation and seed yield in soybean. *Plant Biotechnology Journal, 20*(1), 103.

Ye, H., Song, L., Chen, H., Valliyodan, B., Cheng, P., Ali, L., Vuong, T., Wu, C., Orlowski, J., Buckley, B., & Chen, P. (2018). A major natural genetic variation associated with root system architecture and plasticity improves waterlogging tolerance and yield in soybean. *Plant, Cell & Environment, 41*(9), 2169–2182.

Youn, J. T., Van, K. J., Lee, J. E., Kim, W. H., Yun, H. T., Kwon, Y. U., Ryu, Y. H., & Lee, S. H. (2008). Waterlogging effects on nitrogen accumulation and N_2 fixation of super nodulating soybean mutants. *Journal of Crop Science and Biotechnology, 11*(2), 111–118.

Yu, G. H., Jiang, L. L., Ma, X. F., Xu, Z. S., Liu, M. M., Shan, S. G., et al. (2014). A soybean C2H2-type zinc finger gene *GmZF1* enhanced cold tolerance in transgenic *Arabidopsis*. *PLoS One, 9*(10), e109399.

Yu, H. H., Xie, W. B., Li, J., Zhou, F. S., & Zhang, Q. F. (2014). A whole-genome SNP array (RICE6K) for genomic breeding in rice. *Plant Biotechnology Journal, 12*, 28–37.

Yu, Z., Chang, F., Lv, W., Sharmin, R. A., Wang, Z., Kong, J., Bhat, J. A., & Zhao, T. (2019). Identification of QTN and candidate gene for seed-flooding tolerance in soybean (*Glycine max*) using genome-wide association study (GWAS). *Genes, 10*(12), 957.

Zhang, D., Cheng, H., Geng, L., Kan, G., Cui, S., & Meng, Q. (2009). Detection of quantitative trait loci for phosphorus deficiency tolerance at soybean seedling stage. *Euphytica, 167*, 313–322.

Zhang, D., Liu, C., Cheng, H., Kan, G., Cui, S., & Meng, Q. (2010). Quantitative trait loci associated with soybean tolerance to low-phosphorus stress based on flower and pod abscission. *Plant Breeding, 129*, 243–249.

Zhang, D., Song, H., Cheng, H., Hao, D., Wang, H., Kan, G., et al. (2014). The acid phosphatase-encoding gene *GmACP1* contributes to soybean tolerance to low-phosphorus stress. *PLoS Genetics, 10*, e1004061.

Zhang, J., McDonald, S. C., Wu, C., Ingwers, M. W., Abdel-Haleem, H., Chen, P., & Li, Z. (2022). Quantitative trait loci underlying flooding tolerance in soybean (*Glycine max*). *Plant Breeding, 141*(2), 236–245.

Zhang, Y., Liu, Z., Wang, X., Li, Y., Gou, Z., Zhao, X., Hong, H., Ren, H., Qi, X., & Qiu, L. (2022). Identification of genes for drought resistance and prediction of gene candidates in soybean seedlings based on linkage and association mapping. *The Crop Journal, 10*(3), 830–839.

Zhong, X., Hong, W., Shu, Y., Li, J., Liu, L., Chen, X., Islam, F., Zhou, W., & Tang, G. (2022). CRISPR/Cas9-mediated gene-editing of *GmHdz4* transcription factor enhances drought tolerance in soybean (*Glycine max*). *Frontiers in Plant Science, 13*, 988505.

Zhou, G., Guan, R., Li, Y., Chang, R., & Qiu, L. (2009). Molecular characterization of *GmNHX2*, a Na+/H+ antiporter gene homolog from soybean, and its heterologous expression to improve salt tolerance in *Arabidopsis*. *Chinese Science Bulletin, 54*(19), 3536–3545.

Zhou, Q. Y., Tian, A. G., Zou, H. F., Xie, Z. M., Lei, G., Huang, J., et al. (2008). Soybean WRKY-type transcription factor genes, *GmWRKY13*, *GmWRKY21*, and *GmWRKY54*, confer differential tolerance to abiotic stresses in transgenic *Arabidopsis* plants. *Plant Biotechnology Journal, 6*(5), 486–503.

CHAPTER 13

Physiological and Biochemical Responses to Biotic and Abiotic Stresses and Adaptations in Soybean

JHILICK BANERJEE,[1] PAWAN K. AMRATE,[1,*] RIYA MISHRA,[1] VENNAMPALLY NATARAJ,[2] M. K. SHRIVASTAVA,[1] AKASH BARELA,[1] and SHIVANI JAWARKAR[1]

[1]*Department of Plant Breeding and Genetics, Jawaharlal Nehru Krishi Vishwa Vidyalaya, Jabalpur, Madhya Pradesh, India*

[2]*ICAR–Indian Institute of Soybean Research, Indore, Madhya Pradesh, India*

*Corresponding author

ABSTRACT

A variety of stresses can harm crops, leading to decreased performance in their environment. The mechanisms plants use to respond to stress involve specific physiological and biochemical changes depending on the metabolic pathways involved. Soybean (*Glycine max* L. Merril) is one of the most economically important and protein-rich legume oilseed crops, ranking as the fifth most essential crop in the world. It is valued for its multipurpose utility. As a crop, soybean is vulnerable to several abiotic and biotic stresses, which ultimately reduce profitability for growers. Plants rely on molecular, physiological, or biochemical responses to develop tolerance to these stresses. During drought and flooding, the plant's responses aim to minimize overall photosynthesis. One typical adaptation in soybean is a change in root morphology, particularly the development of a longer taproot that penetrates deeper soil layers, helping the plant survive under water-deficit conditions. In response to biotic stress, soybean leaves accumulate isoflavone aglycones,

Soybeans, Volume 2: Genome, Biotechnology, Stress, and Disease.
R. Ambika Rajendran, Ayyagari Ramlal, Sanjay Kumar Lal, and Godfree Chigeza (Eds.)

DOI: 10.1201/9781998511730-16

glyceollins, and other compounds. Compared to pathogen-induced responses, insect-inducible metabolites in soybean leaves remain poorly understood. Therefore, it is necessary to investigate and understand the mechanisms behind such responses. This chapter aims to discuss the nature of biotic and abiotic stresses affecting soybean, the related physiological and biochemical responses, and the adaptations that enhance stress tolerance.

13.1 INTRODUCTION

Soybean (*Glycine max* L. Merril) is the most prominent oilseed crop adapted to grow in diverse agroecological regions worldwide (Arya et al., 2021; Ramlal et al., 2023a). The soybean seed is a valuable source of vegetarian protein, oil, and several other elements that make it one of the most essential produced and consumed seed legumes globally (Agarwal et al., 2013; Banerjee et al., 2023). The richest source of protein among food crops and other multipurpose utilities, such as human food, animal feed, fiber, edible oil, and other industrial uses, keeps it in the place of topmost agricultural commodities in the world (Banerjee et al., 2022; Raghuvanshi and Bisht, 2010; Ramlal et al., 2023a; Uikey et al., 2022). The highest production of soybeans, 120.5 MT, occurs in the United States of America, followed by Brazil (117.0 MT), Argentina (55.30 MT), China (15.90 MT), and India (10.93 MT), respectively (SOPA, 2020). These five nations together account for almost 90% of the world's soybean production (Leff et al., 2004). The soybean plant has a symbiotic association with N-fixing bacteria that helps acquire most of the nitrogen required for growth and yield. Soybean is the only one among the most cultivated crops (rice, wheat, maize) that can be associated with rhizobia and arbuscular mycorrhizal (AM) fungi, exhibiting the possibility of further exploitation (Miransari, 2016).

Stress is an external factor that influences the growth of plants adversely and brings down their productivity. Stresses may be biotic or environmental, leading to changes in growth rates and, ultimately, crop yield by altering gene expression and cellular metabolism in the plant. Soybean cultivation worldwide always suffers from many abiotic and biotic stresses. Crop growth is severely affected during stress due to the retardation in nutrient uptake, which includes both macro and micronutrients. The influence of biotic and abiotic stress directly affects the N_2 fixation process. Besides this, there are many other physio-morphological, biochemical, and molecular traits that are influenced. The mechanism that plants follow is the result of stimulation from the stress, which generates specific cellular responses. These responses can be physiological or biochemical, depending on the metabolism involved. Understanding crop responses to such stresses leads to better management of crops under stress, which will ultimately produce better yields even during stress.

The soybean crop is sensitive to being attacked by several insect pests at different growth stages, from seedling to maturity. Approximately 15–20% of the total soybean production is lost directly or indirectly due to the attack of insect pests (Biswas, 2008). The major insect pests that affect soybean production include SBA, Seedcorn Maggot, Stemfly, *Helicoverpa armigera,* Green Vegetable Bug, Whitefly, *Spodoptera litura,* Soybean Stem Borer, Brown Root Stink Bug, White Grubs, Grape Colaspis, Three-Cornered Alfalfa Hopper, and Girdle Beetle (Heinrichs and Muniappan, 2018; Hesler et al., 2018; Sharma et al., 2014). The major diseases that affect soybean production across the world include soybean cyst nematode (SCN), bacterial blight, soybean rust, frog eye leaf spot, yellow mosaic virus, charcoal rot, collar rot, aerial blight, bacterial pustule, and anthracnose (Amrate and Shrivastava, 2021; Amrate et al., 2023a; Haile et al., 1998; Rajput et al., 2021; Uikey et al., 2022; Wrather et al., 2010; Yang et al., 1999; Yorinori et al., 2005). According to the recent report by Wrather et al. (2010), 59.9 million metric tons of soybean yield are reduced due to major diseases in the top eight soybean-producing countries. Soybean is also susceptible to a range of abiotic stresses from drought, salinity, heat, flooding, cold stress, as well as mineral deficiencies and toxicities (Rajendran et al., 2023; Ramlal et al., 2023b; Wei et al., 2018). The incidence of insect pests and diseases in soybean is always influenced by abiotic stresses such as high temperature, high rainfall, high humidity, and other environmental elements prevailing at cropping sites (Amrate et al., 2021a, 2023b; Rajput et al., 2021; Srivastava et al., 2021). Some abiotic stresses, such as exposure to drought and high temperatures, directly influence the incidence of charcoal rot, particularly during the reproductive stages of soybean (Amrate et al., 2023a; Nataraj et al., 2019; Smith and Wyllie, 1999). Overall, soybean crops are susceptible, and several biotic and abiotic stresses affect crucial growth stages that can lead to an enormous negative impact on their performance (Tables 13.1(a) and 13.1(b)).

The researchers have contributed much effort to interpreting the various stress signaling pathways in response to individual stresses. There are many sources identified for various stress tolerance in soybean. However, the physiological and biochemical mechanisms need to be better understood.

For instance, irrigation is an excellent strategy to overcome the problem of drought, but the available water sources are also continuously declining. Therefore, crops adapted to water-limited environments with high water use efficiency can be an immediate step to develop climate-resilient varieties capable of producing more food per unit of water used. It becomes vital to know about the different plant adaptations towards these biotic and abiotic stresses that confer tolerance to the crop. This chapter focuses on the various physiological and biochemical responses or adaptations acquired by soybean crops during various pathogenic and insect pest attacks, as well as abiotic stress.

TABLE 13.1(a) Major Biotic Stress and Their Influence on Soybean Production

Pest and Diseases	Scientific Name/ Causal Organism	Attacking Stage	Yield Loss	References
Bean leaf beetle	*Cerotoma trifurcata* Förster	Vegetative and reproductive	3–20%	Musser et al. (2014, 2015)
Grape colaspis	*Colaspis brunnea*	Vegetative and reproductive	10–50%	Flint (1941); Hesler et al. (2018)
			≤0.3% yield loss	Hesler et al. (2018); Musser et al. (2017)
Soybean aphids	*Aphis glycines*	Reproductive	40%	Hesler et al. (2018); Ragsdale et al. (2007)
			>50%	Wang et al. (1996)
			75% and 48%	Catangui et al. (2009)
Three-cornered alfalfa hoppers	*Spissistilus festinus* (Say)	Vegetative and reproductive	60%	Mueller and Dumas (1975)
			25%	Musser et al. (2014, 2015)
Stem fly	*Melanagromyza sojae*	Vegetative and reproductive	20–30%	Jayappa (2000)
			33%	Singh and Singh (1992)
Girdle beetle	*Obereopsis brevis*	Reproductive	14–42%	Kumawat et al. (2010)
Green stink bug	*Acrosternum hilare* (Say) (syn. *Chinavia hilaris*)	Reproductive	63.9–78.5%	Todd and Turnipseed (1974)
Whitefly	*Bemisia tabaci*	Vegetative and reproductive	15–17%	Latha et al. (2018)
Anthracnose	*Colletotrichum truncatum*	Vegetative and reproductive	2,539,000 metric tons	Wrather et al. (2010)
			16–25%	Sharma et al. (2014)
			64.8%	Amrate et al. (2021b)
Brown spot	*Septoria glycines*	Vegetative	4,259,900 metric tons	Wrather et al. (2010)

TABLE 13.1(a) *(Continued)*

Pest and Diseases	Scientific Name/ Causal Organism	Attacking Stage	Yield Loss	References
Soybean rust	*Phakopsora pachyrhizi*	Reproductive	13,239,000 metric tons or up to 100%	Wrather et al. (2010)
Soybean cyst nematode	*Heterodera glycines*	Vegetative and reproductive	7,192,900 metric tons	Wrather et al. (2010)
Charcoal rot	*Macrophomina phaseolina*	Reproductive	2,505,000 metric tons	Wrather et al. (2010)
			92.6% or up to 100%	Amrate et al. (2019)
			6–33% (under irrigated conditions)	Mengistu et al. (2011, 2018)
Phytophthora root and stem rot	*Phytophthora sojae*	Vegetative	2,318,700 metric tons	Wrather et al. (2010)
Root-knot and other nematodes	*Meloidogyne* spp. and others	Vegetative and reproductive	2,691,900 metric tons	Wrather et al. (2010)
Seedling diseases	*Rhizoctonia solani*, *Pythium* spp., *Fusarium* spp., *Phomopsis* spp.	Early stage	3,365,700 metric tons	Wrather et al. (2010)
Rhizoctonia aerial blight	*Rhizoctonia solani*	Reproductive	1,539,900 metric tons	Wrather et al. (2010)
			41.0%	Amrate et al. (2021b)
Yellow mosaic disease	*Mungbean yellow mosaic India virus* (MYMIV)	Early growth to reproductive	>85%	Amrate et al. (2023b)
			15–75%	Sharma et al. (2014)

TABLE 13.1(b) Major Abiotic Stress and Their Influence on Soybean Production

Stress	Sensitive Stage	Yield Loss	References
Waterlogging	V2 and R3 stages	30–93%	Linkemer et al. (1998)
Salinity	Germination	40%	Nadeem et al. (2019)
Heat	Vegetative and reproductive stages	Yield decreases by 17% for every 1 °C increase	Kucharik and Serbin (2008)
Drought	Flowering and seed development	50% reduction in grain yield	Wei et al. (2018)
	Seed development period	45–88% grain yield loss	Eck et al. (1987)

13.2 MECHANISM BEHIND PLANT'S RESPONSES TO STRESSFUL ENVIRONMENTS

It is a well-known fact that stress may come at any stage in the plant, and the plant also tries to adjust accordingly to survive and reproduce itself. The responses made by plants result from regulated signaling pathways, which start with stimulus perception. During its entire lifespan, a plant has to make numerous decisions for its survival. These decisions are made based on the stimuli received from the environment. Abiotic stress occurs due to extreme weather conditions like temperature hikes, altered water availability, and high/low soil ion content. These can lead to decreased growth and productivity. Biotic stress results from attacks by biological agents like insects, pests, pathogens, and nematodes, and so forth.

These stresses are perceived by primary sensors, which translate these physical and chemical stimuli into a biological signal. However, these primary sensors for various stresses are not clearly identified. According to Vu et al. (2019), an accurate sensor is one in which the stress directly results in the manipulation of the structural properties of any component; the sensor properties should affect the signal transduction pathway, and thirdly, the manipulations caused in the activities should result in physiological or biochemical adaptations of a plant in response to stress conditions. For instance, light receptors perceive high temperature, and GIPC sphingolipids sense salinity. Oxidation of the N-terminus of ERFVIIs senses oxygen, and the response is primed by ethylene (Lamers et al., 2020). Therefore, the basic mechanism behind all responses made by plants against stress is a result of stimuli created by the stressful environment, which is sensed or perceived by a sensor that coordinates the cellular signaling pathway that, in turn, produces the physiological or biochemical responses.

13.3 RESPONSES TO ABIOTIC STRESS AND ADAPTATION IN SOYBEAN

Abiotic stresses such as drought, floods, cold, heat, and salinity are likely to occur due to climate change. These stresses are affecting plant growth and development, leading to crop failure globally. However, it is possible to develop tolerant cultivars with enhanced tolerance under stress by incorporating biotechnological and molecular techniques (Miransari, 2016). Therefore, to develop such tolerance, it becomes essential to examine their responses under stress. Some of the most important and evident responses of soybean under significant abiotic stresses are presented below.

13.3.1 DROUGHT STRESS

Drought stress may hit the crop during the entire growth cycle due to climatic variation and agronomic factors (Murtaza, 2016). Drought severely affects the performance of crop plants in terms of growth, production, and productivity compared to any other environmental factor (Shao et al., 2009). Plants cope with water deficit conditions by adaptive changes in plant physiological and biochemical processes; for example, changes in plant structure, growth rate, photosynthesis, respiration, water relations, hormonal metabolism, tissue osmotic potential, and antioxidant defense mechanisms (Bhargava and Sawant, 2012). In soybean, the flowering stage and seed development are considered the most sensitive stages to drought (Eck et al., 1987). Wei et al. (2018) also indicated the significance of drought in soybean cultivation by reporting more than a 50% reduction in grain yield. Due to water deficiency under arid conditions, the leaves of soybean plants started dropping and wilting, which also led to a decrease in cell water potential and turgor (Zhang et al., 2022). Long-term drought stress in soybeans at reproductive stages reduces biomass allocation to reproductive organs, resulting in seed weight reduction in soybeans (Du et al., 2020). Under drought conditions, the poor growth and low yield are also attributed to a disturbance in the activity of nitrogenase, which leads to improper symbiotic nitrogen fixation in soybeans (Arya et al., 2021; Kunert et al., 2016).

To overcome the effect of drought, plants use diverse mechanisms. One of the primary physiological responses to drought in soybean plants is a decrease in stomatal opening coupled with a reduction in photosynthesis (Liu et al., 2003; Mak et al., 2014). Another significant response in soybean cultivars under drought stress is the effect on chlorophyll and relative water content (RWC). The RWC in leaves shares a close connection with drought resistance. The chlorophyll and RWC decrease at each growth stage during drought stress in soybeans (Dong et al., 2019).

Soybeans also responded with biochemical changes that included an increase in the number of plastoglobulins and an increase in total sugar concentration, mainly fructose and sucrose, as well as a decrease in stachyose (Mutava et al., 2015). The abscisic acid content (ABA) increases during drought stress, while jasmonic acid decreases. ABA plays a vital role in reducing water loss by inducing the closure of stomatal openings under dry conditions (Wilkinson and Davies, 2002). For example, higher ABA levels were observed in drought-tolerant compared to drought-susceptible varieties of soybean (Mutava et al., 2015). The accumulation of osmoprotectants such as soluble and complex sugars, sugar alcohols, organic acids, and free amino acids (AAs) due to biochemical changes under drought conditions is another crucial adjustment in soybean, which helps in retaining water and maintaining the integrity of the cell membrane (Ashraf and Iram, 2005; Kido et al., 2013; Silvente et al., 2012).

Soybean plants usually adopt three significant mechanisms to counteract the adverse effects of drought, that is, escape, tolerance, and avoidance (Zhao et al., 2018). In the mechanism of escape, the crop tries to complete its life cycle before the onset of drought. This results in less seed production. In the avoidance mechanism, plants maintain a high-water potential. The plant makes some adaptations to avoid water stress by growing roots deeper in the soil, controlling transpiration through stomata, and reducing water loss. In the tolerance mechanism, the plant maintains good turgor through osmotic adjustments using less tissue water. Soybean plants have evolved mechanisms by undergoing physiological and biochemical changes to cope with drought-like harsh environments and complete their life cycle.

Osmoregulation is one such mechanism in plants wherein the content of malondialdehyde (MDA), which is a product of membrane lipid peroxidation, is produced under drought-stress conditions. Zhou et al. (2022) analyzed osmotic adjustment and antioxidant activity in two spring soybean cultivars at different levels of soil moisture. It was observed that the MDA content increased with decreasing moisture in the soil in the sensitive variety. This resulted in high membrane lipid peroxidation in soybean leaves, which led to a severe degree of injury. Whereas in the case of resistant genotypes, the content of MDA was lower. In terms of osmotic adjustment, proline content, soluble sugars, and proteins reduce cell permeability under stress. Increasing proline and soluble sugar content during drought stress reduces cell permeability while maintaining normal metabolism. Thus, it improves drought tolerance in crops. Adjustment of stomatal conductance to prevent excessive water loss is another mechanism through which soybeans survive an extended drought period (Stolf-Moreira et al., 2010). After 30 days of water stress, the drought-tolerant soybean variety MG/BR46 exhibited a higher degree of reduction (65%) in stomatal conductance in comparison to the drought-sensitive cultivar BR16 (50% reduction) (Stolf-Moreira et al., 2010).

Further, soybean genotypes with a significant rise in proline contents have been considered drought-tolerant (Mishra et al., 2021). An increased level of proline content helps maintain cell water levels under drought conditions (Choudhary et al., 2021; Mishra et al., 2021). However, Sheteiwy et al. (2021) reported that the use of biofertilizers of arbuscular mycorrhizal fungi (AMF) and *Rhizobium japonicum* exhibits improvement in proline content and secondary metabolism in soybean, while simultaneously decreasing MDA, which supports better plant growth under drought stress.

13.3.2 WATER LOGGING STRESS

Most crops are sensitive to waterlogging or flooding conditions in the field at different growth stages of their life. Considering the present scenario of climatic variabilities, this may lead to low production or failure of the crop due to sudden excess rainfall in a given area. In the near future, such problems will be crucial and threaten the food security of an increasing global population (Ramlal et al., 2023b). In the case of soybean, the problem already exists and will become more severe as it is highly prone to waterlogging at the germination, emergence, and grain-filling stages (Rajendran et al., 2023; Ramlal et al., 2023b).

Water logging is the excess of moisture/water in crop-growing areas. Initially, water logging stress causes hypoxic (O_2 deficient) conditions, and then it switches to an anoxic (O_2 absent) condition that leads to limitations of energy, deposition of noxious compounds, loss of carbon, and inhibition of respiration (Tamang et al., 2014; Wegner, 2010). Water logging in the root system causes anaerobic conditions, affecting nodulation and, ultimately, nitrogen metabolism (Sigua et al., 2012). In soybean, about 30% and 93% of grain yield loss due to waterlogging stress at the V2 stage and R3 stage has been estimated (Linkemer et al., 1998). Waterlogging reduced the number and weight of nodules in soybean, and its effects were more severe on varieties with a comparatively smaller number of nodules, root length, total biomass, and slower adventitious formation (Miura, 2012). Sakazono et al. (2014) revealed that waterlogging initially affects the root system and then, after that, the shoot system of soybean. Even short-term waterlogging during early growth may inhibit root elongation and root branching in most waterlogging-sensitive soybean genotypes (Rajendran et al., 2022; Ramlal et al., 2023b).

Because of the lower activity of the roots to absorb water and nutrients under waterlogging conditions, leaf area decreases in affected and more sensitive genotypes (Fatimah and Nurhidayati, 2020; Guang et al., 2012). It is reported that under waterlogging conditions, the accumulation of ethanol increases at various levels compared to standard conditions (Hasanuzzaman et al., 2017). An increase in ethanol harms auxin and cytokinin synthesis and other processes (Fatimah and Nurhidayati, 2020; Hasanuzzaman et al., 2017). Sathi et al. (2022) reported that soybean genotypes exposed to waterlogging exhibited a reduction in leaf RWC, a decrease in SPAD values (an indicator of the chlorophyll content of the leaf), and an increase in electrolyte leakage. It was reported by Souza et al. (2016) that flooding is responsible for the decrease in asparagine, whereas there is an accumulation of gamma-aminobutyric acid (GABA) in soybeans. The AAs inside the nodules of soybeans reported a loss of 15N due to flooding. The enzyme activity of nitrogenase decreases significantly (Souza et al., 2016). It was reported by Kausar et al. (2012) that flooding stress in soybean causes antioxidant activity, which results in the decrease of ascorbate peroxidase (tAPX), an antioxidant enzyme, over time due to waterlogging. Waterlogging, as with any other abiotic stress, also reduces chlorophyll content, which leads to the yellowing of leaves (Puiatti and Sodek, 1999).

Maintaining good root growth with proper elongation and formation of the rapid and high number of adventitious roots are morphological acclimation of soybean genotypes to water logging conditions (Loc et al., 2015; Miura, 2012; Thomas et al., 2005; Sakazono et al., 2014; Sathi et al., 2022). A study also showed that higher root growth, high nitrogen accumulation and total biomass are essential characteristics of soybean cultivars for better performance under water logging conditions (Loc et al., 2015).

13.3.3 SALINITY STRESS

Salt stress is also another significant abiotic stress that affects the yield of soybean. High soil salinity content significantly affects the growth and development of plants in various ways (Hasegawa et al.; 2000; Shu et al., 2017). Salt stress has a direct impact on nodulation as it decreases the number of root nodules, thereby reducing the efficiency of nitrogen fixation (Delgado et al., 1994; Elsheikh and Wood, 1995; Singleton and Bohlool, 1984). The saline conditions delay or hinder the germination of soybean seeds (Nadeem et al., 2019). Salt, like NaCl, delays soybean seed germination by decreasing gibberellin (GA) and increasing ABA biogenesis, hence leading to a decrease in the GA/ABA ratio (Shu et al., 2017). The protein content in soybean seeds is reported to be reduced; however, the effect on oil seeds is still under question. In a study on two soybean genotypes, with one being tolerant and the other susceptible, it was found that the chlorophyll content was on the diminishing side in the susceptible genotype. Lu et al. (2009) reported suppression of the activity of Rubisco with increasing salinity, accompanied by a reduction in the net photosynthetic rate and a decrease in stomatal conductance and intercellular CO_2 concentration. During salt stress, plants adapt by accumulating antioxidant enzymes like ascorbate, catalase (CAT), peroxidase, superoxide dismutase (SOD), and osmolytes like proline, glycine betaine, and total soluble sugars. The toxicity of ions like sodium and chloride in saline soils induces the activation of these antioxidant enzyme machineries. Apart from these, certain methyl esters are involved in soybean salt tolerance, as reported by Shelke et al. (2019).

13.3.4 HEAT STRESS

High temperature has always been linked to a decrease in photosynthetic efficiency and, ultimately, decreased crop yield (Jianing et al., 2022). Soybean is relatively less sensitive to heat than other crops. However, the pollination and seed development stage is prone to high temperatures; a relatively low 22–24°C is optimal during reproductive stages (Chebrolu et al., 2016; Hatfield et al., 2011). Heat stress in soybean also induces adverse effects on the reproductive parts, a significant reason for yield loss. Another significant response seen in soybeans during heat stress is the accumulation of reactive oxygen species (ROS), which include peroxide radicals (O_2) and hydrogen peroxide (H_2O_2), causing oxidative DNA damage (Jianing et al., 2022). Yang et al. (2023) indicated that high night temperatures may result in comparatively lower yields of soybean due to a reduction in seed size, weight, pods, and the number of seeds per plant. Antioxidants produced can impart tolerance with the help of physiological responses like the improvement of gaseous exchange (Sabagh et al., 2020). A series of biochemical responses, such as heat shock protein accumulation, metabolites, antioxidants, and hormones, help control heat stress. Antioxidants like tocopherols, flavonoids, phenylpropanoids, and ascorbate precursors accumulation were observed in the seeds of tolerant genotypes. These antioxidants also help reduce ROS and prevent chloroplast blasts due to increased oxidation. Osmoprotectants are also a necessary mechanism used to cope with abiotic stress. Prolines are considered to be the most effective osmoprotectant in plants. There are many similar biochemical mechanisms, such as osmoregulation, ion homeostasis, and antioxidant and hormonal systems, which induce heat tolerance in soybean (Jianing et al., 2022).

13.4 MECHANISM BEHIND BIOCHEMICAL RESPONSES AGAINST BIOTIC STRESS IN SOYBEAN

Plants have various physiological and biochemical responses to protect themselves from biotic stress. These biochemical responses involve the production of plant secondary metabolites to counteract biotic agents such as insects, pathogens, and nematodes. These metabolites play both direct and indirect roles in combating biotic stress. In a direct role, the metabolites are produced to attack the agents. Indirectly, the compounds produced by attacking pests trigger systemic resistance in the host plant, leading to several biochemical responses and the production of secondary metabolites. Phytohormones such as salicylic acid (SA), ethylene, jasmonic acid, and brassinosteroids mediate the stress-induced signaling of the synthesis of metabolites for plant tolerance against biotic stress.

13.5 RESPONSES TO BIOTIC STRESS AND ADAPTATION IN SOYBEAN

13.5.1 RESPONSES TO INSECT PEST

More than 700 species of insects are associated with soybean worldwide; out of these, only a few insect pests are severe and can cause economic losses (Rupe and Luttrell, 2008). Insects damage plants by root-feeding, stem-feeding, leaf-feeding, and pod-feeding (Higley, 1994). Insect feeding may result in reductions of plant stand, leaf and photosynthetic rate, assimilate removals, and plant structures and phenological activities. Soybean seed and grain quality are influenced by pod or seed feeders such as stink bugs, corn earworms, bean leaf beetles, grasshoppers, armyworms, loopers, and caterpillars (Rupe and Luttrell, 2008). Soybean has an incredible ability to recover from early-stage insect infestation. It has been reported that soybean can withstand up to 40% defoliation before flowering and 25% after flowering without any reduction in yield (Lorenz et al., 2000). However, the yield reduction is high in the case of pod-feeding insects. If insects attack during the early stages of pod set, soybean can still reproduce well, with up to 80% of pods being produced (Smith and Bass, 1972).

Stink bugs are a major problem for soybean crops, as they can cause delayed maturity and reduce seed viability and germination (Boethel et al., 2000; Rupe and Luttrell, 2008). Timbo et al. (2014) conducted a study to better understand how soybean plants respond biochemically to brown stink bug injuries. They analyzed the redox metabolism and proteomic 2DE profiles of susceptible and resistant soybean varieties. The results showed that in susceptible varieties, the injury from the brown stink bug increases the roles of lipoxygenase three and guaiacol peroxidase while decreasing CAT activity. On the other hand, no changes were seen in tAPX. No alterations in resistant varieties were observed in any of these enzymes. The identified changes in protein profiles for susceptible genotypes were related to general stress responses to plant defense, while in resistant genotypes, they were related to Calvin cycle enzymes.

The soybean aphid (SBA) (*Aphis glycines*) is a serious insect pest that causes significant yield losses of up to 40%. This occurs because it reduces the plant's ability to photosynthesize and alters the AA profile of the host plant (Chiozza et al., 2010; Ragsdale et al., 2007). Li et al. (2007) conducted a study on the response of two soybean cultivars—the susceptible cultivar *Williams 82* and the resistant *Rag1* (Dowling)—to aphid feeding. They found that the resistant plant exhibited twice as many differentially expressed (DE) genes related to cell wall metabolism, the phenylpropanoid pathway, and the activation of the salicylic acid (SA) and jasmonic acid (JA) signaling pathways compared to the susceptible cultivar. Hohenstein et al. (2019) also discovered that resistant soybean varieties have a stronger and faster defense response to aphid infestation compared to susceptible ones. Their study also indicated that susceptible plants, especially those under long-term aphid feeding, induce isoflavonoid biosynthesis. Accumulation of daidzein, formononetin, genistein, and glyceollins was reported in aphid-treated susceptible plants. Daidzein might dampen aphid feeding. Finally, long-term SBA colonization in susceptible soybean plants can eventually lead to the development of an effective defense response.

13.5.2 RESPONSES TO PLANT PATHOGEN

Numerous biotic factors are known to decrease soybean yield, such as bacteria, viruses, fungi, and nematodes (Nazarov et al., 2020). The infection of these pathogens in the plant triggers a range of physiological and biochemical responses. Here, we present a summary of the cellular responses demonstrated by soybeans when faced with biotic stresses.

Macrophomina phaseolina, a fungus that causes charcoal rot, is a significant threat to soybean crops worldwide (Amrate et al., 2023a). During infection, chlorophyll and carotenoid levels decrease, while phenolic content and antioxidant enzymes such as peroxidase, polyphenol oxidase, and CAT increase in susceptible genotypes. The levels of superoxide and H_2O_2 radicals also increase throughout the growth stages, peaking during the final grain-filling stage (Navabpour et al., 2014). Genotypes resistant to charcoal rot contain high levels of phenolics, isoflavones, and total boron, contributing to their resistance mechanism (Bellaloui et al., 2012).

Stem and root rot caused by *Phytophthora sojae* is a disease that causes significant loss in productivity worldwide. As already observed, the antioxidant enzyme machinery is quite prominent in plants during stress conditions. One such enzyme, phenylalanine ammonia-lyase (PAL), is studied in plants as a response to biotic and abiotic stresses. Zhang et al. (2017) reported that the infection of *P. sojae* induces the PAL gene family GmPAL2.1, and RNA interference (RNAi) analysis suggested that this enhances resistance to the fungus in soybean transgenic plants. These plants showed an increase in isoflavones like daidzein and genistein, as well as the osmolyte SA and relative glyceollin content. Glyceollin is a soybean-derived phytoalexin that is synthesized via the phenylpropanoid pathway in soybeans and accumulated in seeds in response to stress (Lyne et al., 1976; Nwachukwu et al., 2013). These glyceollins are supposed to protect the tissues of plants from damage caused by oxidation induced by stress. Moreover, glyceollin is necessary to inhibit the pathogen species and growth of *Phytophthora megasperma* var. sojae (Kim et al., 2012; Yoshikawa et al., 1978).

Sclerotinia sclerotiorum, a fungus that causes white mold, can lead to significant economic losses worldwide due to its wide range of hosts. However, researchers have found a potential solution in secondary metabolites produced by the antibiotic Streptomyces. Specifically, a secondary metabolite called "wuyiencin" was shown in a recent report by Yang et al. (2022) to trigger physiological and biochemical defense responses in soybean against *S. sclerotiorum*. Wuyiencin was found to increase the activity of ROS scavenging enzymes such as SOD, CAT, and peroxidase, as well as stimulate callose deposition, phenylalanine ammonia lyase activity, chitinase activity, and beta-1,3-glucanase activity. These responses suggest that Wuyiencin may be an effective defense against *S. sclerotiorum* infection.

Yellow mosaic disease, caused by the mungbean yellow mosaic India virus (MYMIV) and transmitted by whitefly, is a problematic disease in soybean. It significantly reduces the chlorophyll content in infected plants compared to healthy ones, thereby lowering the photosynthesis rate in symptomatic plants (Dantre et al., 1996; Kaur et al., 1991; Kumar et al., 2022). Ghuge et al. (2018) reported that protein content was higher in healthy plants of resistant genotypes than in susceptible ones at 15 and 30 days after sowing (DAS). It increased in infected plants compared to healthy plants of both resistant and susceptible genotypes at 30 DAS. The total phenol content was also highest in resistant plants, followed by moderately resistant and susceptible ones. Yellow mosaic-affected soybean showed an increasing trend in both protein and phenol content (Kumar et al., 2022).

13.5.3 RESPONSES TO NEMATODES

Soybean is affected by several nematodes across the world. Yield loss from nematode-soybean association varies depending on several factors, including nematode species, population levels, cultivars, cropping systems, soil factors, weather conditions, and so forth (Lima et al., 2017). In some reported cases, yield loss can reach 30–100% (Barker, 1998; Schmitt and Barker, 1981). Root-knot nematodes (RKN), Meloidogyne spp., are essential plant-parasitic nematodes worldwide. Infection of RKN in roots leads to changes in plant nutrient and water uptake that ultimately decrease yield (Lima et al., 2017). Likewise, SCN, *Heterodera glycines*, is one of the most prominent soybean pests worldwide. Constantino et al. (2021) revealed that *Heterodera glycines*-infested plants exhibited no significant stunting until 10 days after infestation; prior to the stunting of the plants, it causes elevation of styrene, D-limonene, tetradecane, 2,6-di-tert-butyl-4-methylene-2,5-cyclohexadiene-1-one, butylated hydroxytoluene, and suppression of ethylhexyl benzoate in infested plants. Infested plants yielded certain volatile organic compounds over time that were significantly higher in comparison to mock-infested plants.

13.6 ALTERNATIVE APPROACHES TO COMBAT STRESS IN SOYBEAN

Soybean production can only be sustainable by lowering the stress and its influences on cropping areas of any region worldwide. The prevailing weather conditions always influence the attack and severity of biotic stress, such as insect pests and pathogens (Amrate et al., 2023b; Rajput et al., 2021). Whereas abiotic stresses can significantly limit production without biotic stresses. In soybean, in the cultivating field, the impact of drought stress can be minimized by applying need-based irrigation, especially at the sensitive stage. However, selecting the genotypes with better adaptation features under drought conditions would be best. To address the longer dry spell-related problem or to grow soybeans in areas with lower water availability, these selected genotypes should be included in the development of a high-yielding drought tolerance varietal development program. Marker-assisted breeding can be employed for the accurate and faster development of drought tolerance cultivars. Modifications in cultivation practices, such as early sowing of crops or early maturing varieties, would be intensified to meet pre-determined drought conditions. AMF and *Rhizobium, B. japonicum* increase proline content and other secondary metabolites and decrease MDA, which helps soybean plants thrive under drought conditions (Sheteiwy et al., 2021). Hence, biofertilizers would be an excellent alternative for successfully managing soybean under drought-adverse conditions. Similarly, waterlogging can be minimized by using ridge-furrow cultivation and a suitable drainage system in the field. Some essential criteria, such as the Testa integrity of soybean seeds and delayed germination, can be utilized to select genotypes and further breeding to improve germination waterlogging tolerance (Rajendran et al., 2022). Soybean seed is sensitive to waterlogging. However, some soybean accessions have been identified that can tolerate eight days of waterlogging at the seed germination stage. These genotypes should be evaluated under different field conditions and utilized directly for sowing and crop improvement against waterlogging stress (Rajendran et al., 2023). In order to protect crops from waterlogging stress during germination, it is essential to focus on developing varieties that can germinate under anaerobic conditions. This approach will be critical in the current era of climate change (Ramlal et al., 2023b). It is important to identify soybean genotypes with salt, heat, and cold tolerance and transfer these specific traits to high-yielding genotypes to overcome region- and stage-specific challenges.

It is expected to control biotic stresses like insect pests and diseases with the help of chemicals. However, this method is expensive and harmful to the environment. Therefore, we must develop varieties resistant to multiple biotic stresses at different locations. Molecular breeding technology has become efficient in transferring genes into the desired cultivar. These technologies are precise and quick; plus, another benefit that these technologies provide is selection at a very early stage of plant growth. Identification of QTLs helps us by providing basic information about the marker-trait association of specific molecular markers. It is also helpful in understanding the distance between genes of interest and flanking markers. There are several genotypes reported to be resistant or have moderate resistance to many vital diseases worldwide in soybean (Amrate et al., 2023a, b; Kofsky et al., 2021; Langenbach et al., 2016; Mengistu et al., 2018). All these identified genotypes need to be validated for their resistance to particular diseases biochemically and molecularly. Further, they should be incorporated into intensive breeding programs to reduce the impact of pathogenic stresses. Good cultural practices such as adopting suitable crop rotation, taking care of soil health, using bioagents as seed inoculants, treating seeds with effective fungitoxicants, and adopting the right sowing windows, and so forth, should be followed to reduce the incidence of seed and soil-borne diseases and other diseases.

In today's world, organic farming is gaining popularity. As a result, using different bioagents in various formulations during field preparation, sowing, and at the onset of insect pests and diseases is well documented in various crops (Kumar et al., 2021). These bioagents and chemicals also effectively reduce the severity of various diseases in soybeans (Amrate et al., 2018; Reznikov et al., 2016). The efficacy of this biological agent, with specific modifications through recent biotechnological innovations/tools such as genetic engineering, RNAi technology, and nanotechnology, can be enhanced by the production of secondary metabolites, biofertilizers, bioenergy, and processed products (Kumar et al., 2021). This would be an essential narrow-spectrum alternative that targets specific insect pests and diseases in soybeans, with potential for future use. Nanotechnology exhibits substantial potential for mitigating stress factors in soybean agriculture. It encompasses nanosensors that enable real-time crop monitoring, nano fertilizers designed for precise and efficient nutrient delivery, nano pesticides tailored for specific pest management, and gene delivery techniques for crop enhancement (Mishra et al., 2023). These cutting-edge nanotechnological innovations can bolster soybeans' adaptability and agricultural output when confronted with environmental and biotic adversities.

13.7 FUTURE PROSPECTS

Climate change has been a significant cause of declining crop productivity. The increasing demand for food to feed a growing population requires significant improvements to combat stress. As we discussed above, reactive oxygen species (ROS) are released under abiotic stress; at high concentrations of ROS, cells are damaged, whereas, at low and medium concentrations, they act as secondary messengers in intracellular signaling cascades. Therefore, future research should be directed in a way that exploits ROS-related pathways and identifies other ROS-related metabolic pathways. Another direction for future research is better usage of weather forecasts and improving the irrigation systems for soybean cultivation. Molecular breeding and genomics should play a crucial role in developing soybean varieties that are resistant or tolerant to stress based on their physiological and biochemical responses. Technology and genetic modification could make crops more resistant to abiotic stresses, potentially mitigating the risk of crop failure. In this context, omics approaches, including genomics, transcriptomics, proteomics, ionomics, and phenomics, would play a crucial role in addressing abiotic stress in soybean. These approaches provide valuable insights into the genetic, molecular, and physiological mechanisms underlying stress tolerance. By integrating multi-omics data, researchers can identify candidate genes, proteins, and metabolic pathways associated with stress response and develop breeding and crop improvement strategies to enhance soybean's resilience to abiotic stress (Deshmukh et al., 2014). Likewise, utilizing host genetic resistance against diseases is their most economical and environmentally friendly management mode (Amrate et al., 2023a; Rajput et al., 2021). Identifying and mapping the genes and QTL responsible for various diseases in soybean through biparental mapping populations and GWAS can help enhance soybean productivity. By introgressing these genes/QTL into otherwise susceptible mega varieties, we can effectively reduce the risk of disease outbreaks and increase yields (Nataraj et al., 2023). In the current era, remote sensing has become increasingly important. As a result, proximal remote sensing technologies such as RGB imaging, thermal imaging, hyperspectral imaging, multispectral imaging, and fluorescence imaging provide a non-destructive and rapid way to gather detailed phenotypic data, offering valuable insights into various plant traits at different growth stages (Anand et al., 2023). These would be helpful in the quantitative assessment of various biotic and abiotic stresses and their effective management.

13.8 CONCLUSION

Plants display a range of physiological responses, such as stomatal conductance, osmoregulation, reduced photosynthetic rate, reduced chlorophyll content, and homeostasis, as well as biochemical responses, including an increase in antioxidant enzymes, such as peroxidase and CAT, an increase in osmoprotectants like SA, jasmonic acid, brassinosteroids, proline, and changes in isoflavones, phytohormones, and secondary metabolites in response to both biotic and abiotic stresses. Phytohormones, such as jasmonates, gibberellins, ABA, and nitric oxide, have evolved to regulate plant physiological processes, such as seed germination, photosynthesis, senescence, pollen growth, and defense responses, which also aid in controlling stress. Many phytohormones are thought to provide specific adaptations to variable climates, but their effects have been underreported. It is possible to conduct research to identify dependable sources of tolerance using the suggested responses. This can enable plants to independently handle the harmful effects of biotic and abiotic stress through exogenous applications of these biochemical responses, thus reducing their dependency on chemical management.

KEYWORDS

- **abiotic and biotic stress**
- **antioxidant enzymes**
- **biochemical adaptations**
- **physiological responses**
- **phytohormones**
- **secondary metabolites**
- **soybean stress tolerance**

REFERENCES

Agarwal, D. K., Billore, S. D., & Sharma, A. N. (2013). Soybean: Introduction, improvement, and utilization in India—Problems and prospects. *Agricultural Research, 2*, 293–300.

Amrate, P. K., & Shrivastava, M. K. (2021). Yield response and pathological characterization of promising genotypes of soybean against major diseases in Madhya Pradesh. *Journal of Oilseeds Research, 38*(4), 380–384.

Amrate, P. K., Pancheshwar, D. K., & Shrivastava, M. K. (2018). Application of fungicides and *Trichoderma viride* in management of soybean pod blight complex. *International Journal of Chemical Studies, 6*(4), 3208–3210.

Amrate, P. K., Shrivastava, M. K., & Bhale, M. S. (2019). Resistance in soybean varieties against charcoal rot disease caused by *Macrophomina phaseolina*. *Plant Disease Research, 34*(2), 124–128. https://doi.org/10.5958/2249-8788.2019.00021.0.

Amrate, P. K., Shrivastava, M. K., & Pancheshwar, D. K. (2021a). Crop-weather based relation and severity prediction of aerial blight incited by *Rhizoctonia solani* Kuhn in soybean. *Journal of Agrometeorology, 23*(1), 66–73.

Amrate, P. K., Shrivastava, M. K., & Singh, G. (2021b). Identification of sources of resistance and yield loss assessment for aerial blight and anthracnose/pod blight diseases in soybean. *Legume Research.* https://doi.org/10.18805/LR-4452.

Amrate, P. K., Shrivastava, M. K., Bhale, M. S., Agrawal, N., Kumawat, G., Shivakumar, M., & Nataraj, V. (2013a). Identification and genetic diversity analysis of high-yielding charcoal rot-resistant soybean genotypes. *Scientific Reports, 13*, 8905. https://doi.org/10.1038/s41598-023-35688-2.

Amrate, P. K., Shrivastava, M. K., Borah, M., Routhu, G. K., Sharma, S., Nataraj, V., Pancheshwar, D. K., & Singh, G. (2023b). Molecular characterization of *soybean yellow mosaic virus* isolates and identification of stable resistance sources in central India. *Australasian Plant Pathology, 52*, 165–179. https://doi.org/10.1007/s13313-022-00902-8.

Anand, K. J., Nagre, S. P., Shrivastava, M. K., Amrate, P. K., Patel, T., & Katara, V. K. (2023). Enhancing crop improvement through synergistic integration of advanced plant breeding and proximal remote sensing techniques: A review. *International Journal of Plant & Soil Science, 35*(19), 121–138. https://doi.org/10.9734/ijpss/2023/v35i193533.

Arya, H., Singh, M. B., & Bhalla, P. L. (2021). Towards developing drought-smart soybeans. *Frontiers in Plant Science, 12*, 750664. https://doi.org/10.3389/fpls.2021.750664.

Ashraf, M., & Iram, A. (2005). Drought stress-induced changes in some organic substances in nodules and other plant parts of two potential legumes differing in salt tolerance. *Flora: Morphology, Distribution, Functional Ecology of Plants, 200*, 535–546. https://doi.org/10.1016/j.flora.2005.06.005.

Banerjee, J., Shrivastava, M. K., Amrate, P. K., Singh, Y., Upadhyay, A., & Soni, M. (2022). Genetic variability and association of yield-contributing traits in advanced breeding lines of soybean. *Electronic Journal of Plant Breeding, 13*(2), 597–607.

Banerjee, J., Shrivastava, M. K., Singh, Y., & Amrate, P. K. (2023). Estimation of genetic divergence and proximate composition in advanced breeding lines of soybean (*Glycine max* (L.) Merrill). *Environment & Ecology, 41*(3C), 1960–1968.

Barker, K. R. (1998). Introduction. In K. R. Barker, G. A. Pederson, & G. L. Winhan (Eds.), *Plant Nematode Interactions* (pp. 1–120). American Society of Agronomy.

Bellaloui, N., Mengistu, A., Zobiole, L. H. S., & Shier, W. T. (2012). Resistance to toxin-mediated fungal infection: Role of lignins, isoflavones, other seed phenolics, sugars, and boron in the mechanism of resistance to charcoal rot disease in soybean. *Toxin Reviews, 31*(1, 2), 16–26.

Bhargava, S., & Sawant, K. (2012). Drought stress adaptation: Metabolic adjustment and regulation of gene expression. *Plant Breeding, 132*, 21–32.

Biswas, G. C. (2008). Insect pests and their management of soybean crop in Bangladesh. In M. A. Bakr (Ed.), *Proceedings of the Workshop on Prospects and Performance of Soybean in Bangladesh* (pp. 67). ORC, BARI, Gazipur.

Boethel, D. J., Russin, J. S., Weir, A. T., Layton, M. B., Mink, J. S., & Boyd, M. L. (2000). Delayed maturity associated with southern green stink bug (*Nezara viridula*) injury at various stink bug phenological stages. *Journal of Economic Entomology, 93*, 707–712.

Catangui, M. A., Beckendorf, E. A., & Riedell, W. E. (2009). Soybean aphid population dynamics, soybean yield loss, and development of stage-specific economic injury levels. *Agronomy Journal, 101*(5), 1080–1092.

Chebrolu, K. K., Fritschi, F. B., Ye, S., Krishnan, H. B., Smith, J. R., & Gillman, J. D. (2016). Impact of heat stress during seed development on soybean seed metabolome. *Metabolomics, 12*, 28. https://doi.org/10.1007/s11306-015-0941-1.

Chiozza, M. V., O'Neal, M. E., & MacIntosh, G. C. (2010). Constitutive and induced differential accumulation of amino acids in leaves of susceptible and resistant soybean plants in response to the soybean aphid (*Aphis glycines*). *Environmental Entomology, 39*, 856–864. https://doi.org/10.1603/en09338.

Choudhary, M. L., Tripathi, M. K., Tiwari, S., Pandya, R. K., Gupta, N., Tripathi, N., & Parihar, P. (2021). Screening of pearl millet (*Pennisetum glaucum* (L.) R. Br.) germplasm lines for drought tolerance based on morpho-physiological traits and SSR markers. *Current Journal of Applied Science and Technology, 40*, 46–63.

Constantino, N., Oh, Y., Sennik, E., Andersen, B., Warden, M., Oralkan, Ö., & Dean, R. A. (2021). Soybean cyst nematodes influence aboveground plant volatile signals prior to symptom development. *Frontiers in Plant Science, 12*, 749014. https://doi.org/10.3389/fpls.2021.749014.

Dai Vu, L., Gevaert, K., & De Smet, I. (2019). Feeling the heat: Searching for plant thermosensors. *Trends in Plant Science, 24*(3), 210–219.

Dantre, R. K., Keshwal, R. L., & Khare, M. N. (1996). Biochemical changes induced by yellow mosaic virus in resistant and susceptible cultivars of soybean (*Glycine max* (L.) Merrill.). *Indian Journal of Virology, 12*, 47–49.

Delgado, M. J., Ligero, F., & Liuch, C. (1994). Effects of salt stress on growth and nitrogen fixation by pea, faba bean, common bean, and soybean plants. *Soil Biology and Biochemistry, 26*, 371–376.

Deshmukh, R., Sonah, H., Patil, G., Chen, W., Prince, S., Mutava, R., Vuong, T., Valliyodan, B., & Nguyen, H. T. (2014). Integrating omic approaches for abiotic stress tolerance in soybean. *Frontiers in Plant Science, 5*, 1–12. https://doi.org/10.3389/fpls.2014.00244.

Dong, S., Jiang, Y., Dong, Y., Wang, L., Wang, W., Ma, Z., & Liu, L. (2019). A study on soybean responses to drought stress and rehydration. *Saudi Journal of Biological Sciences, 26*(8), 2006–2017.

Du, Y., Zhao, Q., Chen, L., Yao, X., & Xie, F. (2020). Effect of drought stress at reproductive stages on growth and nitrogen metabolism in soybean. *Agronomy, 10*, 302. https://doi.org/10.3390/agronomy10020302.

Eck, H. V., Mathers, A. C., & Musick, J. T. (1987). Plant water stress at various growth stages and growth and yield of soybeans. *Field Crops Research, 17*(1), 1–16.

Elsheikh, E. A. E., & Wood, M. (1995). Nodulation and N_2 fixation by soybean inoculated with salt-tolerant rhizobia or salt-sensitive bradyrhizobia in saline soil. *Soil Biology and Biochemistry, 27*, 657–661.

Fatimah, V. S., & Nurhidayati, T. (2020). Morphophysiological characteristic responses of soybean (*Glycine max* L.) Grobogan variety in waterlogging stress. *Ecology, Environment & Conservation, 26*, S132–S138.

Flint, W. P. (1941). Found—An insect enemy of soybeans. *Soybean Digest, 1*, 4.

Ghuge, A. S., Gupta, V. R., Patel, S. E., & Zade, S. B. (2018). Biochemical changes induced by mungbean yellow mosaic virus (MYMV) in mungbean (*Vigna radiata* (L.) Wilczek). *International Journal of Chemical Studies, 6*(5), 2156–2159.

Guang, C., Xiugui, W., Yu, L., & Wenbing, L. (2012). Effect of waterlogging stress on cotton leaf area index and yield. *Procedia Engineering, 28*, 202–209. https://doi.org/10.1016/j.proeng.2012.01.706.

Haile, F. J., Higley, L. G., & Specht, J. E. (1998). Soybean cultivars and insect defoliation: Yield loss and economic injury levels. *Agronomy Journal, 90*(3), 344–352.

Hasanuzzaman, M., Al Mahmud, J., Nahar, K., Anee, T. I., Oku, H., & Fujita, M. (2017). Responses, adaptation, and ROS metabolism in plants exposed to waterlogging stress.

In M. I. R. Khan, N. A. Khan, & A. M. Ismail (Eds.), *Reactive Oxygen Species and Antioxidant Systems: Role and Regulation under Abiotic Stress*. Springer.

Hasegawa, P. M., Bressan, R. A., Zhu, J. K., & Bohnert, H. J. (2000). Plant cellular and molecular responses to high salinity. *Annual Review of Plant Physiology and Plant Molecular Biology, 51*, 463–499. https://doi.org/10.1146/annurev.arplant.51.1.463.

Hatfield, J. L., Boote, K. J., Kimball, B. A., Ziska, L. H., Izaurralde, R. C., Ort, D., Thomson, A. M., & Wolfe, D. (2011). Climate impacts on agriculture: Implications for crop production. *Agronomy Journal, 103*, 351–370.

Heinrichs, E. A., & Muniappan, R. (2018). Integrated pest management for tropical crops: Soybeans. *CAB Reviews, 13*, 55. https://doi.org/10.1079/PAVSNNR201813055.

Hesler, L. S., Allen, K. C., Luttrell, R. G., Sappington, T. W., & Papiernik, S. K. (2018). Early-season pests of soybean in the United States and factors that affect their risk of infestation. *Journal of Integrated Pest Management, 9*(1), 19. https://doi.org/10.1093/jipm/pmx028.

Higley, L. G. (1994). Insect injury to soybean. In L. G. Higley & D. J. Boethel (Eds.), *Handbook of Soybean Insect Pests* (pp. 11–13). Entomology Society of America.

Hohenstein, J. D., Studham, M. E., Klein, A., Kovinich, N., Barry, K., Lee, Y. J., & MacIntosh, G. C. (2019). Transcriptional and chemical changes in soybean leaves in response to long-term aphid colonization. *Frontiers in Plant Science, 10*, 310. https://doi.org/10.3389/fpls.2019.00310.

James, T. W., & Turnipseed, S. G. (1974). Effects of southern green stink bug damage on yield and quality of soybeans. *Journal of Economic Entomology, 67*(3), 421–426. https://doi.org/10.1093/jee/67.3.421.

Jayappa, A. H. (2000). *Studies on the Soybean Stem Fly M. sojae (Zehnt) with Reference to Crop Loss Assessment, Screening of Varieties and Management* (Master's thesis, University of Agricultural Science, Bangalore, India).

Jianing, G., Yuhong, G., Yijun, G., Rasheed, A., Qian, Z., Zhiming, X., Mahmood, A., Shuheng, Z., Zhuo, Z., Xiaoxue, W., & Jian, W. (2022). Improvement of heat stress tolerance in soybean (*Glycine max* L.) by using conventional and molecular tools. *Frontiers in Plant Science, 13*, 993189. https://doi.org/10.3389/fpls.2022.993189.

Jun, T. H., Mian, M. A. R., Kang, S. T., & Michel, A. P. (2012). Genetic mapping of the powdery mildew resistance gene in soybean PI 567301B. *Theoretical and Applied Genetics, 125*, 1159–1168.

Kaur, G., Gill, C. K., Rataul, H. S., & Raheja, R. K. (1991). Biochemical changes in soybean (*Glycine max* L.) cultivars infected with yellow mosaic virus. *Biochemie und Physiologie der Pflanzen, 187*, 357–371.

Kausar, R., Hossain, Z., Makino, T., & Komatsu, S. (2012). Characterization of ascorbate peroxidase in soybean under flooding and drought stresses. *Molecular Biology Reports, 39*(12), 10573–10579.

Kido, E. A., Ferreira Neto, J. R. C., Silva, R. L. O., Belarmino, L. C., Bezerra Neto, J. P., & Soares-Cavalcanti, N. M. (2013). Expression dynamics and genome distribution of osmoprotectants in soybean: Identifying important components to face abiotic stress. *BMC Bioinformatics, 14*, 1–11. https://doi.org/10.1186/1471-2105-14-S1-S7.

Kim, H. J., Lim, J. S., Kim, W. K., & Kim, J. S. (2013). Soybean glyceollins: Biological effects and relevance to human health. *Proceedings of the Nutrition Society, 71*, 166–174.

Kofsky, J., Zhang, H., & Song, B. H. (2021). Novel resistance strategies to soybean cyst nematode (SCN) in wild soybean. *Scientific Reports, 11*, 7967.

Kucharik, C. J., & Serbin, S. P. (2008). Impacts of recent climate change on Wisconsin corn and soybean yield trends. *Environmental Research Letters, 3*(3), 034003.

Kumar, J., Ramlal, A., Mallick, D., & Mishra, V. (2021). An overview of some biopesticides and their importance in plant protection for commercial acceptance. *Plants, 10*, 1185. https://doi.org/10.3390/plants10061185.

Kumar, S., Amrate, P. K., Upadhyay, A., Paroda, M., & Mohare, S. (2022). Influence of yellow mosaic disease on biochemical and yield attributing traits in soybean. *Plant Disease Research, 37*(2), 192–197. https://doi.org/10.5958/2249-8788.2022.00032.4.

Kumawat, M. M., Kumar, A., & Bunker, G. K. (2010). Screening of soybean cultivars against girdle beetle. *Annals of Plant Protection Sciences, 18*(1), 246, 247.

Kunert, K. J., Vorster, B. J., Fenta, B. A., Kibido, T., Dionisio, G., & Foyer, C. H. (2016). Drought stress responses in soybean roots and nodules. *Frontiers in Plant Science, 7*, 1015. https://doi.org/10.3389/fpls.2016.01015.

Lamers, J., Van der Meer, T., & Testerink, C. (2020). How plants sense and respond to stressful environments. *Plant Physiology, 182*(4), 1624–1635. https://doi.org/10.1104/pp.19.01464.

Langenbach, C., Schultheiss, H., Rosendahl, M., Tresch, N., Conrath, U., & Goellner, K. (2016). Interspecies gene transfer provides soybean resistance to a fungal pathogen. *Plant Biotechnology Journal, 14*, 699–708.

Latha, G. S., Das, S. B., Swathi, P. S., Neelesh, R., & Marabi, R. S. (2018). Biology of whitefly, *Bemisia tabaci*, on soybean cultivars. *Journal of Entomology and Zoology, 6*(5), 2351–2355.

Leff, B., Ramankutty, N., & Foley, J. A. (2004). Geographic distribution of major crops across the world. *Global Biogeochemical Cycles, 18*(1), GB1009.

Li, Y., Hill, C. B., Carlson, S. R., Diers, B. W., & Hartman, G. L. (2007). Soybean aphid resistance genes in the soybean cultivars Dowling and Jackson map to linkage group M. *Molecular Breeding, 19*, 25–34. https://doi.org/10.1007/s11032-006-9039-9.

Lima, F. S. O., Valdir, C. R., Regina, N. S., & Patrícia, S. R. R. (2017). Nematodes affecting soybean and sustainable practices for their management. In M. Kasai (Ed.), *Soybean–The Basis of Yield, Biomass and Productivity.* http://dx.doi.org/10.5772/67030.

Linkemer, G., Board, J. E., & Musgrave, M. E. (1998). Waterlogging effect on growth and yield components of late-planted soybean. *Crop Science, 38*, 1576–1584.

Liu, F., Jensen, C. R., & Andersen, M. N. (2003). Hydraulic and chemical signals in the control of leaf expansion and stomatal conductance in soybean exposed to drought stress. *Functional Plant Biology, 30*, 65–73. https://doi.org/10.1071/FP02170.

Loc, N. V., Binh, V. T., Hong, D. H., & Mochizuki, T. (2015). Genotypic variation in morphological and physiological response of soybean to waterlogging at flowering stage. *International Journal of Agricultural Science Research, 4*(8), 150–157.

Lorenz, G., Johnson, D., Studebaker, G., Alle, C., & Young, S. (2000). Insect pest management in soybean. In *Arkansas Soybean Handbook* (pp. 84–94). Arkansas Soybean Promotion Board and Arkansas Cooperative Extension Service.

Lu, K. X., Cao, B. H., Feng, X. P., He, Y., & Jiang, D. A. (2009). Photosynthetic response of salt-tolerant and sensitive soybean varieties. *Photosynthetica, 47*(3), 381–387.

Lyne, R. L., Mulheirn, L. J., & Leworthy, D. P. (1976). New pterocarpinoid phytoalexins of soybean. *Journal of the Chemical Society, Chemical Communications, 13*, 497–498.

Mak, M., Babla, M., Xu, S. C., O'Carrigan, A., Liu, X. H., Gong, Y. M., Holford, P., & Chen, Z. H. (2014). Leaf mesophyll K+, H+, and Ca2+ fluxes are involved in drought-induced decrease in photosynthesis and stomatal closure in soybean. *Environmental and Experimental Botany, 98*, 11, 12. https://doi.org/10.1016/j.envexpbot.2013.10.003.

Mengistu, A., Ray, J. D., Smith, J. R., Arelli, P. R., Bellaloui, N., Chen, P., Shannon, G., & Boykin, D. (2018). Effect of charcoal rot on selected putative drought-tolerant soybean genotypes and yield. *Crop Protection, 105*, 90–10.

Mengistu, A., Smith, J. R., & Ray, J. D. (2011). Seasonal progress of charcoal rot and its impact on soybean productivity. *Plant Disease, 95*, 1159–1166.

Michael, C. A., Eric, B. A., & Riedell, W. R. E. (2009). Soybean aphid population dynamics, soybean yield loss, and development of stage-specific economic injury levels. *Agronomy Journal, 101*, 1080–1092. https://doi.org/10.2134/agronj2008.0233x.

Miransari, M. (2016). Enhancing soybean response to biotic and abiotic stresses. In M. Miransari (Ed.), *Abiotic and Biotic Stresses in Soybean Production* (pp. 53–77). Academic Press.

Mishra, N., Tripathi, M. K., Tripathi, N., Tiwari, S., Gupta, N., Sharma, A., & Shrivastav, M. K. (2021). Changes in biochemical and antioxidant enzyme activities play a significant role in drought tolerance in soybean. *International Journal of Agricultural Technology, 17*(4), 1425–1446. Available online: http://www.ijat-aatsea.com (accessed on 25 March 2025).

Mishra, R., Singh, Y., Shrivastava, M. K., & Amrate, P. K. (2023). Nanotechnology: A milestone in agriculture. In *Advances in Biological Sciences and Biotechnology* (Vol. 5, pp. 13–24). Integrated Publications. https://doi.org/10.22271/int.book.316.

Miura, K., Ogawa, A., Matsushima, K., & Morita, H. (2012). Root and shoot growth under flooded soil in wild groundnut (*Glycine soja*) and as a genetic resource of waterlogging tolerance for soybean (*Glycine max*). *Pakistan Journal of Weed Science Research, 18*, 427–423.

Mueller, A. J., & Dumas, B. A. (1975). Effects of stem girdling by the three-cornered alfalfa hopper on soybean yields. *Journal of Economic Entomology, 68*, 1731–1733.

Murtaza, G., Rasool, F., Habib, R., Javed, T., Sardar, K., Ayub, M. M., & Rasool, A. (2016). A review of morphological, physiological, and biochemical responses of plants under drought stress conditions. *Imperial Journal of Interdisciplinary Research, 2*(12), 1600–1606.

Musser, F. R., Catchot, A. L., Davis, J. A., Herbert, D. A., Lorenz, J. G. M., Reed, T., Reisig, D. D., & Stewart, S. D. (2014). 2013 soybean insect losses in the southern US. *Midsouth Entomologist, 7*, 15–28.

Musser, F. R., Catchot, A. L., Davis, J. A., Herbert, D. A., Lorenz, J. G. M., Reed, T., Reisig, D. D., & Stewart, S. D. (2015). 2014 soybean insect losses in the southern US. *Midsouth Entomologist, 8*, 35–48.

Musser, F. R., Catchot, A. L., Jr., Davis, J. A., Lorenz, G. M., Reed, T., Reisig, D. D., Stewart, S. D., & Taylor, S. (2017). 2016 soybean insect losses in the southern US. *Midsouth Entomologist, 10*, 1–13.

Mutava, R. N., Prince, S. J. K., Syed, N. H., Song, L., Valliyodan, B., & Chen, W. (2015). Understanding abiotic stress tolerance mechanisms in soybean: A comparative evaluation of soybean response to drought and flooding stress. *Plant Physiology and Biochemistry, 86*, 109–120. https://doi.org/10.1016/j.plaphy.2014.11.010.

Nadeem, M., Li, J., Yahya, M., Wang, M., Ali, A., Cheng, A., & Ma, C. (2019). Grain legumes and fear of salt stress: Focus on mechanisms and management strategies. *International Journal of Molecular Sciences, 20*(4), 799.

Nataraj, V., Kumar, S., Kumawat, G., Shivakumar, M., Rajput, L. S., Ratnaparkhe, M. B., Ramteke, R., Gupta, S., Satpute, G. K., Rajesh, V., Kamble, V., & Chandra, S. (2019). Charcoal rot resistance in soybean: Current understanding and future perspectives. In S. Wani (Ed.), *Disease Resistance in Crop Plants* (pp. 241–259). Springer. https://doi.org/10.1007/978-3-030-20728-1_10.

Nataraj, V., Rajput, L. S., Shivakumar, M., Kumawat, G., Kumar, S., Maheshwari, H. S., Gupta, S., Amrate, P. K., Tripathi, R., Agrawal, N., & Ratnaparkhe, B. (2023). Crop improvement against *Colletotrichum truncatum* using molecular breeding approaches. In *QTL Mapping in Crop Improvement* (pp. 45–56). Academic Press.

Navabpour, S., Mirkarimi, S., & Mazandarani, A. (2014). Evaluation of enzymatic and non-enzymatic defense mechanisms in response to charcoal rot disease during growth stage in soybean. *Crop Biotechnology, 3*(5), 63–73.

Nazarov, P. A., Baleev, D. N., Ivanova, M. I., Sokolova, L. M., & Karakozova, M. V. (2020). Infectious plant diseases: Etiology, current status, problems, and prospects in plant protection. *Acta Naturae, 12*, 46–59.

Nisar, D. N., Modi, S., & Lone, I. R. (2019). Study of physiological responses in soybean (*Glycine max*) under salt stress conditions. *Journal of Medicinal Plants, 7*(4), 254–259.

Nwachukwu, I. D., Luciano, F. B., & Udenigwe, C. C. (2013). The inducible soybean glyceollin phytoalexins with multifunctional health-promoting properties. *Food Research International, 54*, 1208–1216.

Puiatti, M., & Sodek, L. (1999). Waterlogging affects nitrogen transport in the xylem of soybean. *Plant Physiology and Biochemistry, 37*(10), 767–773. https://doi.org/10.1016/S0981-9428(00)86690-5.

Raghuvanshi, R. S., & Bisht, K. (2010). Uses of soybean: Products and preparation. In G. Singh (Ed.), *The Soybean: Botany, Production, and Uses* (pp. 404–426). CAB International.

Ragsdale, D. W., McCornack, B. P., Venette, R. C., Potter, B. D., MacRae, I. V., Hodgson, E. W., O'Neal, M. E., Johnson, K. D., O'Neil, R. J., & Hunt, T. E. (2007). Economic threshold for soybean aphid (*Hemiptera: Aphididae*). *Journal of Economic Entomology, 100*, 1258–1267.

Rajendran, A., Lal, S. K., Raju, D., & Ramlal, A. (2022). Associations of direct and indirect selection for pregermination anaerobic stress tolerance in soybean (*Glycine max*). *Plant Breeding, 141*(5), 634–643. https://doi.org/10.1111/pbr.13048.

Rajendran, A., Lal, S. K., Raju, D., Mallikarjun, B. P., Ramlal, A., & Sharma, D. (2023). Waterlogging tolerance evaluation methods for soybean (*Glycine max* (L.) Merr.) at the pregermination stage. *Genetic Resources and Crop Evolution, 70*, 2451–2461. https://doi.org/10.1007/s10722-023-01573-0.

Rajput, L. S., Nataraj, V., Kumar, S., Amrate, P. K., Jahagirdar, S., Huilgol, S. N., Chakruno, P., Singh, A., Maranna, S., Ratnaparkhe, M. B., Borah, M., Singh, K. P., Gupta, S., & Khandekar, N. (2021). WAASB index revealed stable resistance sources for soybean anthracnose in India. *Journal of Agricultural Science, 159*(9, 10), 710–720. https://doi.org/10.1017/S0021859622000016.

Ramlal, A., Lal, S. K., & Sathuvalli, V. (2023b). Editorial: Advances in breeding for waterlogging tolerance in crops. *Frontiers in Plant Science, 14*, 1284730. https://doi.org/10.3389/fpls.2023.1284730.

Ramlal, A., Nautiyal, A., Lal, S. K., & Chigeza, G. (2023a). Editorial: A wonder legume, soybean: Prospects for improvement. *Frontiers in Plant Science, 14*, 1294185. https://doi.org/10.3389/fpls.2023.1294185.

Reznikov, S., Vellicce, G. R., Gonzalez, V., Lisi, V. D., Castagnaro, A. P., & Ploper, L. D. (2016). Evaluation of chemical and biological seed treatments to control charcoal rot of soybean. *Journal of Plant Pathology, 82*, 273–280.

Rupe, J. C., & Luttrell, R. L. (2008). Effect of pests and diseases on soybean quality. In L. A. Johnson, P. J. White, & R. Galloway (Eds.), *Soybeans: Chemistry, Production, Processing, and Utilization*. American Oil Chemists' Society.

Sabagh, A. E., Hossain, A., Islam, M. S., Iqbal, M. A., Fahad, S., Ratnasekera, D., & Llanes, A. (2020). Consequences and mitigation strategies of heat stress for sustainability of soybean (*Glycine max* L. Merr.) production under the changing climate. *Plant Stress Physiology*, 119–140.

Sakazono, S., Nagata, T., Matsuo, R., Kajihara, S., Watanabe, M., Ishimoto, M., Shimamura, S., Harada, K., Takahashi, R., & Mochizuki, T. (2014). Variation in root development response to flooding among 92 soybean lines during early growth stages. *Plant Production Science, 17*, 228–236.

Sathi, K. S., Masud, A. A. C., Anee, T. I., Rahman, K., Ahmed, N., & Hasanuzzaman, M. (2022). Soybean plants under waterlogging stress: Responses and adaptation mechanisms. In *Managing Plant Production Under Changing Environment* (pp. 103–134). Springer Nature Singapore.

Schmitt, D. P., & Barker, K. R. (1981). Damage and reproduction potentials of *Pratylenchus brachyurus* and *P. penetrans* on soybean. *Journal of Nematology, 13*(3), 327–332.

Shao, H. B., Chu, L. Y., Jaleel, C. A., Manivannan, P., Panneerselvam, R., & Shao, M. A. (2009). Understanding water deficit stress-induced changes in the basic metabolism of higher plants—Biotechnologically and sustainably improving agriculture and the Eco environment in arid regions of the globe. *Critical Reviews in Biotechnology, 29*(2), 131–151.

Sharma, A. N., Gupta, G. K., Verma, R. K., Sharma, O. P., Bhagat, S., Amaresan, N., Saini, M. R., Chattopadhyay, C., Sushil, S. N., Asre, R., Kapoor, K. S., Satyagopal, K., & Jeyakumar, P. (2014). *Integrated Pest Management for Soybean*. ICAR–National Center for Integrated Pest Management.

Shelke, D. B., Nikalje, G. C., Chambhare, M. R., Zaware, B. N., Penna, S., & Nikam, T. D. (2019). Na+ and Cl− induce differential physiological, biochemical responses and metabolite modulations *in vitro* in contrasting salt-tolerant soybean genotypes. *Biotechnology, 9*(3), 1–15.

Sheteiwy, M. S., Ali, D. F. I., & Xiong, Y. C. (2021). Physiological and biochemical responses of soybean plants inoculated with arbuscular mycorrhizal fungi and *Bradyrhizobium* under drought stress. *BMC Plant Biology, 21*, 195. https://doi.org/10.1186/s12870-021-02949-z.

Shoaib, A., Awan, Z. A., & Khan, K. A. (2018). Involvement of antioxidants and total phenolics in *Glycine max* resistance and susceptibility to charcoal rot. *International Journal of Biology and Biotechnology, 15*, 655–660.

Shu, K., Qi, Y., Chen, F., Meng, Y., Luo, X., Shuai, H., Zhou, W., Ding, J., Du, J., Liu, J., Yang, F., Wang, Q., Liu, W., Yong, T., Wang, X., Feng, Y., & Yang, W. (2017). Salt stress represses soybean seed germination by negatively regulating GA biosynthesis while positively mediating ABA biosynthesis. *Frontiers in Plant Science, 8*, 1372. https://doi.org/10.3389/fpls.2017.01372.

Sigua, G. C., Williams, M., Grabowski, J., Chase, C., & Kongchum, M. (2012). Effect of flooding duration and nitrogen fertilization on yield and protein content of three forage species. *Agronomy Journal, 104*(3), 791–798.

Silvente, S., Sobolev, A. P., & Lara, M. (2012). Metabolite adjustments in drought-tolerant and sensitive soybean genotypes in response to water stress. *PLoS One, 7*(6), e38554.

Singh, K. J., & Singh, O. P. (1992). Influence of stem tunneling by the maggots of *Melanagromyza sojae* (Zehn.) on yield of soybean. *Journal of Insect Science, 5*(2), 198–200.

Singleton, P. W., & Bohlool, B. B. (1984). Effect of salinity on nodule formation by soybean. *Plant Physiology, 74*, 72–76.

Smith, G. S., & Wyllie, T. D. (1999). Charcoal rot. In G. L. Hartman, J. B. Sinclair, & J. C. Rupe (Eds.), *Compendium of Soybean Disease* (4th ed., pp. 29–31). American Phytopathological Society.

Smith, R. H., & Bass, M. H. (1972). Relationship of artificial pod removal to soybean yields. *Journal of Economic Entomology, 63*, 606–609.

Souza, S. C., Mazzafera, P., & Sodek, L. (2016). Flooding of the root system in soybean: Biochemical and molecular aspects of N metabolism in the nodule during stress and recovery. *Amino Acids, 48*(5), 1285–1295.

Soybean Processors Association of India (SOPA). (2020). *World Soybean Production Statistics*. Retrieved from: http://www.sopa.org/statistics/world-soybean-production (accessed on 25 March 2025).

Srivastava, A. K., Marabi, R. S., & Bal, L. M. (2021). Weather-based rules for yellow mosaic disease prediction on soybean in Madhya Pradesh. *Indian Journal of Biochemistry and Biophysics, 58*, 486–497.

Stolf-Moreira, R., Medri, M. E., Neumaier, N., Lemos, N. G., Pimenta, J. A., Tobita, S., Brogin, R. L., Marcelino-Guimarães, F. C., Oliveira, M. C. N., Farias, J. R., Abdelnoor, R. V., & Nepomuceno, A. L. (2010). Soybean physiology and gene expression during drought. *Genetics and Molecular Research, 9*, 1946–1956.

Sultana, S. K., Chowdhury, M. A. A., Rahman, F. M., Naznin, A., Khussboo, R., & Mirza, H. (2022). Screening of soybean genotypes for waterlogging stress tolerance and understanding the physiological mechanisms. *Advances in Agriculture, 2022*, Article ID 5544665. https://doi.org/10.1155/2022/5544665.

Tamang, B. G., Magliozzi, J. O., Maroof, M. A., & Fukao, T. (2014). Physiological and transcriptomic characterization of submergence and reoxygenation responses in soybean seedlings. *Plant, Cell & Environment, 37*, 2350–2365.

Thomas, A., Gueffeiro, S., & Sodek, L. (2005). Aerenchyma formation and recovery from hypoxia of the flooded root system of nodulated soybean. *Annals of Botany, 96*, 1191–1198.

Timbo, R. V., Hermes-Lima, M., Silva, L. P., Mehta, A., & Moraes, M. C. B. (2014). Biochemical aspects of the soybean response to herbivory injury by the brown stink bug *Euschistus heros* (Hemiptera: Pentatomidae). *PLoS One, 9*(10), e109735. https://doi.org/10.1371/journal.pone.0109735.

Todd, J. W., & Turnipseed, S. G. (1974). Effects of southern green stink bug damage on yield and quality of soybeans. *Journal of Economic Entomology, 67*(3), 421–426.

Uikey, S., Sharma, S., Amrate, P. K., & Shrivastava, M. K. (2022). Identification of rich oil-protein and disease resistance genotypes in soybean. *International Journal of Bio-Resource and Stress Management, 13*(5), 497–506. https://doi.org/10.23910/1.2022.2478.

Wang, S., Bao, X., Sun, Y., Chen, R., & Zhai, B. (1996). Study on the effect of population dynamics of soybean aphid (*Aphis glycines*) on both growth and yield of soybean. *Soybean Science (China), 15*, 243–247.

Wegner, L. H. (2010). Oxygen transport in waterlogged plants. In S. Mancuso & S. Shabala (Eds.), *Waterlogging Signaling and Tolerance in Plants* (pp. 3–22). Springer.

Wei, Y., Jin, J., Jiang, S., Ning, S., & Liu, L. (2018). Quantitative response of soybean development and yield to drought stress during different growth stages in the Huaibei plain, China. *Agronomy, 8*, 97. https://doi.org/10.3390/agronomy8070097.

Wilkinson, S., & Davies, W. J. (2002). ABA-based chemical signaling: The coordination of responses to stress in plants. *Plant, Cell & Environment, 25*, 195–210. https://doi.org/10.1046/j.0016-8025.2001.00824.x.

Wrather, A., Shannon, G., Balardin, R., Carregal, L., Escobar, R., Gupta, G. K., Morel, M. A. W., Ploper, D., & Tenuta, A. (2010). Effect of diseases on soybean yield in the top eight producing countries in 2006. *Plant Health Progress, 11*(1). https://doi.org/10.1094/PHP-2010-0125-01-RS.

Wrather, J. A., Stienstra, W. C., & Koenning, S. R. (2001). Soybean disease loss estimates for the United States from 1996 to 1998. *Canadian Journal of Plant Pathology, 23*(2), 122–131.

Yang, L., Song, W., Xu, C., Sapey, E., Jiang, D., & Wu, C. (2023). Effects of high night temperature on soybean yield and compositions. *Frontiers in Plant Science, 14*, 1065604. https://doi.org/10.3389/fpls.2023.1065604.

Yang, M., Zhang, W., Lv, Z., Shi, L., Zhang, K., & Ge, B. (2022). Induced defense response in soybean to *Sclerotinia sclerotiorum* using wuyiencin from *Streptomyces albulus* CK-15. *Plant Disease, 107*(1), 107–115. https://doi.org/10.1094/PDIS-03-22-0582-RE.

Yang, X. B., Lundeen, P., & Uphoff, M. D. (1999). Soybean varietal response and yield loss caused by *Sclerotinia sclerotiorum*. *Plant Disease, 83*(5), 456–461.

Yorinori, J. T., Paiva, W. M., Frederick, R. D., Costamilan, L. M., Bertagnolli, P. F., Hartman, G. E., Godoy, C. V., & Nunes, J. J. (2005). Epidemics of soybean rust (*Phakopsora pachyrhizi*) in Brazil and Paraguay from 2001 to 2003. *Plant Disease, 98*, 675–677.

Yoshikawa, M., Yamauchi, K., & Masago, H. (1978). Glyceollin: Its role in restricting fungal growth in resistant soybean hypocotyls infected with *Phytophthora megasperma* var. *sojae*. *Physiological and Molecular Plant Pathology, 12*, 73–82.

Zhang, C., Xin, W., Feng, Z., Lidong, D., Junjiang, W., Qun, C., & Dongyue, Q. (2017). *Phenylalanine ammonia-lyase* 2.1 contributes to the soybean response towards *Phytophthora sojae* infection. *Scientific Reports, 7*(1), 1–13.

Zhang, M., Duan, L., Tian, X., He, Z., Li, J., Wang, B., & Li, Z. (2007). Uniconazole-induced tolerance of soybean to water deficit stress in relation to changes in photosynthesis, hormones, and antioxidant system. *Journal of Plant Physiology, 164*(6), 709–717.

Zhang, Y., Liu, Z., Wang, X., Li, Y., Li, Y., Gou, Z., & Qiu, L. (2022). Identification of genes for drought resistance and prediction of gene candidates in soybean seedlings based on linkage and association mapping. *Crop Journal, 10*(3), 830–839.

Zhang, Z. N., Hao, J. J., Yuan, J. Z., Song, Q. J., Hyten, D. L., Cregan, P. B., Zhang, G. R., Gu, C. H., Ming, L., & Wang, D. C. (2014). *Phytophthora* root rot resistance in soybean E00003. *Crop Science, 54*, 492–499.

Zhao, T., Aleem, M., & Sharmin, R. A. (2018). Adaptation to water stress in soybean: Morphology to genetics. In *Plant, Abiotic Stress and Responses to Climate Change* (pp. 33–68).

Zhou, Q., Li, Y., Wang, X., Yan, C., Ma, C., Liu, J., & Dong, S. (2022). Effects of different drought degrees on physiological characteristics and endogenous hormones of soybean. *Plants, 11*, 2282. https://doi.org/10.3390/plants11172282.

PART IV

Pharmacological and Therapeutic Applications of Soybean

CHAPTER 14

Soybean in Medicine

SUMIT KUMAR KASHYAP,[1] RAJA SEKHAR REDDY NORU,[2] AMOORU HARIKA,[3] RAJ VRAJESH KOTHARI,[4] AYYAGARI RAMLAL,[5,6,*] AMBIKA RAJENDRAN,[5,*] and SANJAY KUMAR LAL[5]

[1]*Department of Molecular and Cellular Engineering, Jacob Institute of Biotechnology and Bioengineering, Sam Higginbottom University of Agriculture, Technology, and Sciences, Naini, Prayagraj, Uttar Pradesh, India*

[2]*Kerala Agricultural University, College of Agriculture Vellayani, Thiruvananthapuram, Kerala, India*

[3]*Division of Plant Physiology, ICAR–Indian Agricultural Research Institute (IARI), Pusa Campus, New Delhi, Delhi, India*

[4]*Department of Plant Biotechnology, MGM College of Agricultural Biotechnology, Gandheli, Aurangabad, Maharashtra, India*

[5]*Division of Genetics, Indian Council of Agricultural Research (ICAR)–Indian Agricultural Research Institute (IARI), Pusa Campus, New Delhi, India*

[6]*School of Biological Sciences, Universiti Sains Malaysia (USM), Georgetown, Penang, Malaysia*

[*]Corresponding author

ABSTRACT

Soybean (*Glycine max* (L.) Merr.), a widely consumed legume, is drawing attention for its potential in healthcare by unveiling a wealth of biologically active compounds, including isoflavones, proteins, saponins, and phytic acid. These compounds have exhibited significant potential as

Soybeans, Volume 2: Genome, Biotechnology, Stress, and Disease.
R. Ambika Rajendran, Ayyagari Ramlal, Sanjay Kumar Lal, and Godfree Chigeza (Eds.)

DOI: 10.1201/9781998511730-18

therapeutic agents for various health conditions. Soybean is a critical link between its bioactive components and its versatile pharmacological properties, encompassing antioxidant, anti-inflammatory, and anti-cancer attributes. The chapter delves into the historical foundations of soybean in traditional medicine (TM) and highlights its resurgence in modern healthcare. It explores the diverse health benefits of soybeans, encompassing cardiovascular health, bone health, menopausal symptom relief, cancer prevention, diabetes management, obesity control, and skin care.

14.1 INTRODUCTION

Soybean (*Glycine max* (L.) Merr.) has a long history of use in traditional medicine (TM) for various health conditions, including menopause symptoms, bone health, and heart health. With advances in modern medicine, soy has increasingly been incorporated into various forms of medication. Soy contains isoflavones, which have estrogen-like effects on the body. These compounds possess potential health benefits, such as reducing night sweats along with other associated symptoms of menopause (Ahmad et al., 2014; Ramlal et al., 2022a, b, 2023; Song et al., 2007; Swallah et al., 2023). Aside from its potential health benefits, soy is also commonly used as a dietary supplement and a component in health products, such as protein bars and shakes (Mishra et al., 2021). Soy protein is complete as it contains all the essential amino acids (AAs) the body requires for good health. As a result, it is a popular ingredient in protein supplements for athletes and bodybuilders (Singh et al., 2021). Furthermore, soy may be used in prescription medications to enhance their effectiveness or reduce specific side effects. Soy also offers a range of pharmacological benefits thanks to its bioactive compounds. While soybeans are primarily recognized for their nutritional content, they also have phytochemicals with potential health-promoting properties (Afrin et al., 2020; Mahmoud et al., 2014). Additionally, soy may also be used in hormone replacement therapy to alleviate menopause symptoms or in chemotherapy drugs to reduce specific side effects (Aguiar and Barbosa, 2014). Despite having beneficial properties, there are also some safety considerations to keep in mind. For example, high doses of soy supplements may negatively affect thyroid function in some individuals, and there may be interactions with certain medications (Barnes, 1998; Duda-Chodak and Tarko, 2023).

Soybeans have a rich history in TM and have also found applications in modern medicine and research due to their various bioactive compounds (Figure 14.1). Soybean-based products, including miso and natto, are traditional Japanese foods believed to support digestive health. Miso, for instance, is used as a probiotic food due to its content of beneficial microorganisms (Dukariya et al., 2020; Isanga and Zhang, 2008). They have been used in traditional remedies for various ailments, such as sore throats and skin conditions, and even as a poultice for wound healing. These remedies are often based on their nutrient content and the belief in their healing properties, whereas modern research has shown that soy can contribute to heart health (Dukariya et al., 2020; Messina et al., 2021; Song et al., 2007). Although soy is generally considered safe for consumption as part of a balanced diet for many, some people may experience side effects or have concerns regarding soy consumption, as soybeans have some potential side effects, which we will see further in this chapter (Anthony et al., 1996).

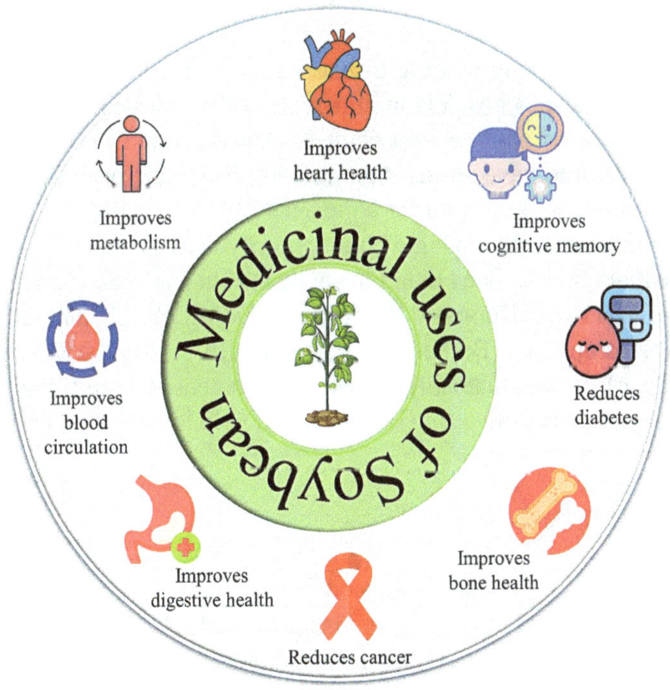

FIGURE 14.1 Medicinal importance of soybean.

14.2 BIOACTIVE COMPONENTS OF SOYBEAN

Soybeans contain biologically active components, including isoflavones, proteins, saponins, and phytosterols. Among these components, isoflavones are perhaps the most extensively studied for their potential health benefits (Isanga and Zhang, 2008; Ramlal et al., 2022a, b, 2023; Swallah et al., 2023).

14.2.1 ISOFLAVONES

The isoflavones found in soybeans include genistein, daidzein, and glycitein. These isoflavones are present in varying amounts depending on the soybean variety, growing conditions, and processing methods used. Genistein is the most abundant isoflavone in soybeans and has been studied for its potential anticancer effects (Perabo et al., 2008; Spagnuolo et al., 2015). Breast, prostate, and colon cancer cells are cancer cells found to be inhibited in laboratory studies. In addition, genistein has been found to have anti-inflammatory properties, which may be beneficial in preventing and treating chronic diseases such as cardiovascular diseases (CVDs) and diabetes (Anderson and Wolf, 1995; Messina and Messina, 2000). Daidzein is another important isoflavone found in soybeans and has been shown to have antioxidant and anti-inflammatory effects (Yu et al., 2016). It is also beneficial for bone health by stimulating the growth of bone-forming cells and reducing bone loss. Glycitein is the least studied of the three major isoflavones found in soybeans but has been shown to have similar effects to genistein and daidzein, including anti-inflammatory and anti-cancer properties (Barnes and Peterson, 1995). Similarly, Ramlal et al. (2022a, b) showed that isoflavones can be potential ligands that can act against angiotensin-converting enzymes and help lower blood pressure.

14.2.2 SAPONINS

Saponins are compounds widely distributed in the plant kingdom, including soybeans. These compounds are named after their characteristic foaming properties, as they can form stable froths or soapy solutions when mixed with water (Anthony et al., 1997; Timilsena et al., 2023). Saponins comprise a sugar molecule (glycone) and a hydrophobic or lipophilic group (aglycone). They are classified into two groups based on their chemical structure: triterpenoid and steroid saponins. Triterpenoid saponins are mainly found in plants such as ginseng and liquorice, while steroid saponins are found in plants such as yucca and soybeans (Singh et al., 2020). Saponins are primarily found in the seed coat, the bean's outer layer. The amount of soybean saponins varies depending on the variety and processing methods used. Typically, raw soybeans contain higher levels of saponins than processed soy products. Saponins have been studied for their potential health benefits, particularly in cholesterol reduction (Bodoira et al., 2021). They have been found to bind to bile acids, preventing their reabsorption and, in turn, promoting excretion. This process can reduce low-density lipoprotein (LDL) cholesterol levels in the blood, thus lowering the risk of CVDs. In addition to their cholesterol-lowering effects, saponins have been shown to have a range of other potential health benefits (Isanga and Zhang, 2008). For example, they have been found to have anti-inflammatory and antioxidant effects and may also play a role in regulating immune function. Saponins are also studied for their potential anti-cancer properties (Kerwin, 2004; Man et al., 2010). Some studies have shown that saponins can inhibit the growth of cancer cells and induce apoptosis (cell death) in cancer cells. While saponins have many potential health benefits, they can also have some adverse effects. In high doses, saponins can be toxic to humans and animals. They sometimes also interfere with the absorption of elements such as iron and zinc and may cause gastrointestinal upset in some individuals (Xiao et al., 2017).

14.2.3 SOY PROTEINS

Soy proteins have gained popularity recently as a healthful alternative to animal-based proteins, particularly for vegetarians (Qin et al., 2022). Soy protein comes in different forms, including soy concentrate, flour, and isolate. Soy isolate contains the highest concentration of protein, up to 90%, while soy flour and concentrate typically contain between 50% and 70% protein (Lephart, 2016). Soy proteins also help in weight loss. Studies have shown that soy protein can lower LDL levels and decrease the risk of heart disease. Soy protein has also been shown to have a satiating effect, helping reduce hunger and supporting weight loss efforts (Merritt and Jenks, 2004). Choosing soy protein over animal-based proteins can help reduce one's carbon footprint. Additionally, some people may have allergies or sensitivities to soy protein (Barnes and Peterson, 1995; Merritt and Jenks, 2004).

14.2.4 PHYTOSTEROLS

Phytosterols are a group of compounds that occur in plant-based foods. They are structurally similar to cholesterol and can show similar effects on the body (Moreau et al., 2018). However, phytosterols have been shown to have numerous health benefits, particularly for heart health. Phytosterols are found in many plant-based foods, including fruits, vegetables, nuts, and grains. They are particularly abundant in certain oils, such as soybean and corn. Additionally, some fortified foods, such as margarine and orange juice, may be enriched with phytosterols (Lephart et al., 2013). In addition to their cholesterol-lowering effects, phytosterols may also have anti-inflammatory and antioxidant properties. Research suggests they reduce inflammation and improve immune function (Singh et al., 2021). Phytosterols have numerous health benefits; they should be consumed in moderation. Consuming large amounts of phytosterols may interfere with the absorption of specific fat-soluble vitamins (D and E) (Fardet et al., 2017).

14.3 PHARMACOLOGICAL PROPERTIES OF SOYBEAN

Soybean is a versatile plant with numerous health benefits due to its various pharmacological properties (Dukariya et al., 2020; Modgil et al., 2021). Its beneficial effects on the body are derived from its various compounds, such as isoflavones, saponins, and phytosterols, which have been studied for their pharmacological effects.

14.3.1 ANTIOXIDANT PROPERTY

One of the most well-known pharmacological properties of soybean is its antioxidant activity. Soybean isoflavones have potent antioxidant activity (Chen et al., 2021; Kim, 2021). These compounds can help to neutralize free radicals in the body, reducing oxidative stress and inflammation (Song et al., 2007). This may help to prevent chronic diseases, such as heart disease, diabetes, and cancer. In addition, soybean also contains other antioxidant compounds, such as vitamin E and selenium, which may further enhance its antioxidant effects. These effects have been attributed to the presence of various bioactive compounds such as isoflavones, saponins, phytosterols, and essential FAs. Antioxidants help to protect the body from damage caused by free radicals, which are unstable molecules. The isoflavones in soybean, especially genistein, have been shown to have strong antioxidant properties. Studies have found that genistein can scavenge free radicals and inhibit lipid peroxidation, a process that produces harmful compounds that damage cells and tissues.

Soybean saponins have also been shown to have antioxidant effects. In vitro studies have found that soy saponins can scavenge free radicals and inhibit the production of reactive oxygen species (ROS), which are molecules that can damage cells and contribute to the development of chronic diseases (Zha et al., 2014). Phytosterols, another group of bioactive compounds found in soybeans, have also been shown to have antioxidant properties, protecting cells and tissues from oxidative damage (Bhatia and Greer, 2008). Finally, essential FAs, such as linoleic and linolenic acids in soybeans, have shown antioxidant effects. These FAs are essential components of cell membranes and play a role in maintaining their integrity, which protects cells from oxidative damage.

14.3.2 ANTI-INFLAMMATORY PROPERTIES

Soybeans have been known to have anti-inflammatory properties due to their high content of phytochemicals such as isoflavones, saponins, and phytic acid. Various studies have shown these compounds to have anti-inflammatory effects (Ahsan et al., 2018; Anthony et al., 1996; Isanga and Zhang, 2008). Isoflavones, a type of phytoestrogen found in soybeans, have been shown to reduce inflammation by inhibiting the production of pro-inflammatory cytokines such as interleukin-1 beta and tumor necrosis factor-alpha (Mijiti et al., 2021). Saponins, another type of phytochemical found in soybeans, have been shown to have anti-inflammatory effects by inhibiting the activity of enzymes such as cyclooxygenase-2 and lipoxygenase that are involved in the production of inflammatory mediators (Zhu et al., 2018). Phytic acid, a natural antioxidant found in soybeans, has also been shown to have anti-inflammatory effects by reducing oxidative stress and inflammation (Upadhyay et al., 2022).

14.3.3 ANTICANCER PROPERTIES

Soybeans have been studied extensively for their potential anti-cancer properties. Some of the compounds in soybeans, such as isoflavones, protease inhibitors (PIs), and phytic acid, have been shown to have anti-cancer effects in various types of cancer (Swallah et al., 2023). Isoflavones have been shown to have anti-cancer effects by inhibiting the growth of cancer cells, inducing cancer cell death (apoptosis), and reducing inflammation (Sheehan, 1998). Studies have suggested that isoflavones may be particularly effective in reducing the risk of breast and prostate cancer. PIs, another compound found in soybeans, have been shown to inhibit the growth of cancer cells by preventing the breakdown of proteins needed for cancer cell growth and survival (Merritt and Jenks, 2004). Phytic acid, a natural antioxidant found in soybeans, has also been shown to have anti-cancer effects by reducing oxidative stress and inflammation in the body, which are known risk factors for cancer. Overall, the anti-cancer properties of soybeans are likely due to the combination of these and other compounds found in soybeans and may be particularly effective when consumed as part of a balanced and healthy diet (Lephart et al., 2013).

14.4 SOYBEAN IN TRADITIONAL MEDICINE (TM)

Soybean has been used for its medicinal properties for centuries in TM systems worldwide. In traditional Chinese medicine (TCM), soybeans are believed to have a cooling effect on the body, which can help alleviate symptoms of heat-related illnesses such as fever and inflammation (Hu et al., 2023). Soybean is often used to treat constipation, diarrhea, coughs, and skin disorders. Soybean milk is also believed to benefit the kidneys and is often recommended for individuals with kidney problems (Gorzkiewicz et al., 2021). In Japan, soybeans have been used in TM for centuries. For instance, natto, a fermented soybean dish that is rich in nutrients and enzymes, is believed to have a range of health benefits, including improving digestion, boosting the immune system, and reducing the risk of CVDs (Wang et al., 2023). Also, it is used as a dietary supplement to support overall health and well-being. Soybean products like tofu and miso are commonly consumed in Japan and are believed to have anti-aging and longevity-promoting effects (Lephart et al., 2013). In Korea, soybeans are believed to have a nourishing effect on the body and are often used to treat anemia and other conditions related to blood deficiency (Singh et al., 2020). Soybeans are believed to have a nourishing effect on the spleen and stomach and are often used to improve digestion and treat digestive disorders. Soybeans are also used in Korean medicine to treat respiratory conditions such as coughs and asthma and are believed to benefit the kidneys (Das et al., 2021).

Nonetheless, some of the bioactive compounds found in soybeans have been the subject of scientific study and have been shown to have potential health benefits. Many studies have examined the effects of isoflavone supplements on menopausal symptoms, with mixed results. Some studies have found that isoflavone supplements can reduce hot flashes and other symptoms, while others have found no significant effects (Messina, 2016). The evidence for the use of isoflavones to reduce the risk of breast cancer is also mixed. Some studies suggest that consuming soy foods may reduce the risk of breast cancer, while others have found no significant association. In addition to their estrogenic effects, isoflavones have anti-inflammatory and antioxidant properties, which may contribute to their potential health benefits (Gorzkiewicz et al., 2021). In Ayurveda, the TM of India, soybeans are believed to have a cooling effect on the body. They treat conditions associated with excess heat, such as skin rashes, fever, and inflammation. Soybeans are also used in Ayurveda to support digestive health and treat related conditions such as constipation and diarrhea (Suzuki et al., 2008). Moreover, they are believed to balance the body's three energies and are often used to improve digestion and promote overall health. Soybeans have also been used in TM in several medications and formulations by others. For example, in Mexican TM, soybeans are used to treat respiratory and digestive conditions, while in African TM, they are used to treat fever and skin infections. Studies have suggested that consuming soy products may help reduce the risk of CVDs, osteoporosis, and certain types of cancer (Li et al., 2019).

14.5 SOYBEAN IN MODERN MEDICINE

Soybeans have been widely studied for their potential health benefits in modern medicine (Sanjukta and Rai, 2016). The bioactive components (isoflavones, saponins, and phytosterols) have been linked to various health benefits, including reducing the risk of chronic diseases and menopausal symptoms, promoting weight loss, improving bone health, and potentially reducing the risk of certain types of cancer (Ahmad et al., 2014). One of the most significant potential benefits of soybeans in modern medicine is their ability to reduce the risk of CVDs. Additionally, some studies have suggested that soy protein and isoflavones can improve blood vessel function, lower blood pressure, and reduce inflammation, all of which can contribute to better cardiovascular health (Singh et al., 2021). However, some studies have produced conflicting results, and more research is needed to determine the optimal dose and duration of soy protein or isoflavone consumption for cardiovascular health. Studies have suggested that soy protein can promote weight loss and improve insulin sensitivity, particularly in individuals with type 2 diabetes (Gougeon, 2013). Additionally, soy protein may help to reduce the risk of developing type 2 diabetes in the first place. However, other studies have found no significant effects of soy protein consumption on weight loss or glycemic control, and more research is needed to determine the optimal role of soy protein in weight management and diabetes prevention and treatment. Studies have found that soy isoflavone supplements can prevent bone loss and may even increase bone density in postmenopausal women (Levis et al., 2011). Additionally, soy protein may benefit bone health, although more research is needed to determine the mechanisms behind these effects. Soybeans have also been studied for their potential role in cancer prevention and treatment. Some studies have suggested that consuming soy isoflavones may reduce the risk of breast cancer, particularly in premenopausal women (Dong and Qin, 2011). However, other studies have produced conflicting results, and more research is needed to fully understand the potential role of soybeans in cancer prevention and treatment. Soy-based supplements have been investigated for their potential use in the treatment of prostate cancer, with mixed results (Suzuki et al., 2008). Soy isoflavones may have anti-inflammatory and antioxidant effects that could benefit skin health. Soy products can be easily incorporated into a healthy diet and lifestyle and may offer potential health benefits as part of a balanced and varied diet. Lastly, soybeans have been studied for their potential benefits for skin health. One study found that a topical soy extract reduced hyperpigmentation and improved overall skin tone and texture in women with facial hyperpigmentation (Bhatia and Greer, 2008).

14.6 FDA-APPROVED SOY SUPPLEMENTS

Soybean products have been extensively studied for their medicinal properties, and several have been approved by the US Food and Drug Administration (FDA) (Kim et al., 2021). Soy protein is approved as a food ingredient and is used in meat substitutes, protein bars, and dietary supplements. Soy isoflavone supplements, approved as dietary supplements, are marketed for health benefits, including menopausal symptom relief and cardiovascular health. Lunasin, a soy protein-derived peptide approved by the FDA, is a dietary supplement and has shown potential in cancer prevention and treatment due to its anti-inflammatory and antioxidant effects (Singh et al., 2020). It has also demonstrated cholesterol-lowering effects in preclinical studies. Additionally, soybean oil, an FDA-approved product, is widely used in cooking and food production. It is a source of unsaturated fats, beneficial for heart health when consumed in moderation (Bhatia and Greer, 2008).

14.6.1 OTHER SOY SUPPLEMENTS

In addition to the FDA-approved soy products, other soy supplements are being investigated for their potential medicinal properties in modern medicine. Soy phosphatidylcholine (PC) is a type of phospholipid derived from soybeans and has been studied for its potential to improve liver function (Mak and Shekhar, 2024). In preclinical studies, soy PC has been shown to have antioxidant and anti-inflammatory effects and to protect against liver damage caused by toxins or disease. While more research is needed, soy PC shows promise as a potential therapeutic agent for liver diseases such as non-alcoholic fatty liver disease and alcoholic liver disease. Soy germ extract (SGE) is another soy supplement that has been studied for its potential health benefits. SGE is rich in bioactive compounds such as isoflavones, saponins, and phytosterols, and has been shown to have antioxidant and anti-inflammatory effects (Song et al., 2003). In preclinical studies, SGE has been investigated for its potential to prevent or treat various health conditions, including cancer, CVDs, and osteoporosis. While more research is needed, SGE shows promise as a potential therapeutic agent for these and other health conditions (Choi et al., 2018; Kim et al., 2019). Soy-derived peptides are also being investigated for their potential medicinal properties. Soy-derived peptides have been shown to have various bioactive properties, including antioxidant, anti-inflammatory, and immunomodulatory effects. In preclinical studies, soy-derived peptides have been investigated for their potential to prevent or treat various health conditions, including CVDs, diabetes, and neurodegenerative diseases. While more research is needed, soy-derived peptides show promise as potential therapeutic agents for these and other health conditions (Shu et al., 2009). Therefore, it is recommended that other supplements should also be explored and investigated for their potential therapeutic properties as alternatives for various diseases.

14.7 POTENTIAL USES OF SOYBEAN IN MEDICINE

Soybean has been found to have beneficial effects on blood glucose and lipid levels, making it a promising natural alternative for managing diabetes. Soybean has been shown to reduce cholesterol levels and improve blood vessel function, making it beneficial for preventing and managing CVDs (Ramdath et al., 2017). The isoflavones in soybeans have been found to have anti-cancer properties and may help prevent the development of certain types of cancer, including breast, prostate, and colon cancer. Soybean contains compounds that can help improve bone density and reduce the risk of osteoporosis. Soybean has been found to benefit menopausal symptoms due to its estrogen-like effects (Hooper et al., 2009; Xiao, 2008).

Another potential use of soybean in medicine is in the prevention and treatment of neurodegenerative diseases such as Alzheimer's disease (Lu et al., 2018). Soybean contains neuroprotective compounds and may help prevent the formation of beta-amyloid plaques associated with Alzheimer's disease. Additionally, soybeans may benefit cognitive function and memory, making them a potential natural alternative for managing age-related cognitive decline (Chatterjee et al., 2015; Lu et al., 2018). Soybean also has potential uses in wound healing and skin care. Soybeans contain compounds that can help promote collagen synthesis, which is essential for wound healing and skin regeneration. Soybean oil has also been found to have moisturizing and anti-inflammatory properties, making it beneficial for treating various skin conditions such as eczema and psoriasis (Poljšak and Kočevar Glavač, 2022). Soybean-based products have been used in various cosmetics and personal care products, indicating the potential of soybeans in the skincare industry.

14.7.1 CARDIOVASCULAR HEALTH

Soy isoflavones in CVDs improve lipid profiles (Li and Zhang, 2017). In addition to inhibiting the production of cholesterol in the liver, isoflavones can also increase cholesterol excretion from the body through several mechanisms, such as stimulation of bile acid synthesis and inhibition of intestinal cholesterol absorption. This is because isoflavones can bind to bile acids and prevent them from being reabsorbed in the intestine. This results in the excretion of more cholesterol from the body (Dong et al., 2011).

Isoflavones can also reduce triglyceride levels, another type of fat in the blood that can increase the risk of heart disease. Isoflavones have been shown to inhibit the synthesis of FAs in the liver (Mariani et al., 2005). Isoflavones can also stimulate the activity of lipoprotein lipase (LPL), an enzyme that breaks down triglycerides in the blood into FAs and glycerol. This can lead to an increase in the clearance of triglycerides from the blood. Overall, these effects on lipid profiles can help reduce the risk of developing heart disease and other cardiovascular conditions. However, it is essential to note that the benefits of isoflavones may vary depending on individual factors such as age, gender, and overall health status. Additionally, soybeans may have anti-inflammatory effects and may help to improve endothelial function, which is an essential factor in maintaining healthy blood vessels (Rivas et al., 2002).

14.7.2 BONE HEALTH

Soybean also improves bone health (Ishimi, 2009; Messina, 2016). Soy isoflavones have been shown to positively affect bone health (George et al., 2020). This may be because phytates can bind to harmful substances in the gut, reducing their impact on mineral absorption. Reducing bone turnover can protect bone health by decreasing the rate at which bone is broken down and replaced. This can help to prevent bone loss and maintain bone density, reducing the risk of fractures and osteoporosis (Lydeking-Olsen et al., 2004). Soy foods with isoflavones can prevent lumbar spine bone loss in postmenopausal women, who may otherwise lose 1.5–3% of bone per year (Ewe and Yeo, 2015). If continued into old age, this prevention of bone loss could translate into a decrease in the lifetime risk of osteoporosis and a lowering of fracture rates (Papandreou et al., 2019). However, the evidence suggests that consuming soybeans as part of a balanced diet may positively impact bone health, particularly in postmenopausal women.

14.7.3 MENOPAUSAL SYMPTOMS

Soybean isoflavones are phytoestrogens and have a similar structure to estrogen hormones produced in the human body (Pabich and Materska, 2019). This property of soybean has been studied extensively for its potential to alleviate menopausal symptoms, such as hot flashes, night sweats, and mood changes caused by a decline in estrogen levels during menopause (Xiao et al., 2017). Isoflavones can bind to estrogen receptors (ER) because isoflavones are similar in structure to 17-β-oestradiol (Muthyala et al., 2004). ERs have two forms, ERα and ERβ, distributed among different tissues (Ishimi, 2009). One of the most well-studied isoflavones found in soybeans is called genistein. Genistein has a higher affinity for estrogen receptor beta (ERβ) than for estrogen receptor alpha (ERα). Therefore, isoflavones can exhibit estrogenic or anti-estrogenic behavior based on the number and type of ER present and the level of endogenous estradiol. For instance, if there are more ERβ than ERα present in a particular tissue, isoflavones can exhibit estrogenic behavior by binding to ERβ and activating estrogenic signaling pathways, thereby helping to alleviate menopausal symptoms by compensating for the decrease in endogenous estrogen levels during menopause (Messina, 2016).

14.7.4 CANCER PREVENTION

Cancer is a significant health concern worldwide, and researchers have been investigating various foods that may have cancer-preventive properties. Soybeans are found to be associated with a reduced risk of CVDs and different types of cancer, including breast, prostate, endometrial, lung, and bladder cancer (Fan et al., 2022; Mai et al., 2007). Soybeans contain a variety of biologically active compounds or phytochemicals, such as genistein, daidzein, and glycitein, which can regulate critical proteins involved in different stages of carcinogenesis, including cellular proliferation, differentiation, apoptosis, angiogenesis, and metastasis. These soy isoflavonoids exert hormonal and non-hormonal actions in cancer prevention (Amaral et al., 2017; Hernández-Ledesma et al., 2011; Messina, 2016; Perabo et al., 2008). Plant proteases such as soybean proteases, bowman-Birk protease inhibitor (BBI), KTI, and lectins have demonstrated promising chemopreventive properties as an alternative to traditional chemotherapy (Srikanth and Chen, 2016). These biologically active compounds have been shown to specifically target cancer cells while avoiding normal cells, minimizing toxicity and side effects, making them effective in preventing cancer. They can also reduce cell proliferation and induce apoptosis (Hernández-Ledesma et al., 2011).

14.7.5 DIABETES MANAGEMENT

Soybean and soy-based products have been found to have potential benefits in improving insulin sensitivity in individuals with diabetes (Kwon et al., 2010). One of the main components of soybeans studied for its effects on insulin sensitivity is isoflavones. Insulin sensitivity refers to the ability of the body's cells to respond to insulin, a hormone that regulates blood sugar levels. In individuals with type 2 diabetes, the cells become less responsive to insulin, leading to high blood sugar levels. This is known as insulin resistance (Hernández-Ledesma et al., 2011). Inflammation and oxidative stress have been implicated in the development of insulin resistance. Oxidative stress, an imbalance between the production of ROS and the body's ability to detoxify them, has also been implicated in the development of insulin resistance. ROS are harmful molecules that can damage cells and tissues and have been shown to contribute to inflammation and insulin resistance (Messina, 2016). Soybeans are a low-glycemic index (GI) food, which means they cause a slower and more gradual rise in blood sugar levels compared to high-GI foods. This can help regulate blood sugar levels and reduce the risk of insulin resistance and type 2 diabetes, often associated with obesity. The GI measures how quickly and how much food raises blood sugar levels.

14.7.6 OBESITY MANAGEMENT

Obesity is a significant health concern, as it increases the risk of several chronic diseases, such as type 2 diabetes, CVDs, and certain types of cancer. Soybeans are a low-calorie food that can help promote weight loss and improve overall health in individuals with obesity (Flegal et al., 2016; Velasquez and Bhathena, 2007). Soybeans also contain bioactive compounds, such as isoflavones, which have been found to have potential benefits for obesity management. Furthermore, soybeans are rich in protein and essential for building and maintaining lean muscle mass (Messina, 2016). Maintaining muscle mass is vital for individuals with obesity, as muscle burns more calories at rest than fat, which can help increase metabolism and promote weight loss (Westerterp et al., 2018). In addition, soybeans are rich in dietary fiber, which can induce satiety and decrease energy intake by promoting feelings of fullness and reducing hunger sensations.

14.7.7 SKIN HEALTH

Soybean is a versatile and nutrient-dense legume shown to have potential benefits for skin health. One of the main ways soybeans can promote skin health is through their high content of isoflavones, a type of phytoestrogen that can mimic the effects of estrogen on the body. Estrogen plays a vital role in maintaining skin health, particularly in women, by promoting collagen production and reducing the appearance of fine lines and wrinkles (Patisaul et al., 2010). Studies have also suggested that soybean consumption or the application of soy-based products can help improve skin health and reduce the signs of aging. Moreover, its potential benefits for aging skin and soybeans may also have applications for managing hyperpigmentation (Leyden and Wallo, 2011) and inflammation (Ishimi, 2009). Soybean extract has been shown to inhibit melanin production in the skin, which can help reduce the appearance of dark spots and uneven skin tone. Soybeans are also rich in antioxidants, such as vitamin E and isoflavones, that can help reduce inflammation and protect the skin against oxidative stress, which is a critical factor in the development of various skin conditions, including acne, rosacea, and psoriasis (Kim et al., 2016; Wang et al., 2013).

14.8 POTENTIAL ALLERGENS AND ALLERGIES

Soybean consumption can potentially cause allergies in some individuals, particularly those with a history of soy allergies or other legume allergies. The potential allergens in soybean that can cause an allergic reaction are proteins known as Gly m 4 and Gly m Bd 30K (Holzhauser et al., 2009; Tsai et al., 2017). These proteins are resistant to heat and digestion, making it challenging to eliminate them from soy-based products. Soy allergies occur when the immune system mistakenly identifies soy proteins as harmful, triggering an allergic reaction (Galvez and de Lumen, 1999). The symptoms of soy allergies can range from mild to severe. They may include hives or rash, itching or tingling in the mouth, swelling of the lips, face, tongue, or throat, difficulty breathing, wheezing, abdominal pain, nausea, vomiting, or diarrhea. In rare cases, soy allergies can lead to anaphylaxis, a life-threatening condition (Jayachandran and Xu, 2019). It is essential for individuals with known soy allergies or legume allergies to avoid soybeans and soy-based products. Soybeans are an ingredient in many food products due to their protein content and emulsifying properties. Soy lecithin, a byproduct of soybean processing, is also commonly used as an emulsifier in food products and may trigger an allergic reaction in sensitive individuals (Gu et al., 2001; Müller et al., 1998).

14.9 CHALLENGES AND LIMITATIONS

Despite the numerous potential uses of soybeans in medicine, some challenges must be addressed. One significant challenge is the variability in the composition of soybean bioactive compounds, which can vary depending on the variety, cultivation conditions, and processing methods (Messina and Messina, 2000). This variability can significantly impact the quality and efficacy of soybean-based products. Developing standardized methods for extracting and quantifying bioactive compounds may be challenging. For example, studies have shown that the levels of isoflavones, one of the essential bioactive compounds in soybeans, can vary significantly depending on the cultivar and growing conditions. Similarly, the processing method used to extract isoflavones can affect the final product's quality and bioactivity (Messina, 2016). Inconsistencies in the composition of bioactive compounds can lead to variability in the effectiveness of soybean-based products, making it challenging to determine optimal dosages and ensure consistent quality. Addressing these challenges will require further research to identify the optimal cultivars and processing methods for creating high-quality soybean-based products for medical use and a comprehensive understanding of the factors that influence soybean bioactive compound composition.

Additionally, the effectiveness of soybean-based products may vary depending on the individual's genetic makeup and co-existing medical conditions. For example, a clinical trial was conducted to investigate the effects of soybean isoflavones on menopausal symptoms in postmenopausal women (Hu et al., 2023). The study found that soybean isoflavones reduced the frequency and severity of hot flashes in these women, suggesting that soybean-based products may be a safe and effective natural alternative to conventional hormone replacement therapy. Clinical trials can also help identify potential interactions between soybean-based products and other medications, as well as potential side effects in specific populations, such as pregnant or breastfeeding women (Plamada et al., 2023). For example, a clinical trial was conducted to evaluate the safety of soybean-based infant formula in premature infants. The study found that premature infants well tolerated soybean-based formula, suggesting that it may be a suitable alternative to cow's milk-based formula in this population (Heinig et al., 1993). However, further research is needed to fully evaluate the safety and efficacy of soybean-based products. Future research directions for soybean in medicine include exploring the mechanisms by which soybean bioactive compounds exert their health benefits, identifying the optimal dosages and potential side effects of soybean-based products, and evaluating the safety and efficacy of soybean-based products in clinical trials.

14.10 CONCLUSION AND FUTURE PERSPECTIVES

In recent years, soybeans have emerged as promising plants because of their potential therapeutic benefits. Further research is needed to fully understand the mechanisms behind these benefits and determine optimal doses for different populations. Nonetheless, the evidence to date suggests that soybeans and their derivatives have the potential to offer a range of health benefits and should be considered part of a healthy diet. Soybeans contain several bioactive compounds, including isoflavones, saponins, and phytosterols, which have been reported to have various health benefits. These compounds can potentially be used in the treatment and prevention of chronic diseases such as diabetes, CVD, and cancer. However, there are also challenges to using soybeans in medicine. One significant challenge is the limited availability of high-quality soybean-based products, which could limit the potential applications of soybeans in medicine. More extensive research is needed to understand soybean-based products' optimal dosages and potential side effects. Additionally, soybean allergies are prevalent, and more research is needed to develop soybean products that people can safely consume. Future research directions could focus on developing new soybean-based products tailored to specific healthcare needs.

KEYWORDS

- **chronic disease prevention**
- **functional foods**
- **isoflavones**
- **phytochemicals**
- **soybean bioactive compounds**
- **therapeutic properties**
- **traditional medicine**

REFERENCES

Afrin, S., Giampieri, F., Gasparrini, M., Forbes-Hernandez, T. Y., Cianciosi, D., Reboredo-Rodriguez, P., et al. (2020). Dietary phytochemicals in colorectal cancer prevention and treatment: A focus on the molecular mechanisms involved. *Biotechnology Advances, 38*, 107322.

Aguiar, P. M., & Barbosa, P. (2014). Use of soy isoflavones on hormone replacement therapy during climacteric. *African Journal of Pharmacy and Pharmacology, 8*(42), 1071–1078.

Ahmad, A., Hayat, I., Arif, S., Masud, T., Khalid, N., & Ahmed, A. (2014). Mechanisms involved in the therapeutic effects of soybean (*Glycine max*). *International Journal of Food Properties, 17*(6), 1332–1354.

Ahsan, F., Imran, M., Gilani, S. A., Bashir, S., Khan, A. A., Khalil, A. A., et al. (2018). Effects of dietary soy and its constituents on human health: A review. *Biomedical Journal, 1*(6), 0.

Amaral, C., Toloi, M. R. T., Vasconcelos, L. D., Fonseca, M. J. V., Correia-da-Silva, G., & Teixeira, N. (2017). The role of soybean extracts and isoflavones in hormone-dependent breast cancer: Aromatase activity and biological effects. *Food & Function, 8*(9), 3064–3074.

Anderson, R. L., & Wolf, W. J. (1995). Compositional changes in trypsin inhibitors, phytic acid, saponins, and isoflavones related to soybean processing. *Journal of Nutrition, 125*, 581S–588S.

Anthony, M. S., Clarkson, T. B., Bullock, B. C., & Wagner, J. D. (1997). Soy protein versus soy phytoestrogens in the prevention of diet-induced coronary artery atherosclerosis of male cynomolgus monkeys. *Arteriosclerosis, Thrombosis, and Vascular Biology, 17*(11), 2524–2531.

Anthony, M. S., Clarkson, T. B., Hughes Jr, C. L., Morgan, T. M., & Burke, G. L. (1996). Soybean isoflavones improve cardiovascular risk factors without affecting the reproductive system of peripubertal rhesus monkeys. *Journal of Nutrition, 126*(1), 43–50.

Barnes, S. (1998). Evolution of the health benefits of soy isoflavones. *Proceedings of the Society for Experimental Biology and Medicine, 217*(3), 386–396.

Barnes, S., & Peterson, T. G. (1995). Biochemical targets of the isoflavone genistein in tumor cell lines. *Proceedings of the Society for Experimental Biology and Medicine, 208*(1), 103–108.

Bhatia, J., & Greer, F. (2008). Use of soy protein-based formulas in infant feeding. *Pediatrics, 121*(5), 1062–1068.

Chatterjee, G., Roy, D., Khemka, V. K., Chattopadhyay, M., & Chakrabarti, S. (2015). Genistein, the isoflavone in soybean, causes amyloid beta peptide accumulation in human neuroblastoma cell line: Implications in Alzheimer's disease. *Aging and Disease, 6*(6), 456.

Chen, Q., Wang, X., Yuan, X., Shi, J., Zhang, C., Yan, N., & Jing, C. (2021). Comparison of phenolic and flavonoid compound profiles and antioxidant and α-glucosidase inhibition properties of cultivated soybean (*Glycine max*) and wild soybean (*Glycine soja*). *Plants, 10*(4), 813. https://doi.org/10.3390/plants10040813.

Choi, C. W., Choi, S. W., Kim, H. J., Lee, K. S., Kim, S. H., Kim, S. L., et al. (2018). Germinated soy germ with increased soyasaponin Ab improves BMP-2-induced bone formation and protects against *in vivo* bone loss in osteoporosis. *Scientific Reports, 8*(1), 12970. https://doi.org/10.1038/s41598-018-31341-3.

Das, G., Heredia, J. B., de Lourdes Pereira, M., Coy-Barrera, E., Oliveira, S. M. R., Gutiérrez-Grijalva, E. P., et al. (2021). Korean traditional foods as antiviral and respiratory disease prevention and treatments: A detailed review. *Trends in Food Science & Technology, 116*, 415–433.

Dong, J. Y., & Qin, L. Q. (2011). Soy isoflavones consumption and risk of breast cancer incidence or recurrence: A meta-analysis of prospective studies. *Breast Cancer Research and Treatment, 125*, 315–323.

Dong, J. Y., Tong, X., Wu, Z. W., Xun, P. C., He, K., & Qin, L. Q. (2011). Effect of soya protein on blood pressure: A meta-analysis of randomised controlled trials. *British Journal of Nutrition, 106*(3), 317–326.

Duda-Chodak, A., & Tarko, T. (2023). Possible side effects of polyphenols and their interactions with medicines. *Molecules, 28*(6), 2536. https://doi.org/10.3390/molecules28062536.

Dukariya, G., Shah, S., Singh, G., & Kumar, A. (2020). Soybean and its products: Nutritional and health benefits. *Journal of Nutritional Science and Healthy Diet, 1*(2), 22–29.

Ewe, J. A., & Yeo, S. K. (2015). Fermented soymilk as a nutraceutical. In *Beneficial Microorganisms in Food and Nutraceuticals* (pp. 133–159).

Fan, Y., Wang, M., Li, Z., Jiang, H., Shi, J., Shi, X., et al. (2022). Intake of soy, soy isoflavones, and soy protein and risk of cancer incidence and mortality. *Frontiers in Nutrition, 9*, 847421. https://doi.org/10.3389/fnut.2022.847421.

Fardet, A., Morise, A., Kalonji, E., Margaritis, I., & Mariotti, F. (2017). Influence of phytosterol and phytostanol food supplementation on plasma liposoluble vitamins and provitamin A carotenoid levels in humans: An updated review of the evidence. *Critical Reviews in Food Science and Nutrition, 57*(9), 1906–1921.

Flegal, K. M., Kruszon-Moran, D., Carroll, M. D., Fryar, C. D., & Ogden, C. L. (2016). Trends in obesity among adults in the United States, 2005 to 2014. *JAMA, 315*(21), 2284–2291.

Galvez, A. F., & De Lumen, B. O. (1999). A soybean cDNA encoding a chromatin-binding peptide inhibits mitosis of mammalian cells. *Nature Biotechnology, 17*(5), 495–500.

George, K. S., Muñoz, J., Akhavan, N. S., Foley, E. M., Siebert, S. C., Tenenbaum, G., et al. (2020). Is soy protein effective in reducing cholesterol and improving bone health? *Food & Function, 11*(1), 544–551.

Gorzkiewicz, J., Bartosz, G., & Sadowska-Bartosz, I. (2021). The potential effects of phytoestrogens: The role in neuroprotection. *Molecules, 26*(10), 2954.

Gougeon, R. (2013). Insulin resistance of protein metabolism in type 2 diabetes and impact on dietary needs: A review. *Canadian Journal of Diabetes, 37*(2), 115–120.

Gu, X., Beardslee, T., Zeece, M., Sarath, G., & Markwell, J. (2001). Identification of IgE-binding proteins in soy lecithin. *International Archives of Allergy and Immunology, 126*(3), 218–225.

Heinig, M. J., Nommsen, L. A., Peerson, J. M., Lonnerdal, B., & Dewey, K. G. (1993). Energy and protein intakes of breast-fed and formula-fed infants during the first year of life and their association with growth velocity: The DARLING Study. *American Journal of Clinical Nutrition, 58*(2), 152–161.

Hernández-Ledesma, B., Hsieh, C. C., Dia, V., de Mejia, E. G., & de Lumen, B. O. (2011). Lunasin, a cancer preventive seed peptide. *Soybean & Health*, 145.

Holzhauser, T., Wackermann, O., Ballmer-Weber, B. K., Bindslev-Jensen, C., Scibilia, J., Perono-Garoffo, L., et al. (2009). Soybean (*Glycine max*) allergy in Europe: Gly m 5 (β-conglycinin) and Gly m 6 (glycinin) are potential diagnostic markers for severe allergic reactions to soy. *Journal of Allergy and Clinical Immunology, 123*(2), 452–458.

Hooper, L., Ryder, J. J., Kurzer, M. S., Lampe, J. W., Messina, M. J., Phipps, W. R., & Cassidy, A. (2009). Effects of soy protein and isoflavones on circulating hormone concentrations in pre-and post-menopausal women: A systematic review and meta-analysis. *Human Reproduction Update, 15*(4), 423–440.

Hu, X., Liu, W., He, M., Qiu, Q., Zhou, B., Liu, R., et al. (2023). Comparison of the molecular mechanisms of Fuzi Lizhong Pill and Huangqin decoction in the treatment of the cold and heat syndromes of ulcerative colitis based on network pharmacology. *Computational Biology and Medicine, 159*, 106870.

Hu, X., Zhang, Q., Ding, J., Liu, Y., & Qin, W. (2023). An updated review of functional properties, debittering methods, and applications of soybean functional peptides. *Critical Reviews in Food Science and Nutrition, 63*(27), 8823–8838.

Isanga, J., & Zhang, G. N. (2008). Soybean bioactive components and their implications to health—A review. *Food Reviews International, 24*(2), 252–276.

Ishimi, Y. (2009). Soybean isoflavones in bone health. *Food Factors & Health Promotion, 61*, 104–116.

Jayachandran, M., & Xu, B. (2019). An insight into the health benefits of fermented soy products. *Food Chemistry, 271*, 362–371.

Kerwin, S. M. (2004). Soy saponins and the anticancer effects of soybeans and soy-based foods. *Current Medicinal Chemistry–Anti-Cancer Agents, 4*(3), 263–272.

Kim, B., Hong, V. M., Yang, J., Hyun, H., Im, J. J., Hwang, J., et al. (2016). A review of fermented foods with beneficial effects on brain and cognitive function. *Preventive Nutrition and Food Science, 21*(4), 297.

Kim, H. J., Choi, E. J., Kim, H. S., Choi, C. W., Choi, S. W., Kim, S. L., et al. (2019). Germinated soy germ extract ameliorates obesity through beige fat activation. *Food & Function, 10*(2), 836–848.

Kim, I. S. (2021). Current perspectives on the beneficial effects of soybean isoflavones and their metabolites for humans. *Antioxidants, 10*(7), 1064.

Kim, I. S., & Kim, C. H., & Yang, W. S. (2021). Physiologically active molecules and functional properties of soybeans in human health—A current perspective. *International Journal of Molecular Sciences, 22*(8), 4054.

Kwon, D. Y., Daily III, J. W., Kim, H. J., & Park, S. (2010). Antidiabetic effects of fermented soybean products on type 2 diabetes. *Nutrition Research, 30*(1), 1–13.

Lephart, E. D. (2016). Skin aging and oxidative stress: Equol's anti-aging effects via biochemical and molecular mechanisms. *Ageing Research Reviews, 31*, 36–54.

Levis, S., Strickman-Stein, N., Ganjei-Azar, P., Xu, P., Doerge, D. R., & Krischer, J. (2011). Soy isoflavones in the prevention of menopausal bone loss and menopausal symptoms: A randomized, double-blind trial. *Archives of Internal Medicine, 171*(15), 1363–1369.

Leyden, J., & Wallo, W. (2011). The mechanism of action and clinical benefits of soy for the treatment of hyperpigmentation. *International Journal of Dermatology, 50*(4), 470–477.

Li, Y., & Zhang, H. (2017). Soybean isoflavones ameliorate ischemic cardiomyopathy by activating Nrf2-mediated antioxidant responses. *Food & Function, 8*(8), 2935–2944.

Lu, Y., An, Y., Lv, C., Ma, W., Xi, Y., & Xiao, R. (2018). Dietary soybean isoflavones in Alzheimer's disease prevention. *Asia Pacific Journal of Clinical Nutrition, 27*(5), 946–954.

Lydeking-Olsen, E., Beck-Jensen, J. E., Setchell, K. D., & Holm-Jensen, T. (2004). Soymilk or progesterone for prevention of bone loss: A 2-year randomized, placebo-controlled trial. *European Journal of Nutrition, 43*, 246–257.

Mahmoud, A. M., Yang, W., & Bosland, M. C. (2014). Soy isoflavones and prostate cancer: A review of molecular mechanisms. *Journal of Steroid Biochemistry and Molecular Biology, 140*, 116–132.

Mai, Z., Blackburn, G. L., & Zhou, J. R. (2007). Soy phytochemicals synergistically enhance the preventive effect of tamoxifen on the growth of estrogen-dependent human breast carcinoma in mice. *Carcinogenesis, 28*(6), 1217–1223.

Mak, K. M., & Shekhar, A. C. (2024). Soybean polyenyl phosphatidylcholine (PPC) is beneficial in liver and extrahepatic tissue injury: An update in experimental research. *The Anatomical Record, 307*(6), 2162–2186.

Man, S., Gao, W., Zhang, Y., Huang, L., & Liu, C. (2010). Chemical study and medical application of saponins as anti-cancer agents. *Fitoterapia, 81*(7), 703–714.

Mariani, E., Polidori, M. C., Cherubini, A., & Mecocci, P. (2005). Oxidative stress in brain aging, neurodegenerative and vascular diseases: An overview. *Journal of Chromatography B, 827*(1), 65–75.

Merritt, R. J., & Jenks, B. H. (2004). Safety of soy-based infant formulas containing isoflavones: The clinical evidence. *Journal of Nutrition, 134*(5), 1220S–1224S.

Messina, M. (2016). Soy and health update: Evaluation of the clinical and epidemiologic literature. *Nutrients, 8*(12), 754.

Messina, M., & Messina, V. (2000). Soyfoods, soybean isoflavones, and bone health: A brief overview. *Journal of Renal Nutrition, 10*(2), 63–68.

Messina, M., Shearer, G., & Petersen, K. (2021). Soybean oil lowers circulating cholesterol levels and coronary heart disease risk, and has no effect on markers of inflammation and oxidation. *Nutrition, 89*, 111343.

Mijiti, N., Someya, A., & Nagaoka, I. (2021). Effects of isoflavone derivatives on the production of inflammatory cytokines by synovial cells. *Experimental and Therapeutic Medicine, 22*(5), 1–8.

Mishra, A., Chandel, A. K., Bhalani, D. V., & Shrivastava, R. (2021). Importance of dietary supplements to health. *Current Nutrition & Food Science, 17*(6), 583–600.

Modgil, R., Tanwar, B., Goyal, A., & Kumar, V. (2021). Soybean (*Glycine max*). In B. Tanwar & A. Goyal (Eds.), *Oilseeds: Health Attributes and Food Applications* (pp. 1–46). Springer.

Moreau, R. A., Nyström, L., Whitaker, B. D., Winkler-Moser, J. K., Baer, D. J., Gebauer, S. K., & Hicks, K. B. (2018). Phytosterols and their derivatives: Structural diversity, distribution, metabolism, analysis, and health-promoting uses. *Progress in Lipid Research, 70*, 35–61.

Müller, U., Weber, W., Hoffmann, A., Franke, S., Lange, R., & Vieths, S. (1998). Commercial soybean lecithins: A source of hidden allergens? *Zeitschrift für Lebensmitteluntersuchung und -Forschung A, 207*, 341–351.

Muthyala, R. S., Ju, Y. H., Sheng, S., Williams, L. D., Doerge, D. R., Katzenellenbogen, B. S., et al. (2004). Equol, a natural estrogenic metabolite from soy isoflavones: Convenient preparation and resolution of R- and S-equols and their differing binding and biological activity through estrogen receptors alpha and beta. *Bioorganic & Medicinal Chemistry, 12*(6), 1559–1567.

Pabich, M., & Materska, M. (2019). Biological effect of soy isoflavones in the prevention of civilization diseases. *Nutrients, 11*(7), 1660.

Papandreou, C., Becerra-Tomás, N., Bulló, M., Martínez-González, M. Á., Corella, D., Estruch, R., et al. (2019). Legume consumption and risk of all-cause, cardiovascular, and cancer mortality in the PREDIMED study. *Clinical Nutrition, 38*(1), 348–356.

Patisaul, H. B., & Jefferson, W. (2010). The pros and cons of phytoestrogens. *Frontiers in Neuroendocrinology, 31*(4), 400–419.

Perabo, F. G. E., Von Löw, E. C., Ellinger, J., Von Rücker, A., Müller, S. C., & Bastian, P. J. (2008). Soy isoflavone genistein in prevention and treatment of prostate cancer. *Prostate Cancer and Prostatic Diseases, 11*(1), 6–12.

Plamada, D., Teleky, B. E., Nemes, S. A., Mitrea, L., Szabo, K., Călinoiu, L. F., et al. (2023). Plant-based dairy alternatives—A future direction to the milky way. *Foods, 12*(9), 1883.

Poljšak, N., & Kočevar Glavač, N. (2022). Vegetable butters and oils as therapeutically and cosmetically active ingredients for dermal use: A review of clinical studies. *Frontiers in Pharmacology, 13*, 868461.

Qin, P., Wang, T., & Luo, Y. (2022). A review on plant-based proteins from soybean: Health benefits and soy product development. *Journal of Agricultural and Food Research, 7*, 100265.

Ramdath, D. D., Padhi, E. M., Sarfaraz, S., Renwick, S., & Duncan, A. M. (2017). Beyond the cholesterol-lowering effect of soy protein: A review of the effects of dietary soy and its constituents on risk factors for cardiovascular disease. *Nutrients, 9*(4), 324.

Ramlal, A., Bhat, I., Nautiyal, A., Baweja, P., Mehta, S., Kumar, V., et al. (2023). *In silico* analysis of angiotensin-converting enzyme (ACE) inhibitory compounds obtained from soybean (*Glycine max* (L.) Merr.). *Frontiers in Physiology, 14*, 1172684. https://doi.org/10.3389/fphys.2023.1172684.

Ramlal, A., Dey, N., Sharma, U., & Rajendran, A. (2022a). *In-silico* studies to reveal the potential inhibitory capacity of soy isoflavonoids against angiotensin-converting enzyme. In H. Chopra, V. Ghuriani, P. M. Arora, S. Babbar, & P. Baweja (Eds.), *Sustainable Future for Humanity: The New Learning Curve* (pp. 155–167). Imperial Publications.

Ramlal, A., Nautiyal, A., Baweja, P., Kumar, V., Mehta, S., Mahto, R. K., et al. (2022b). Angiotensin-converting enzyme (ACE) inhibitory peptides and isoflavonoids from soybean (*Glycine max* (L.) Merr.). *Frontiers in Nutrition, 9*, 1068388. https://doi.org/10.3389/fnut.2022.1068388.

Rivas, M., Garay, R. P., & Escanero, J. F. (2002). Soy milk lowers blood pressure in men and women with mild to moderate essential hypertension. *Alternative Medicine Review, 7*(5), 432–433.

Sanjukta, S., & Rai, A. K. (2016). Production of bioactive peptides during soybean fermentation and their potential health benefits. *Trends in Food Science & Technology, 50*, 1–10.

Sheehan, D. M. (1998). Herbal medicines, phytoestrogens and toxicity: Risk-benefit considerations. *Proceedings of the Society for Experimental Biology and Medicine, 217*(3), 379–385.

Shu, X. O., Zheng, Y., Cai, H., Gu, K., Chen, Z., Zheng, W., & Lu, W. (2009). Soy food intake and breast cancer survival. *JAMA, 302*(22), 2437–2443.

Singh, A. K., Singh, A., Sarkar, S., Ganapathysubramanian, B., Schapaugh, W., Miguez, F. E., et al. (2021). High-throughput phenotyping in soybean. In J. Zhou & H. T. Nguyen (Eds.), *High-Throughput Crop Phenotyping. Concepts and Strategies in Plant Sciences* (pp. 129–163). Springer.

Singh, G., Dukariya, G., & Kumar, A. (2020). Distribution, importance and diseases of soybean and common bean: A review. *Biotechnology Journal International, 24*(6), 86–98.

Song, T., Lee, S. O., Murphy, P. A., & Hendrich, S. (2003). Soy protein with or without isoflavones, soy germ and soy germ extract, and daidzein lessen plasma cholesterol levels in golden Syrian hamsters. *Experimental Biology and Medicine, 228*(9), 1063–1068.

Song, W. O., Chun, O. K., Hwang, I., Shin, H. S., Kim, B. G., Kim, K. S., et al. (2007). Soy isoflavones as safe functional ingredients. *Journal of Medicinal Food, 10*(4), 571–580.

Spagnuolo, C., Russo, G. L., Orhan, I. E., Habtemariam, S., Daglia, M., Sureda, A., et al. (2015). Genistein and cancer: Current status, challenges, and future directions. *Advances in Nutrition, 6*(4), 408–419.

Srikanth, S., & Chen, Z. (2016). Plant protease inhibitors in therapeutics—Focus on cancer therapy. *Frontiers in Pharmacology, 7*, 470.

Suzuki, T., Matsuo, K., Tsunoda, N., Hirose, K., Hiraki, A., Kawase, T., et al. (2008). Effect of soybean on breast cancer according to receptor status: A case–control study in Japan. *International Journal of Cancer, 123*(7), 1674–1680.

Swallah, M. S., Yang, X., Li, J., Korese, J. K., Wang, S., Fan, H., et al. (2023). The pros and cons of soybean bioactive compounds: An overview. *Food Reviews International, 39*(8), 5104–5131.

Timilsena, Y. P., Phosanam, A., & Stockmann, R. (2023). Perspectives on saponins: Food functionality and applications. *International Journal of Molecular Sciences, 24*(17), 13538.

Tsai, J. J., Chang, C. Y., & Liao, E. C. (2017). Comparison of allergenicity at Gly m 4 and Gly m Bd 30K of soybean after genetic modification. *Journal of Agricultural and Food Chemistry, 65*(6), 1255–1262.

Upadhyay, J., Tiwari, N., Durgapal, S., Jantwal, A., & Kumar, A. (2022). Phytic acid: As a natural antioxidant. In S. M. Nabavi & A. S. Silva (Eds.), *Antioxidants Effects in Health* (pp. 437–450). Elsevier.

Velasquez, M. T., & Bhathena, S. J. (2007). Role of dietary soy protein in obesity. *International Journal of Medical Sciences, 4*(2), 72.

Wang, C., Chen, J., Tian, W., Han, Y., Xu, X., Ren, T., et al. (2023). Natto: A medicinal and edible food with health function. *Chinese Herbal Medicines, 15*(3), 349–359.

Wang, Q., Ge, X., Tian, X., Zhang, Y., Zhang, J., & Zhang, P. (2013). Soy isoflavone: The multipurpose phytochemical. *Biomedical Reports, 1*(5), 697–701.

Westerterp, K. R. (2018). Exercise, energy balance and body composition. *European Journal of Clinical Nutrition, 72*(9), 1246–1250.

Xiao, C. W. (2008). Health effects of soy protein and isoflavones in humans. *Journal of Nutrition, 138*(6), 1244S–1249S.

Xiao, Y., Zhang, S., Tong, H., & Shi, S. (2018). Comprehensive evaluation of the role of soy and isoflavone supplementation in humans and animals over the past two decades. *Phytotherapy Research, 32*(3), 384–394.

Yu, J., Bi, X., Yu, B., & Chen, D. (2016). Isoflavones: Anti-inflammatory benefit and possible caveats. *Nutrients, 8*(6), 361.

Zha, L., Chen, J., Sun, S., Mao, L., & Chu, X., et al. (2014). Soyasaponins can blunt inflammation by inhibiting the reactive oxygen species-mediated activation of PI3K/Akt/NF-κB pathway. *PLOS ONE, 9*(9), e107655.

Zhu, F., Du, B., & Xu, B. (2018). Anti-inflammatory effects of phytochemicals from fruits, vegetables, and food legumes: A review. *Critical Reviews in Food Science and Nutrition, 58*(8), 1260–1270.

PART V

Future Prospects

CHAPTER 15

Soybean: The Path Ahead

AYYAGARI RAMLAL,[1,2] R. AMBIKA RAJENDRAN,[2,*]
SANJAY KUMAR LAL,[2,*] and GODFREE CHIGEZA[3]

[1]*School of Biological Sciences, Universiti Sains Malaysia (USM), Georgetown, Penang, Malaysia*

[2]*Division of Genetics, Indian Council of Agricultural Research (ICAR)–Indian Agricultural Research Institute (IARI), Pusa Campus, New Delhi, India*

[3]*International Institute of Tropical Agriculture (IITA), Lusaka, Zambia*

[*]Corresponding author

ABSTRACT

Soybean (*Glycine max* (L.) Merr.) is a marvelous, versatile legume known for its economic value and unique nutraceutical and medicinal properties. An abundant supply of proteins, fats, and carbohydrates makes it a valuable source of food and feed. Soybean is consumed by both humans and animals. Various soy-based foods—including soy paste, soya milk, miso, and tofu—are widely used in many cuisines and recipes. It harbors a variety of bioactive compounds of pharmaceutical and therapeutic importance, such as inhibitors of angiotensin-converting enzymes, storage proteins, trypsin inhibitors, lectins, lipids, and other minor constituents effective against hypocholesterolemia, cardiovascular diseases (CVDs), and cancer. Although traditionally produced using conventional breeding techniques, rising demand from a growing population is increasing production costs. Breeding efforts also aim to reduce anti-nutritional factors while improving oil and protein quality. Soybean's extensive genetic resources and available physiological tools offer opportunities to study seed development,

Soybeans, Volume 2: Genome, Biotechnology, Stress, and Disease.
R. Ambika Rajendran, Ayyagari Ramlal, Sanjay Kumar Lal, and Godfree Chigeza (Eds.)

DOI: 10.1201/9781998511730-20

polyploidization, and interactions with pathogens and symbionts. It also plays a significant role in genetic mapping, marker development, and genome-wide association studies (GWAS) for identifying genes linked to key traits and advancing breeding. Furthermore, various biotic and abiotic stresses pose external threats that reduce productivity and yield. Thus, understanding these yield-limiting factors is essential for developing superior soybean varieties with desirable traits.

15.1 INTRODUCTION

Soybeans were initially domesticated over 4000 years ago in ancient China when they were used for food. Cultivation extended to Korea and Japan in the sixth and seventh centuries and subsequently to other regions of Asia in the following centuries. In India, soybean cultivation began in the early 20th century and became popular in the 1970s to diversify agriculture. Soybeans were also introduced into Europe and North America as livestock forage in the 18th and 19th centuries, but their cultivation as an oilseed crop became popular in the early 20th century. The domestication of soybeans involved selecting plants with desirable traits, such as larger seeds, improved taste, higher yields, and better resistance to pests and diseases. Traditional breeding methods, such as mass selection and hybridization, were used to develop improved varieties. Using a conventional Chinese fermentation process called "natto" likely facilitated domestication. It is crucial to understand the history and significance of soybean cultivation and domestication and its role in modern agriculture. The field of taxonomy and systematics plays an essential role in understanding the classification and its ancestry. It is classified in the pea family, *Fabaceae*, within the subfamily Papilionoideae, tribe Phaseoleae, and subtribe Glycininae. Taxonomy is also essential to gain insights into the genetics, distribution, and adaptation to different environments of any crop, including soybean. The evolutionary history and domestication incidents of soybean can be traced back to dissect information about its wild ancestor, *Glycine soja* Siebold & Zucc. (thereafter *G. soja*), which is native to East Asia.

Soybean is a leguminous plant cultivated for its protein and oil content and wide usability in the food, feed, and industrial sectors. Globally, soybeans are grown; the United States, Brazil, Argentina, China, and India are the primary producers. A crop that can be cultivated in various climates and soil types is soybean. Soybean also has many industrial uses, such as in biodiesel production and manufacturing various products, including cosmetics and lubricants, further adding value to the crop. Soybeans produce different food commodities, including oil, tofu, soymilk, and soy sauce. Additionally, the crop undergoes processing to create a range of value-added food products, enriching nutritional choices for consumers. The cultivation of soybeans has contributed to economic growth and development in many regions of the world, especially in countries that rely heavily on agriculture.

In ancient China, soybeans were initially cultivated for food and medicinal purposes. However, the modern era of soybean breeding began in the late 1800s when soybean germplasm was introduced from Asia to the United States. Early breeding efforts focused on improving agronomic traits and increasing yield. During the mid-twentieth century, breeding efforts shifted towards developing cultivars resistant to diseases and pests. Subsequently, breeding sought to enhance the nutritional value of goods by lowering antinutritional variables and raising the amount of necessary amino acids (AAs). Likewise, attention was given to developing cultivars with enhanced tolerance against abiotic stress such as drought, among others, as well as improving agronomic traits like oil quality. Genetically modified (GM) soybean varieties with herbicide resistance and other characteristics have also been developed. Soybeans are one of the most extensively grown crops in the world today, and breeding soybeans is still essential to raising their yields, quality, and sustainability.

Soybean breeding faces several challenges and problems that hinder the development of new and improved varieties with desirable traits and more economic value. One such challenge is genetic complexity, which makes it difficult to identify and select the desired genetic combinations that contribute to agronomically important traits such as yield, soil quality, disease resistance, and stress tolerance. Low heritability is another shortcoming in soybean breeding programs, wherein a significant proportion of the observed trait variation is due to non-genetic factors such as environmental variability, making it challenging to accurately predict and select these desired traits. A further obstacle in this area is the minimal genetic variety of soybean germplasm compared to other crops, making expanding the genetic base of soybean cultivars challenging.

Soybean breeding is a crucial aspect of modern agriculture that aims to improve soybean crops' yield, quality, tolerance, and resistance.

Conventional breeding methods have been in use for several decades, involving the selection of desirable traits through controlled crosses between parents. These methods include hybridization, mass selection, pedigree selection, and backcrossing. However, these techniques have limitations such as time consumption, limited precision, and efficiency in identifying and selecting genes associated with complex traits, necessitating modern breeding methods. In recent years, modern breeding techniques such as genetic engineering, marker-assisted selection, genomic selection (GS), and transcriptomics have gained popularity due to their ability to accelerate the breeding process and produce crops with specific traits, thereby overcoming the constraints of conventional methods. However, the successful adoption of these modern soybean breeding techniques, along with the integration of traditional techniques, has the potential to create superior soybean varieties that meet the ever-increasing demand for protein, oil, and industrial products while also promoting sustainable soybean farming practices.

Breeding and product profiles of soybeans and their associated challenges have been discussed in previous chapters. A scientific perspective on the history of soybean breeding, tracing its evolution from traditional practices to modern biotechnology tools, is provided. The key innovations and milestones that have shaped the field are highlighted, along with the challenges and opportunities facing soybean breeders today. In general, the history of soybean breeding demonstrates the significance of plant breeding in ensuring food security and tackling issues related to nutrition, disease resistance, and climate change worldwide. Soybean breeding has been the focus of future research and development opportunities and problems. The authors' recommendations can guide soybean breeders and researchers in developing new and improved soybean varieties that are more productive, sustainable, and resilient to environmental challenges. Along with these, the conventional and modern breeding techniques and their integration into soybean breeding programs to enhance soybean crops' genetic diversity and sustainability have also been covered. Finally, improving the efficiency and effectiveness of soybean breeding programs that can contribute to the development of improved varieties to meet the needs of farmers and consumers has been addressed in the product profile chapter.

Heterosis is crucial in modern agriculture for developing newer varieties with improved agronomical and economic traits. This field has been widely studied and utilized in soybeans to produce hybrid seeds to create high-yielding, disease-resistant, and stress-tolerant hybrids that can improve crop productivity. Soybean hybridization involves crossing two genetically diverse parent lines to produce F1 hybrid seeds. The F1 hybrids typically exhibit heterosis, attributed to the complementary interaction of dominant and recessive genes between the parental lines. However, hybridization is challenging due to the characteristics of the crop, such as self-pollination and low natural outcrossing rates. Different methods used for hybridization in soybean, such as manual crossing, emasculation, and cytoplasmic male sterility, are time-consuming and inefficient. Furthermore, the genetic separation between the parental lines, environmental factors, and management techniques can all affect how heterosis manifests in soybean hybrids. Understanding these factors and their interactions is crucial for successful hybridization and maximizing heterosis expression.

Commercial seed production and marketing in soybeans are essential aspects of the soybean industry. Soybean is a significant oilseed crop that is farmed all over the world. Achieving good yields and profitability depends on the seed quality used for sowing. In this context, seed production and marketing involve the production of high-quality seeds that meet the needs of farmers and other stakeholders, as well as the distribution of these seeds to target markets. Appropriate soybean varieties are chosen to ensure high-quality seed production, and proper crop management practices are implemented. To create high-yielding soybean varieties resistant to biotic and abiotic stressors, seed firms frequently employ genetic and agronomic features in the selection process. Proper seed treatment and storage practices can also improve seed quality. Fungicides and insecticides can prevent damage from pests and diseases, while controlling seed moisture levels can prevent fungal growth. Marketing strategies for soybean seeds involve the development of effective distribution channels to reach target markets, such as farmers, seed dealers, and processors. These strategies may include targeted advertising, promotions, and pricing schemes tailored to specific customer segments.

Soy foods are a diverse array of food products derived from soybeans, a staple food in Asian diets, and have gained popularity in other culinary applications. This nutrient-dense food source appeals to health-conscious consumers looking for a plant-based substitute for animal products because it is high in protein, fiber, and phytochemicals. Soy foods include tofu, tempeh, soy milk, soy sauce, miso, and edamame, each with unique textures and flavors. Soy foods have gained popularity due to their potential health benefits. Isoflavones, a kind of phytoestrogen found in soybeans, have been related to a lower risk of osteoporosis, heart disease, and several types of cancer. A balanced diet necessitates the moderation of soy foods, notwithstanding their possible health benefits. Moreover, individuals with soy allergies or other medical conditions should consult their healthcare provider before consuming soy foods. Additionally, soy foods may benefit vegetarians and vegans as they offer plant-based sources of protein that can be difficult to obtain from other sources.

Soybean is a crop with high nutritional and economic value. However, antinutritional elements such as lectins, phytic acid, and trypsin inhibitors lower soybean seeds' nutritional content and digestibility. Therefore, value addition in soybeans has been a significant focus of research to improve soybean products' nutritional quality and economic value. Value addition in soybeans involves several processes, including dehulling, heat treatment, fermentation, and enzyme treatment. Value addition in soybeans has several benefits, including increased nutritional value, improved digestibility, enhanced flavor and aroma, and increased economic value. Value-added soybean products, such as soy milk, tofu, and soy flour, have become popular in many countries and are consumed as a source of high-quality protein and other nutrients.

Soybeans are used in various industries, such as food, fuel, and pharmaceuticals. The crop can be processed to produce multiple products for industrial and household use. In the food industry, soybeans can produce a wide range of products, such as tofu, soy milk, and various meat substitutes. The crop also makes vegetable and margarine oils for food processing and cooking. Soybean meal (SM), a by-product of soybean processing, is a high-protein feed for livestock and aquaculture. Soybeans can also be used to produce biofuels, including biodiesel and bioethanol. In the pharmaceutical industry, soybeans produce various drugs, including antibiotics and anti-cancer agents. Additionally, it has been discovered that soybean extracts contain anti-inflammatory and antioxidant qualities, making them helpful in treating various illnesses. Besides industrial uses, soybeans can also be used as a household crop. Small-scale gardens can cultivate this crop, which can be utilized for fuel, feed, and domestic food. Soybean products such as tofu and soy milk can be made at home using simple equipment, making soybeans convenient and cost-effective for household processing.

Soybean is a versatile crop rich in protein, essential AAs, and other nutrients, making it an important food and nutritional security crop. Soybean has a balanced nutritional profile, containing macronutrients (such as carbohydrates, proteins, and fats) and micronutrients (such as vitamins and minerals). Soybean protein is of high quality, containing all essential AAs required for human growth and development. Soybean cultivation contributes to food security by providing a reliable source of protein and other nutrients, especially in developing countries where malnutrition is a significant issue. Furthermore, soybeans are an important cash crop for many farmers, providing income and contributing to the local economy.

In recent years, soybeans have become increasingly recognized as necessary for enhancing cropping systems, especially in areas where conventional crop rotations are becoming unsustainable. Soybeans can contribute to sustainable intensification by improving soil fertility, reducing soil erosion, and suppressing weeds and pests, thus enhancing the yield and quality of subsequent crops in the rotation. Soybeans are well suited to intercropping systems, where they are planted alongside other crops, such as maize, cotton, or sorghum, which complement their growth and nutrient requirements. Intercropping soybeans can increase resource-use efficiency and reduce resource competition, leading to higher yields and improved soil health. To further increase their potential for intensifying cropping systems, soybean cultivars with enhanced resilience to abiotic stresses, including drought and salinity, and increased resistance to pests and diseases are now under development. Despite the benefits of soybeans, challenges to their widespread adoption in cropping systems remain. These include limited access to quality seeds and fertilizers, market challenges, and socioeconomic factors. Addressing these challenges requires integrated approaches that involve improving access to inputs, promoting market linkages, and increasing awareness and knowledge of the benefits of soybean-based cropping systems. Overall, soybeans have great potential for intensifying cropping systems and promoting sustainable agriculture, and they should be considered a key component of cropping system diversification strategies.

Soybean is an important crop with a complex genome that has been the focus of genetic research for several decades. The soybean genome is diploid, consisting of 20 pairs of chromosomes with an estimated size of 1.1 Gb. The organization of the soybean genome has been extensively studied, with efforts to develop genetic maps, physical maps, and genomic resources to aid in gene discovery and functional characterization. Soybean chromosomes have been classified into 20 linkage groups based on genetic mapping studies. The availability of high-throughput sequencing technologies has enabled the generation of whole-genome sequences for multiple soybean cultivars, allowing for the identification of genomic variations and structural rearrangements between cultivars. Several studies have also investigated soybean's transcriptome and epigenome to understand gene expression regulation better. The soybean genome contains many essential genes involved in seed development, stress response, and disease resistance. Research has focused on identifying and characterizing these genes to develop soybean cultivars with improved traits. Genome editing (GE) technologies, such as CRISPR/Cas9, have also introduced precise genetic modifications into soybean plants.

Soybean is an important crop due to its nutritional and industrial values. The grain structure of soybean consists of three main components: the seed coat, cotyledon, and embryo. The two layers of the seed coat—the inner parenchyma and the outer epidermis—shield the seed. The bulk of the seed is made up of cotyledon cells, which store fats, proteins, and carbohydrates. The embryo is the developing plant and contains the embryonic shoot and root. Soybean grains also contain a variety of cellular structures, such as protein bodies, oil bodies, and starch granules. Protein bodies are the primary storage form of protein in soybeans and are composed of various proteins, including glycinin and beta-conglycinin. Oil bodies are the primary storage form of lipids and are surrounded by a protein layer. Starch granules are found in the cotyledon cells, providing energy for the developing plant. The composition of the grain structure of soybean has important implications for its nutritional and industrial properties. Soybean is a rich source of protein, with the protein bodies in the cotyledon cells accounting for up to 40% of the grain weight. The oil bodies in soybean grains are also of industrial importance, as they can be extracted and used as a biodiesel supply. The starch granules in soybean grains can produce various products, such as sweeteners and bioplastics.

Plant breeding has two categories of attributes: qualitative and quantitative traits. One or a small number of genes determine qualitative features, which lead to phenotypic differences that are easily distinguished. Qualitative traits in soybeans include seed coat color, plant growth habits, and resistance to specific diseases. These traits are usually inherited in a simple Mendelian manner and can be identified by visual inspection or molecular markers. Quantitative traits, on the other hand, are controlled by multiple genes and are characterized by continuous variation. Examples of quantitative traits in soybeans include seed yield, protein content, and oil content. These traits are affected by both genetic and environmental factors, and interactions between genes and the environment can influence their expression. Quantitative traits are usually measured using statistical methods, and their genetic basis is studied using techniques such as QTL mapping and genome-wide association studies (GWAS). Both qualitative and quantitative traits are essential targets for improvement. Qualitative traits help develop new varieties with specific traits, such as disease resistance, while quantitative traits are essential for increasing yield and improving overall quality. Advances in genomics and molecular breeding techniques have enabled the identification and manipulation of genes underlying both types of traits, making it possible to develop new soybean varieties with improved traits for specific environments and end users.

Anti-nutritional factors can lower protein digestibility and bioavailability, as well as reduce the quantities of some vital AAs, thereby decreasing the quality of soybean protein. Various strategies have been employed to improve soybean protein quality. One strategy is to employ molecular breeding techniques to boost the amounts of vital AAs in soybean protein. Another approach is to use genetic engineering to modify the expression of genes involved in soybean protein synthesis, resulting in soybean plants that produce proteins with improved AA profiles. Processing techniques such as heat treatment, fermentation, and enzyme treatment can also improve soybean protein quality by reducing anti-nutritional factors and increasing digestibility. Improving soybean protein quality is a critical objective for breeding and research.

Vegetable soybean oil is a significant source for industrial and culinary purposes. However, the quality and quantity of soybean oil can be improved through various genetic and agronomic approaches. In this context, several studies have been conducted to enhance the oil content, FA composition, and other quality traits of soybean oil. One of the approaches to improve soybean oil quality is genetic modification, which has been shown to improve oxidative stability, shelf life, and nutritional value. Another approach to improving soybean oil is through agronomic practices. For instance, applying nitrogen fertilizer can increase soybean oil content and yield. PGRs can further increase the amount of oil that accumulates in soybean seeds. Moreover, biotic and abiotic stress control strategies can enhance soybean oil quality and quantity.

Soybean is an essential crop with high protein content and is widely used in human and animal diets. However, anti-nutritional elements like phytic acid and trypsin inhibitors reduce their nutritional value. Molecular interventions have been employed to improve the nutritional quality of soybeans by reducing the levels of these anti-nutritional factors and increasing the levels of desirable nutrients. One approach is to use genetic engineering to reduce the expression of genes involved in the biosynthesis of phytic acid and trypsin inhibitors. Similarly, gene editing tools such as CRISPR/Cas9 have been explored to modify the expression of trypsin inhibitors. Another approach is to increase the expression of genes involved in the biosynthesis of desirable nutrients such as essential AAs, vitamins, and minerals such as iron and zinc, which have been targeted to improve the nutritional quality of soybeans. In addition, molecular markers have been developed to facilitate the selection of soybean lines with desirable nutritional traits.

Soybeans are vulnerable to various biotic stresses caused by pathogens, pests, and weeds. Biotic stresses can significantly reduce the yield and quality of soybeans, resulting in substantial economic losses for growers. Therefore, effective management strategies are essential for mitigating the impact of biotic stresses on soybean production. Biotic stressors, such as soybean cyst nematode (SCN), fungal pathogens, *Phytophthora sojae* and *Fusarium virguliforme*, and insects can affect soybean plants at different growth stages. Soybean biotic stressors are frequently managed by integrated pest management (IPM) tactics, which include chemical, biological, and cultural control techniques. A sustainable and economical way to address biotic pressures is through cultural management techniques such as planting date adjustment, resistant cultivars, and crop rotation. The impact of biotic stressors on soybean production can also be lessened by using biological management techniques, such as beneficial microbes. Chemical management techniques like fungicides, insecticides, and nematicides help control biotic stressors. However, they must be applied carefully to prevent harm to non-target organisms and the environment.

Soybean is an important legume crop worldwide and is crucial to global food security. Various abiotic factors influence soybean yield and productivity, including temperature, water availability, soil nutrients, and photoperiod. Soybean is a warm-season crop, and its optimal temperature range for growth and development is between 20°C and 30°C. High temperatures above 35°C can reduce soybean yield and quality due to heat stress. In addition, soybeans require an adequate and consistent water supply for optimal growth and development. Water stress during the reproductive stage can significantly reduce the yield of soybeans. Soybeans require soil nutrients, including nitrogen, phosphorus, and potassium, for proper growth and development. Nitrogen is the most critical nutrient for soybeans, and its deficiency can lead to reduced yield and quality. Adequate and balanced fertilization could increase the yield and productivity of soybeans.

Soybean is a major oilseed crop frequently exposed to biotic and abiotic stresses that can negatively affect crop productivity. In response to these stresses, soybean plants activate various physiological and biochemical adaptations to enhance their tolerance to these adverse conditions. Biotic stresses caused by pathogenic microorganisms, such as fungi, bacteria, viruses, and nematodes, trigger various defense mechanisms in soybean plants. For example, when the fungal pathogen *P. sojae* attacks, soybean plants produce phytoalexins such as glyceollins and isoflavonoids to restrict pathogen growth. Similarly, soybean plants can activate various signaling pathways and produce defense-related proteins such as chitinases and glucanases in response to bacterial and fungal infections. On the other hand, abiotic stresses, such as drought, salinity, high and low temperatures, and heavy metal toxicity, can also trigger various physiological and biochemical responses in soybean plants. For instance, soybean plants can produce osmolytes such as proline and sugars to maintain cellular water balance and protect against drought and salinity stress. They can also synthesize heat shock proteins, such as Hsp70 and Hsp90, which protect cellular structures from damage caused by high-temperature stress.

Soybean also acts as a potential medicinal crop. It contains several bioactive compounds, such as isoflavones, saponins, and phytosterols, which have been found to have anti-inflammatory, antioxidant, and anti-cancer effects. Isoflavones, specifically genistein and daidzein, are the most studied compounds in soybeans and have been shown to have anti-cancer effects. They prevent the spread of cancerous cells and lower the chance of tumor formation. On the other hand, saponins have anti-inflammatory and antioxidant properties, making them helpful in preventing and treating chronic diseases such as heart disease and diabetes. Soybean also contains phytosterols, which are structurally similar to cholesterol and have been found to lower blood cholesterol levels. This makes soybean a potentially valuable crop for preventing and treating cardiovascular diseases (CVDs). In addition, studies have shown that soybeans have additional health benefits, including lowering the risk of osteoporosis, enhancing cognitive function, and lowering the likelihood that women will experience menopausal symptoms.

Soybean has a long history of medicinal use, and many tools and resources are available to study its potential health benefits. These include genomic and proteomic databases, metabolomics tools, and in vitro and in vivo assays. One of the most essential resources for studying soybeans is the Soybean Knowledge Base (SoyKB), a comprehensive database of genomic, transcriptomic, and proteomic data for soybeans. This resource allows researchers to investigate the molecular mechanisms behind soybean's medicinal properties and to identify genes and proteins of interest. Other databases and resources for soybeans include the Soybean Genome Database, which provides access to the soybean genome sequence, and SoyBase, which provides information on soybean genetic maps, QTLs, and other genetic resources. Metabolomics tools are also crucial for studying soybeans in medicine. For example, liquid chromatography-mass spectrometry (LC-MS) has been used to identify and quantify the bioactive compounds in soybeans, such as isoflavones. These tools have been used to investigate the effects of soybeans on human health, including their potential as a treatment for various diseases. In vitro and in vivo assays are also crucial for studying the medicinal properties of soybeans. For example, cell culture assays have been used to investigate the effects of soybean extracts on cancer cells. In contrast, animal models have been used to study the effects of soybeans on CVD, diabetes, and other conditions. In addition to these tools and resources, many soybean-based products are available for medicinal use. For example, soy protein is commonly used in infant formulas, and soy isoflavones are often used as a dietary supplement for menopausal women.

15.2 CONCLUSION

Recent studies indicate that issues, including disease, pests, and climate change, affect soybean productivity. Consequently, a multidisciplinary strategy integrating multiple domains like genetics, genomics, and breeding is needed for soybean research. While genetic modification of soybeans to improve yield and stress tolerance has shown promising results, the regulatory landscape for GM crops remains challenging. Hence, traditional breeding methods such as marker-assisted selection and GE hold significant potential in soybean improvement. With the advancements in high-throughput sequencing technologies, genomic tools have been developed to aid in identifying genetic variation and functional elements in soybeans. These tools have enabled the identification of essential genes and genomic regions associated with desirable traits. Furthermore, combining genomic and phenotypic data using machine learning techniques has much potential to improve breeding initiatives. These approaches can accelerate the identification of desirable traits, aid in developing predictive models, and facilitate the selection of optimal cultivars. Additionally, the study of the soybean microbiome and its interactions with the plant shows significant potential in developing sustainable production practices. The microbiome can be manipulated to enhance plant growth and stress tolerance while reducing reliance on synthetic fertilizers and pesticides.

KEYWORDS

- **bioactive compounds**
- **genomic tools**
- **machine learning in breeding**
- **marker-assisted selection**
- **soybean microbiome**
- **soybean nutraceutical properties**
- **stress tolerance**

Index

C

H

I

T

U

V

W

X

Y

Z

For Product Safety Concerns and Information please contact our EU representative GPSR@taylorandfrancis.com
Taylor & Francis Verlag GmbH, Kaufingerstraße 24, 80331 München, Germany

www.ingramcontent.com/pod-product-compliance
Lightning Source LLC
LaVergne TN
LVHW020627300926
849161LV00004B/634
9781998511723